# Automotive Electrical and Electronics

SECOND EDITION

**A K BABU**
Associate Professor
Department of Automobile Engineering
SRM Easwari Engineering College
Chennai-600089
Tamilnadu, INDIA

**KHANNA BOOK PUBLISHING CO. (P) LTD.**
Publisher of Engineering and Computer Books
4C/4344, Ansari Road, Darya Ganj, New Delhi-110002
**Phone** : 011-23244447-48 **Mobile:** +91-9910909320
**E-mail** : contact@khannabooks.com
**Website:** www.khannabooks.com

Dear Readers,

To prevent the piracy, this book is secured with HIGH SECURITY HOLOGRAM on the front title cover. In case you don't find the hologram on the front cover title, please write us to at contact@khannabooks.com or whatsapp us at +91-99109 09320 and avail special gift voucher for yourself.

Specimen of Hologram on front Cover title:

Moreover, there is a SPECIAL DISCOUNT COUPON for you with EVERY HOLOGRAM.

How to avail this SPECIAL DISCOUNT:

Step 1: Scratch the hologram
Step 2: Under the scratch area, your "coupon code" is available
Step 3: Logon to www.khannabooks.com
Step 4: Use your "coupon code" in the shopping cart and get your copy at a special discount
Step 5: Enjoy your reading!

Thanks,
Khanna Publishing House

**Price: ₹369.00**

**AUTOMOTIVE ELECTRICAL AND ELECTRONICS**

*by* **AK Babu**

**ISBN:** 978-93-82609-69-8

**Second Edition:** 2018
**Reprint:** 2023, 2024

To view complete list of books, please scan the QR Code:

*Published by:*
**Khanna Book Publishing Co. (P) Ltd.**
CIN: U22110DL1998PTC095547

**Visit us at:** www.khannabooks.com
**Write us at:** contact@khannabooks.com

Printed in India.

# PREFACE

This book is meticulously designed for the core subject "Automotive Electrical and Electronics" critical to the engineering curriculum and offered to the undergraduate and postgraduate students of Automobile Engineering of various engineering colleges and universities in India. The principle objective of this comprehensive book, designed for use as a primary text book, is to provide a clear, broad and thorough understanding of the many systems and component parts that constitute the vehicle electrical and electronics in a detailed manner. The book attempts to address the need of students to obtain a good understanding of basic concepts thereby laying a solid foundation in fundamental principles and serves as a valuable source of information and reference.

The focus of this book is to provide clear explanations of vehicle electrical and electronic components and systems in an easy-to-read style and with unique supportive illustrations, which should be of value both to the students and to experienced faculty members. Each chapter takes the reader systematically through the details of each component system. Key topics are emphasized and are reinforced by numerous illustrations. The subject matter is presented to-the-point and lucid manner; the material is covered in sufficient detail such that the reader may gain a clear understanding of the important principles on which automobile engineering is based.

I hope that students and teachers enjoy reading this book and that they find its contents informative and useful. I am optimistic that the material presented in this book will be helpful in extending the knowledge and understanding of students and facilitate their gaining a thorough understanding of the theory and applications of the subject.

**A K BABU**

सत्यमेव जयते

प्रो. एम.पी. पूनियाँ
उपाध्यक्ष
**Prof. M.P. Poonia**
**Vice-Chairman**

अखिल भारतीय तकनीकी शिक्षा परिषद्
(भारत सरकार का एक सांविधिक निकाय)
(शिक्षा मंत्रालय, भारत सरकार)
नेल्सन मंडेला मार्ग, वसंत कुंज, नई दिल्ली-110070
दूरभाष : 011-26131495
ई-मेल : vcm@aicte-india.org

**ALL INDIA COUNCIL FOR TECHNICAL EDUCATION**
(A STATUTORY BODY OF THE GOVT. OF INDIA)
(Ministry of Education, Govt. of India)
Nelson Mandela Marg, Vasant Kunj, New Delhi-110070
Phone : 011-26131495
E-mail : vcm@aicte-india.org

## Foreword

Much like cell phones are now capable of doing more than making calls, cars can do much more than drive and park. In recent years, the automotive industry has worked hand-in-hand with major technology companies in order to deliver the most advanced, safest and most comfortable vehicles out there. Cars are becoming large smart devices with advanced emergency braking capabilities, mapping technology for autonomous driving, better fuel efficiency and cars as a service as a form of transportation.

With a view to enhance the employability skills and impart deep knowledge in Automobile Sector, AICTE with the assistance of industry-academia experts, has designed a curriculum in this area.

I am happy to know that Dr. A.K. Babu, a well-known author of automobile engineering books, has collaborated with a highly reputed publisher, M/s. Khanna Book Publishing Company Private Limited, to write as series of books on the subject of **"Automobiles"**.

Having written these books in a simple way in accordance with the requirements of AICTE, Dr. Babu has helped our students better understand these topics, as well as capture the plethora of employment opportunities available in this field.

The books are written according to the AICTE guidelines for the subject.

As a dedicated teacher in the field of Automobile Engineering, Dr. Babu's commitment and efforts in writing these series of books are highly appreciated. congratulate him for his contributions towards our higher education.

I wish the author, publisher as well as readers all the best!

(Prof. M.P. Poonia)

# ACKNOWLEDGEMENT

Acknowledgements for this book are rather difficult to make because many books and manuals have influenced me directly or indirectly in its preparation. I am keen to express my grateful thanks to all those authors, publishers and manufacturers.

I wish to express my sincere gratitude to Dr. V. Antony Aroul Raj, Easwari Engineering College and Dr. G. Kumaresan, Institute for Energy Studies, Anna University for their many helpful and important comments, suggestions and advice.

I owe my deepest appreciation to my professor Dr. S. Parthasarathy for his unstinting guidance, insightful discussions, patient advice and immense support throughout the compilation of this book.

I am immensely grateful to Khanna Book Publishing Co (P) Ltd. for providing me with the opportunity to be the lead author for this high-quality book.

# ABOUT THE AUTHOR

**A.K.BABU**, obtained his M.E. in Automobile Engineering from Madras Institute of Technology (M.I.T.), Anna University in 1997. A senior academician, he has over 18 long years of experience in teaching, research and industry, out of which 5 years were spent abroad. A.K.Babu has published a considerable number of research papers in renowned international/national journals and conferences. His research paper, published in SAE Journal of Fuels and Lubricants, received tremendous appreciation and was distinguished as one of the most outstanding papers in the year 2003.

# CONTENTS

# Fundamentals of Electricity

The components of the automobile electrical system are of complicated construction, but all of them are applications of the various actions made possible by the flow of electricity. Before taking up the construction and operation of these various electric components, the fundamentals of electricity will first be reviewed.

## ELECTRICITY

Electricity is a form of energy and is produced by the movement of small atomic particles known as electrons. It can be easily transmitted or converted to other forms, such as mechanical or heat energy. To transmit electricity, conductors such as copper, steel, carbon or an electrolyte are used. Insulators such as rubber, porcelain or bakelite are used to prevent electrical leakage from conductors.

There are two kinds of electricity: static and dynamic. Dynamic electricity can be further divided into direct current (DC) and alternating current (AC).

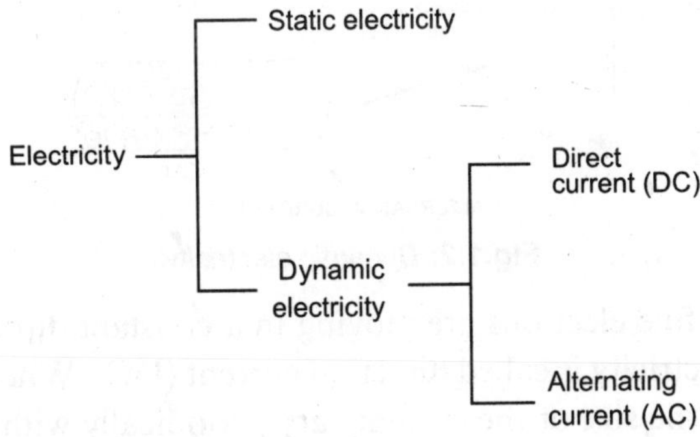


## STATIC ELECTRICITY

When a non-conductive substance such as a glass rod is rubbed with a silk cloth, the surfaces of both the rod and the cloth become charged with electricity, one positively and one negatively. Unless the two materials touch, or are connected by a conductor, the electrical charge will remain in the glass rod and silk cloth. Since it does not move, this type of electricity is called static electricity.

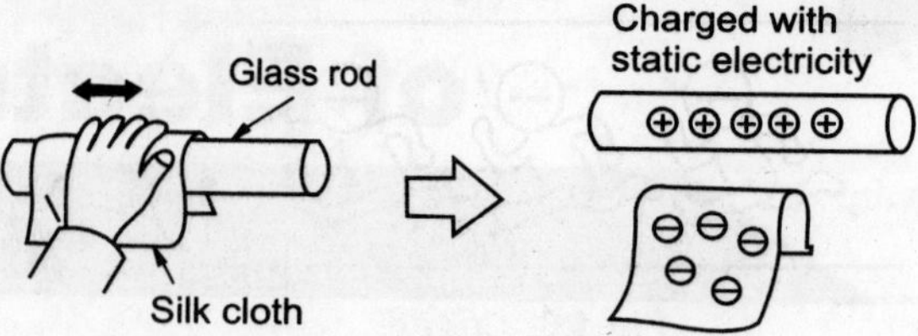


**Fig.1.1:** *Static electricity*

In terms of free electron, static electricity refers to a state in which free electrons are separated from their atoms and do not move about on the surface of the substance.

## DYNAMIC ELECTRICITY

Dynamic electricity refers to a state in which there is a flow of free electrons, i.e. electrons that are separated from their atoms and moving about inside the conductive substance.

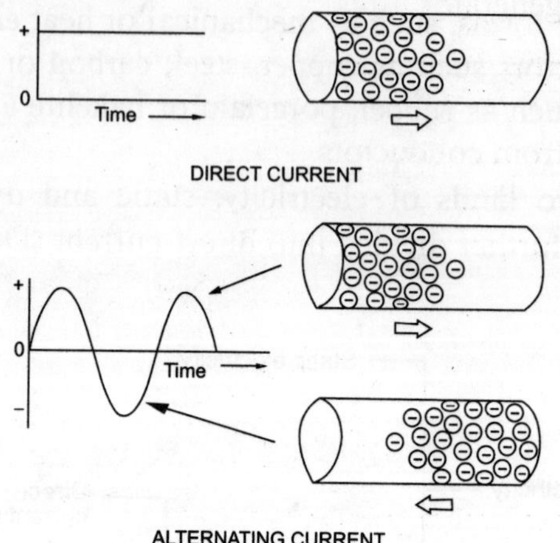


**Fig.1.2:** *Dynamic electricity*

When the free electrons are moving in a constant direction, the state of dynamic electricity is called direction current (DC). When the direction of motion and the size of the current vary periodically with time, the state of dynamic electricity is called alternating current (AC).

## ELEMENTS OF ELECTRICITY

The three elements of electricity are voltage, current and resistance.

## CURRENT

Electric current is a flow of electrons and is measured in amps, (short for amperes). The amp is a measure of flow rate. In other words how much electricity is flowing per second.

**Fig.1.3:** *Current*

1 amp = 6,240,000,000,000,000,000 electrons per second

In electronic circuits a current of one amp is quite large, so often the milli-amp is used instead. A milli-amp (mA) is a thousandth of an amp (A).

## VOLTAGE

Electricity does not flow through a wire unless it is pushed. There has to be an electrical pressure and this pressure is called voltage and is measured in Volts. The symbol is V. It is also called electrical pressure, and electromotive force (emf). This pressure can be provided by a battery, or it can be produced by a generator.

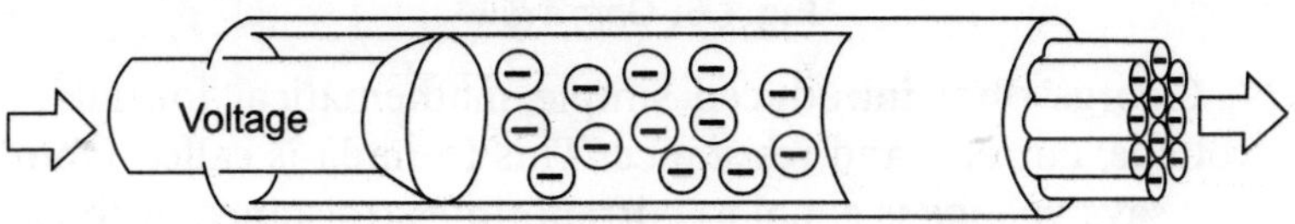


**Fig.1.4:** *Voltage*

## RESISTANCE

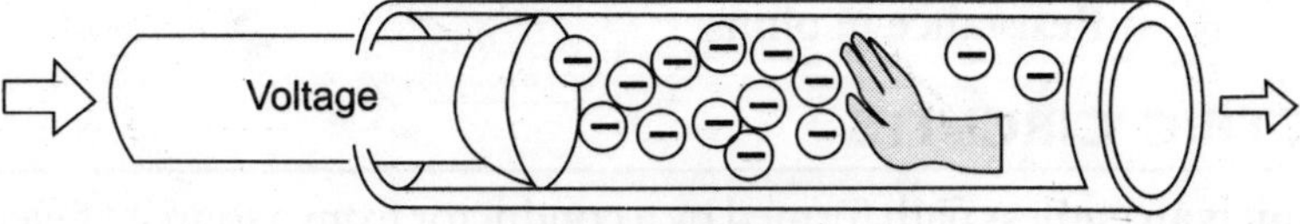


**Fig.1.5:** *Resistance*

The rate that the electrons pass around the circuit (i.e. the amount of current) is dependent not only on the force (voltage) but also by how much resistance there is in the circuit. Resistance is measured in ohms (W).

In the case of electricity, a same conductor will allow better current flow when its diameter becomes larger due to less resistance. When it becomes longer, its resistance increases so that the current gets weaker.

That is, the resistance of a conductor is directly proportional to its length and inversely proportional to its cross-sectional area. Because in a car the same 12V battery powers all circuits, the only way that current can vary is by varying the resistance.

The conductor resistance normally changes with the temperature. The resistances of metal conductors increase when the temperature rises, but resistances of carbon and aqueous solutions decrease. The resistance also varies with the conductor material. A good conductor will have small resistance, whereas the resistance of insulating material will be enormous.

## OHM'S LAW

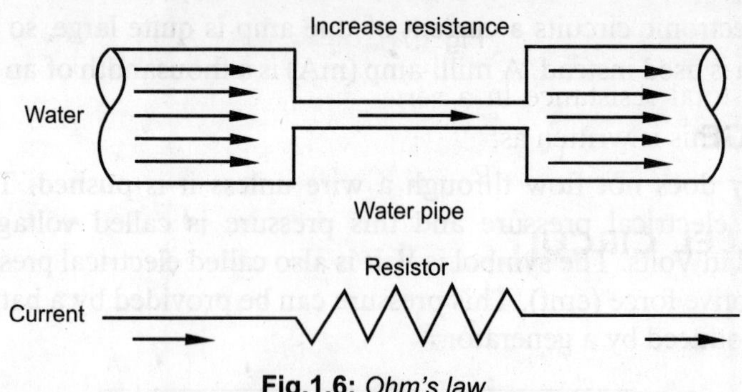


**Fig.1.6:** *Ohm's law*

In 1827, George ohm introduced simple mathematical formula, which relates voltage, current, and resistance. This formula is called Ohm's law. Ohm's law says voltage is equal to current times resistance. A force of 1V is required to push a current of 1A through a resistance of 1W.

$$V = I \times R$$

Where, V = Voltage in volts

I = Current in amperes

R = Resistance in ohms

## ELECTRIC CIRCUITS

A circuit is an endless path formed by a conductor from a source of electrical supply to the control, through the load, and back to the source. There are various types of electrical circuits used in the modern automobile, among them the series, and parallel circuits.

## SERIES CIRCUIT

In a series circuit, there is only one current path. The same battery current flows through each component. A breakage in the path would cease the function of all the equipments.

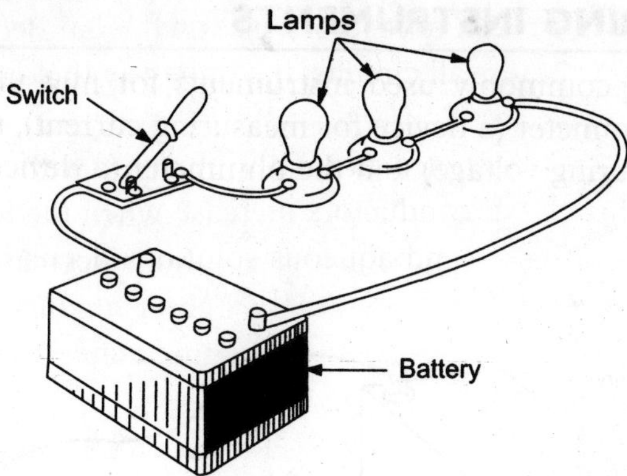


**Fig.1.7:** *Series circuit*

The total resistance in a series circuit is the sum of the individual resistors. This is written as:

$$R_{total} = R_1 + R_2 + R_3$$

## PARALLEL CIRCUIT

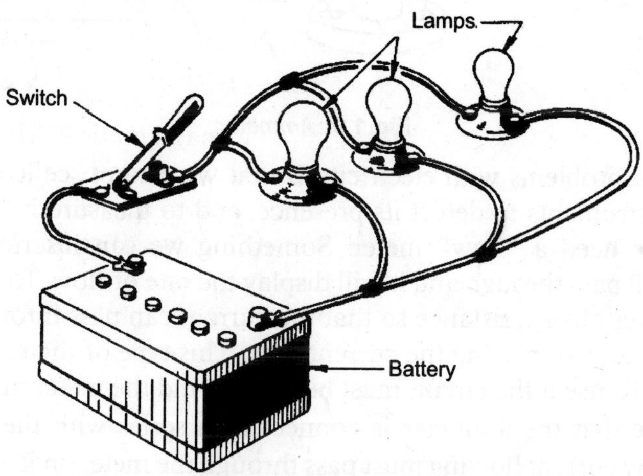


**Fig.1.8:** *Parallel circuit*

In a parallel circuit, there is more than one current path. The current splits up, with the greater current flowing through the smallest resistance. A breakage in any one path would not interfere with the operation of the remainder of the units in the circuit.

The total resistance in a parallel circuit is found by using the formula:

$$R_{total} = \frac{1}{\frac{1}{R_1} + \frac{1}{R_2} + \frac{1}{R_3}}$$

## DC MEASURING INSTRUMENTS

The three most commonly used instruments for measuring DC circuit values are the ammeter (a device for measuring current), the voltmeter (a device for measuring voltage) and the ohmmeter (a device for measuring resistance).

### AMMETER

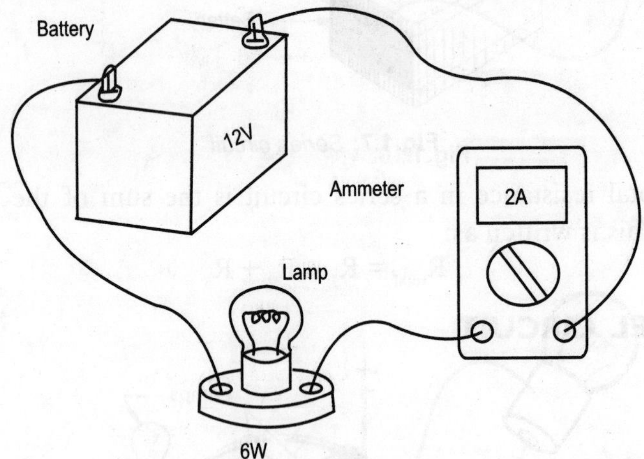


**Fig.1.9:** *Ammeter*

One of the problems with electricity is that we cannot see it. We have to rely on instruments to detect its presence, and to measure it. To measure current we need a "flow" meter. Something we can insert so that the current will pass through and it will display the rate of flow. To work, it has to have a very low resistance so that the current can pass through without the meter itself restricting the current flow. This type of meter is called an ammeter. To use it the circuit must be broken and the meter inserted.

Notice that the ammeter is connected in series with the rest of the circuit. Any current flowing must pass through the meter on its way around the circuit.

### VOLTMETER

When current flows through a resistance some of the supply pressure (voltage) is lost across that resistance. There is a pressure difference across the resistance. Therefore a voltage difference is found across components. To measure it the voltmeter has to be connected across the component (i.e. in parallel with it).

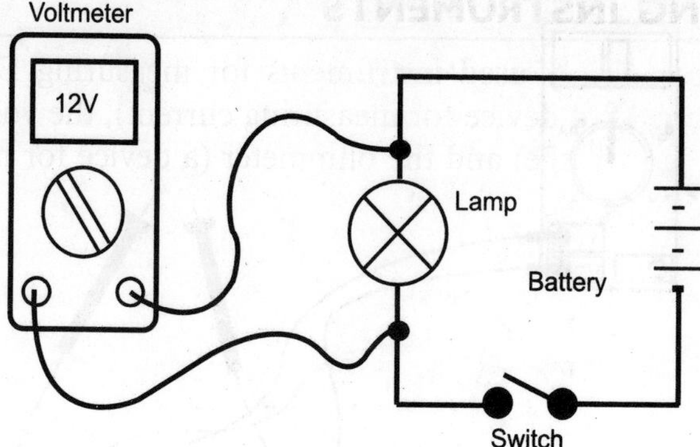


**Fig.1.10:** *Voltmeter*

## OHMMETER

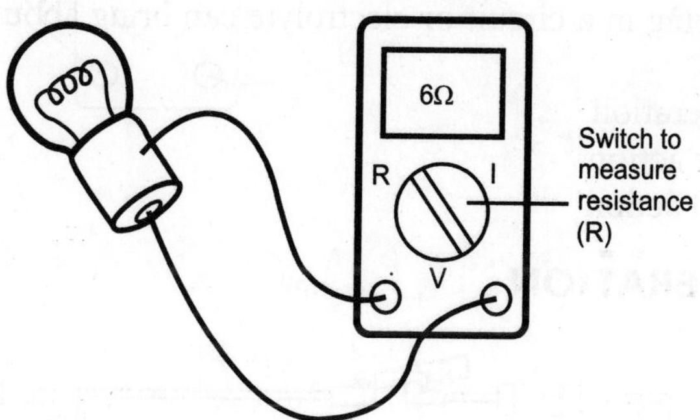


**Fig.1.11:** *Ohmmeter*

When measuring resistance the ohmmeter provides the voltage to pass a current through the components. From the current flowing, the meter works out the resistance. It is therefore important that there are no other currents flowing. The component to be measured must therefore be disconnected from its circuit first.

## MULTIMETER

Most meters used for measuring DC circuit values by technicians are multipurpose meters or multimeters. They are generally designed to measure voltage, resistance, and current. Therefore, a multimeter is often called a volt-ohm-milliammeter (VOM), because it can measure volts, ohms, and milliamperes.

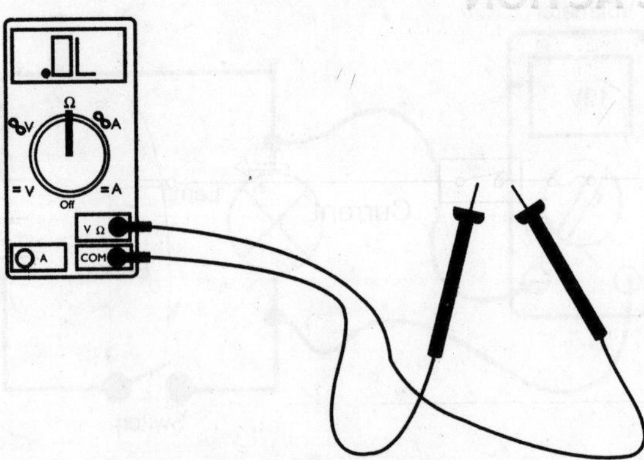


**Fig.1.12:** *Multimeter*

## ACTIONS OF ELECTRIC CURRENT

A current flowing in a circuit or electrolyte can bring about the following circuits:

1. Heat generation
2. Magnetic action
3. Chemical action

## HEAT GENERATION

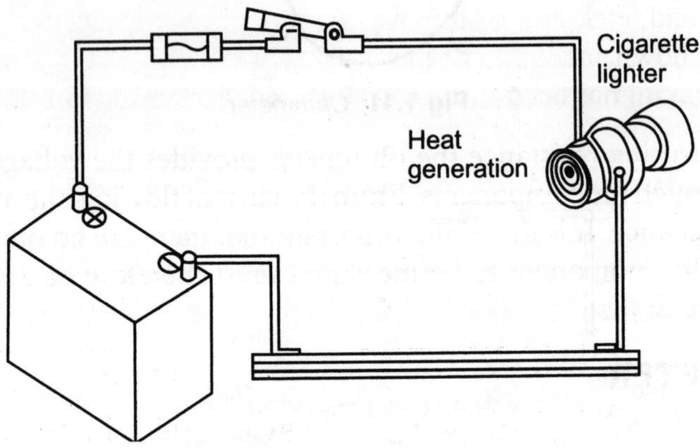


**Fig.1.13:** *Heat generation*

When current is applied to the cigarette lighter in an automobile, the nichrome wire in the cigarette lighter heats up and becomes red. This is because the electric energy is consumed by the nichrome wire's resistance and converted into heat energy. The action which generates heat when current flows in a resistance is called the heat generating action of current.

## MAGNETIC ACTION

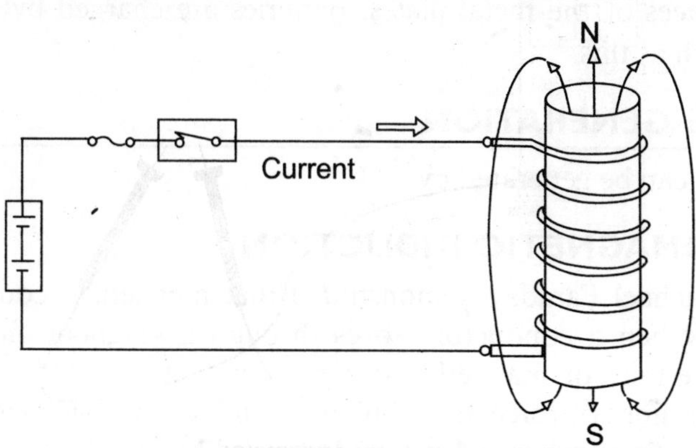


**Fig.1.14:** *Magnetic action*

When current flows in a solenoid as shown in Fig.1.14, the direction of the magnetic flux is such that the S pole is created under the solenoid, and the N pole above it. The number of magnetic lines of force also increases in proportion to the number of turns of the coil. When current flows in a solenoid, the number of magnetic lines of force increases in proportion to the strength of the current.

## CHEMICAL ACTION

When two metal plates are placed in an aqueous solution of salt or sulphuric acid, etc., and a lamp and battery are connected by wires to the plates, as shown in Fig.1.15, the lamp will light up. This proves that an electrical circuit has been formed, and current is flowing through the liquid solution.

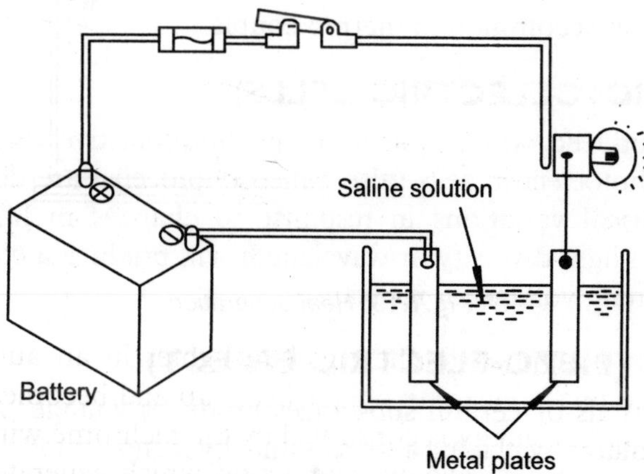


**Fig.1.15:** *Chemical action*

When electricity flows in such a solution, chemical action takes place on the surfaces of the metal plates. Batteries are charged by a chemical reaction such as this.

## VOLTAGE GENERATION

The voltage can be generated by:

### ELECTROMAGNETIC INDUCTION

In 1831, Michael Faraday demonstrated that magnetism could produce electricity. When a conductor moves through a magnetic field at right angles to the lines of magnetic force, the conductor is said to be cutting the magnetic lines of force and a voltage is induced in the conductor. The amount of voltage generated can be increased by increasing the strength of the magnetic field and by increasing the speed of the motion. Induced voltages are created by generators and alternators.

### ELECTROCHEMICAL CELLS

Electrochemical cells are the most compact sources of electric power. They produce electricity as a result of chemical actions inside the cell. When the cell is made, it is packed with chemical compounds that are ready to react together. As the current is drawn from the cell the chemical reaction occurs. Current can be obtained from the cell until the chemicals have completed their reaction and no more of the original compounds are left.

### HEAT (THERMO ELEMENT)

Voltage is produced by heating the joint (junction) where two unlike metals are joined. It may be noted that while doing so, one of the two metals is always maintained at lower temperature. This kind of voltage (emf) generation is common in thermo couple.

### LIGHT (PHOTOELECTRIC CELLS)

Voltage is produced when light striking photosensitive (light sensitive) substances. Photoelectric cells (also called photocells) are devices which produce electrical variations in response to changes in light intensity. Any change in light diversity or wavelength will produce a corresponding change in current, voltage, or resistance.

### PRESSURE (PIEZO-ELECTRIC EFFECT)

Squeezing crystals of certain substances produces voltage. The effect of producing a voltage by placing a stress, either by compression, by expansion or by twisting, on a crystal is called photoelectric effect.

## REVIEW QUESTIONS

1. What is an electric current?
2. What is an electromotive force?
3. What is electrical resistance?
4. Which is the source of electric energy in an automobile?
5. Name and describe two types of electricity.
6. Define a free electron.
7. What is a conductor?
8. What is an insulator?
9. What is the formula for electrical power?
10. How a resistance can be measured?
11. Define Ohm's law.
12. Define the following terms: (a) electric circuit, (b) volt, (c) ampere, (d) short circuit, (e) open circuit.
13. Explain the differences between a series and a parallel circuit.
14. Why is it important that all the lights on a vehicle be wired in parallel?
15. What effects may be produced by an electric current?
16. Explain why it is dangerous for a mechanic to wear rings, metal bracelets or metal watch straps when working on a vehicle.

UNIT 2

# Vehicle Electrical System

The electrical system in the present-day automobiles can be divided up into the storage battery which serves as the source of electric power, the ignition system which ignites the air-fuel mixture drawn in the cylinders, the charging system that charges the storage battery, the starting system that cranks and starts the engine, and the lighting and signalling system.

The five main headings in the electrical system of automobiles are:

1. Charging system
2. Starting system
3. Ignition system
4. Lighting system
5. Accessories

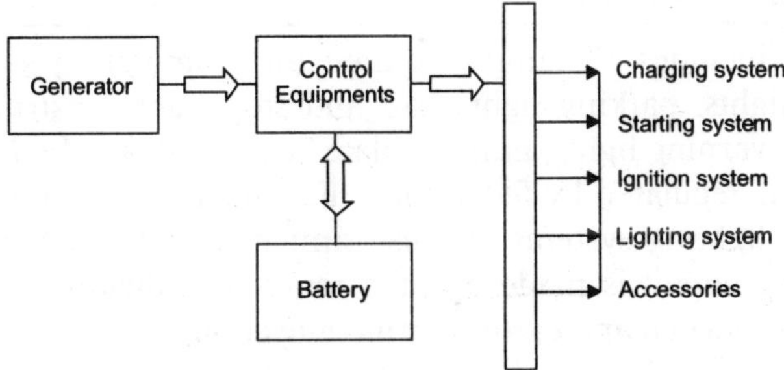


**Fig.2.1:** *Vehicle electrical system*

## CHARGING SYSTEM

The charging system includes the battery, the alternator and the regulator. This system is responsible to keep the battery charged for the regular supply of electric current to the various electrical systems in an automobile. It is not possible for the battery (due to its limited capacity) to supply electric current for long periods. So it is to be kept charged to save it from run down and thus to fulfil the requirements of electric supply to various electrical components. This problem is solved by means of a alternator (generator) which replaces the current used in the various systems of the automobile during its operation. After the engine has started, the alternator charges the battery because it produces a higher voltage than the battery. The regulators control the alternator voltage to prevent overcharging and to maintain the voltage at the proper level.

## STARTING SYSTEM

The internal combustion engines are not capable of self-starting. Early cars were cranked by hand, but as engines became larger, the starting system replaced the hand crank. This system cranks the engine when it is being started. Automotive engines are cranked by a small but powerful electric motor. The battery sends current to the starter motor when the driver turns the ignition switch to start. This causes a pinion gear in the starter motor mesh with teeth on the ring gear around the engine flywheel. Then the starter motor rotates the crankshaft for starting.

## IGNITION SYSTEM

The ignition system in spark ignition engines is used to ignite the air-fuel mixture that is sucked into the cylinders. A very high voltage around 20,000-30,000 volts is required to create the strong spark to ignite the mixture at the proper time. This high voltage is developed by ignition system.

## LIGHTING SYSTEM

This system includes all lamps required for illumination such as head lights, side lights, parking lights, tail and stop lights, instrument panel lights, brake warning lights, number plate lamps, cab and body lights, etc. This system is regulated by federal law. For this reason, all vehicles have standardised lighting systems and use similar circuit arrangements. The entire lighting circuit is made up of a series of individual circuits each having its own switch and earth return connection.

## ACCESSORIES

This system includes warning and safety devices such as horn, trafficator (flasher), windshield wiper, speedometer, etc.

## REVIEW QUESTIONS

1. What are the main circuits in an auto electrical system?
2. Draw the layout of vehicle electrical system and indicate its components.
3. Discuss the importance of electrical system in automobiles.

UNIT 3

# Battery

A storage battery is an electrochemical device that converts chemical energy into electric energy. It does not store electricity. The energy stored is chemical energy, which is transformed into electric energy when a circuit is completed across the terminals of the battery. There are two types of batteries used to produce electric energy: Dry cells (Primary cells) and Wet cells (Secondary cells).

### Dry cell (Primary cell)

The batteries which use electrolyte as paste are called dry cells or primary cells. These are used in flashlights, cameras, portable radios, etc. A major disadvantage of dry cells is that they are not rechargeable. Once the chemical energy is exhausted, the cell must be replaced.

### Wet cell (Secondary cell)

The batteries which use liquid electrolyte are called wet cells or secondary cells. Automotive batteries are of these types. A major advantage of wet cells is that they are rechargeable. Discharged wet cells may be recharged to their original state by forcing an electric current from some other source through the cell in the direction opposite to that of discharge.

## VEHICLE BATTERY

The battery is the heart of the electrical system of an automobile. An automotive battery is an electrochemical secondary cell capable of producing electric energy. The automotive battery produces direct current electricity that flows in only one direction. When discharging (current flowing from the battery), the battery changes chemical energy into electrical energy. It means battery releases stored energy. During charging (current flowing through the battery from the charging system), electrical energy is converted into chemical energy.

The batteries used in automotive applications are lead-acid battery, nickel-alkaline battery, nickel-iron battery, nickel-cadmium battery, aluminium-air battery, zinc-air battery, sodium-nickel chloride battery, sodium-lithium battery and sodium-sulphur battery. Although there are many types of storage batteries, the lead-acid is most commonly used for automotive service because it has been found to be the most practical and economical type. This battery can be charged and discharged continuously throughout its life.

## FUNCTIONS

The main functions of a battery in automobile are as follows:

### Engine is off

Electricity from the battery is used to operate lighting, accessories, or other electrical systems when the engine is not running.

### Engine is starting

Electricity from the battery is used to operate the starter motor and to provide current for the ignition system during engine cranking. Starting the car is the battery's most important function.

### Engine is running

Electricity from the battery may be needed to supplement the charging system when the vehicle's electrical load requirements exceed the charging system's ability to produce electricity. Both the battery and the alternator supply electricity when demand is high.

## PRINCIPLE

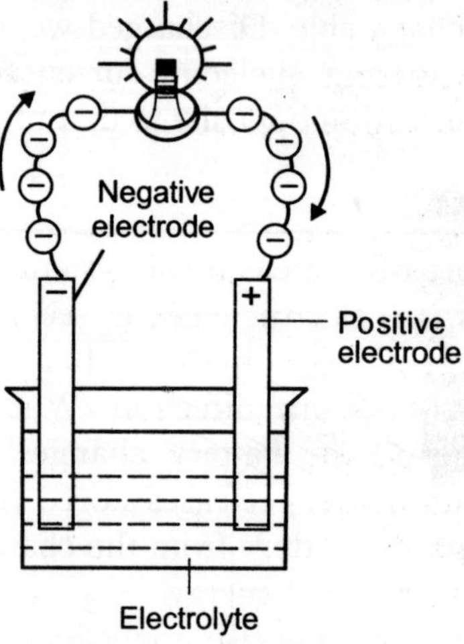


**Fig.3.1:** *Principle of a battery*

Electrons are moved because of the chemical reaction between the electrolyte and the two kinds of plates. The electrolyte allows electrons to move away from one kind of plate and build up on another. As a result, one kind of plate will have more than the normal number of electrons. The other plate is with fewer electrons. In the plate with fewer electrons, the positive protons will predominate giving that plate a positive charge. The plate with more electrons will have a negative charge.

## CONSTRUCTION OF A LEAD-ACID BATTERY

All Lead-acid batteries are similar in construction and operation but their difference lies in the number of cells, the arrangement of the cells, the number of plates in each cell, the size and thickness of the plates and the type of separators used. Automotive batteries consist of:

1. Cells
2. Grid and plates
3. Plate groups
4. Separator
5. Element
6. Electrolyte
7. Case (or) container
8. Cell cover
9. Vent plug
10. Terminal posts

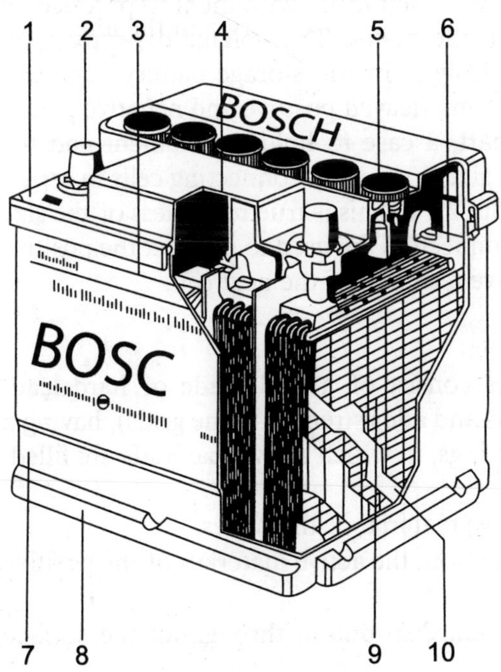


1. Cover
2. Post
3. Electrolyte level mark
4. Intercell connector
5. Vent plug
6. Plate strap
7. Battery case
8. Bottom rail
9. Positive and negative plates
10. Separators

**Fig.3.2:** *Parts of a lead-acid battery*

## CELLS

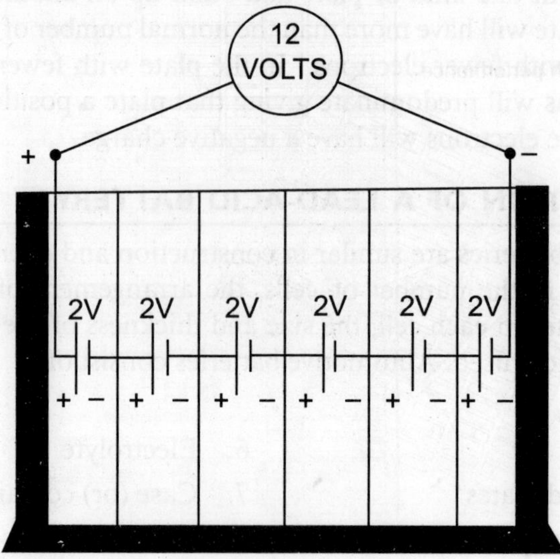


**Fig.3.3:** *Battery cell*

A battery cell basically consists of plates of two different materials immersed in an electrolytic solution that reacts chemically with them to produce an electrical pressure (emf) between the plates. The voltage depends upon the plate materials and the electrolyte. In the storage battery each cell includes an element composed of interleaved positive and negative plates, separators to keep the plates apart, a case to hold the element and the electrolyte. The required battery size is made by connecting cells in series. A fully charged cell will have about 2.1V. This is true regardless of the size or the electrolyte strength. Cell connectors are used to connect the positive terminal of one cell to the negative terminal of the next cell.

## GRIDS

The positive and negative plates consist of a grid made of hard lead-antimony alloy (antimony stiffens and adds strength to the grids), having a large number of rectangular openings, in which active materials are filled. Grid serves as:

1. to hold and support the active material of the plates.
2. to conduct the current to and from the active materials of the positive and negative plates.
3. to maintain a uniform current distribution throughout the surface area of the active materials.

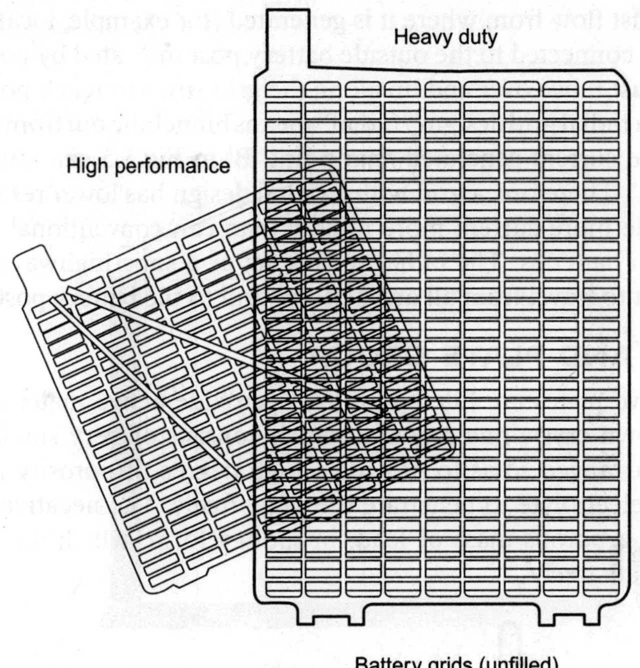


**Fig.3.4:** *Grids*

The active materials, which make chemical reaction to produce electricity, are stored on the plates in the form of paste. The positive plate's grids are filled with a lead peroxide paste, and negative plate grids are filled with spongy lead.

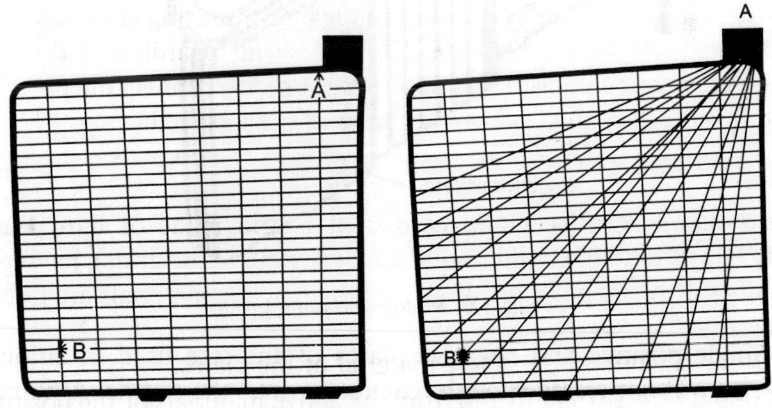


**Fig.3.5:** *Grid design*

Some batteries use a grid design with only vertical and horizontal strips. The battery plate creates electrical energy from chemical energy, and this current must flow from where it is generated (for example, location 'B') to where it is connected to the outside battery post indicated by point A. The current must move over and up along the grid strips to reach point 'A'.

With a radial grid design ("radial" means branching out from a common center), the current generated near point 'B' in Fig.3.5 can travel directly to point 'A'. Therefore, a grid with a radial design has lower resistance and can provide more current more rapidly than can conventional non-radial grid design batteries. The radial spokes act as a superhighway system for the current to travel from all areas of the grid to the battery post.

## PLATES AND PLATE GROUP

The positive plates are filled with lead peroxide active materials. This is a dark brown crystalline material which consists of very small grains or particles, arranged so as to provide a high degree of porosity in order to allow the electrolyte to penetrate the plate freely. The negative plates are filled with a porous mass of lead, in spongy form, which the electrolyte can penetrate freely.

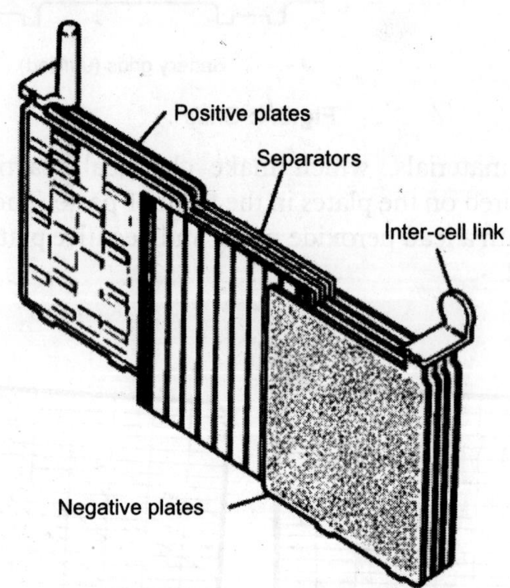


**Fig.3.6:** *Plates and plate group*

Single positive and negative plates of the size used in automobile batteries will not create enough reaction to start the engine. A number of plates are connected together by their lug ends to form a plate group. Each group is made by welding, called *lead burning*, the plate lugs to a lead alloy plate strap. Thus each cell is composed of a positive plate group and a negative plate group. The negative group always contains one more plates than the positive plate group so that there is always a negative plate on both sides of each positive plate. So the chemical action is distributed evenly on both sides of the positive plate and there is less tend for the plate to warp.

## SEPARATORS

Between the positive and negative plates, an inert separator is fixed so that the plates cannot touch one another and give up their stored energy. The functions of separators are to:

- insulate positive and negative plates from physical contact
- act as packaging material
- provide passage for transport of irons

The positive and negative plates are always interleaved and separators are placed between each plate. The plates insulated by the separators electrically and the separators are porous for the passage of electrolyte between them. One face of the separator is grooved vertically to allow for free circulation of electrolyte within the cell and to permit loosen active material to fall to the sediment chamber at the bottom of the battery case.

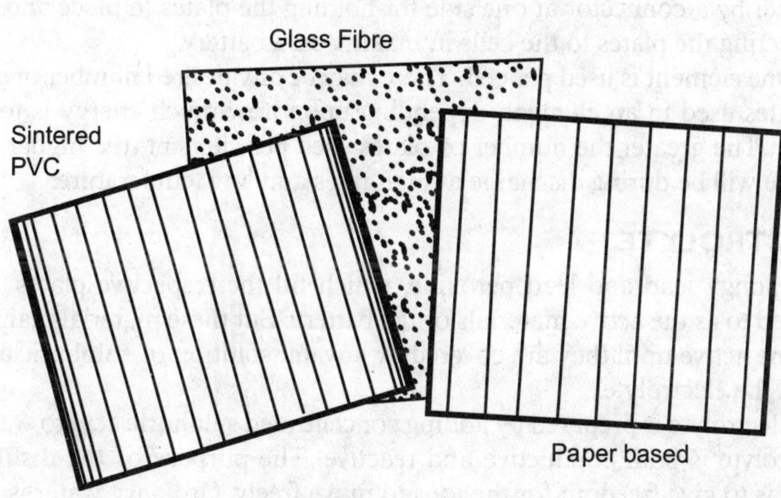


**Fig.3.7:** *Separators*

Separators are non-electrical conducting materials that resist heat and acid. Treated wooden separators were used in the past, however, resin impregnated cellulose fibre types are used nowadays. Micro-porous rubber and plastic are used alone or a combination with glass fibre mats and also flat glass sheets with a micro-porous backing are used.

## ELEMENT

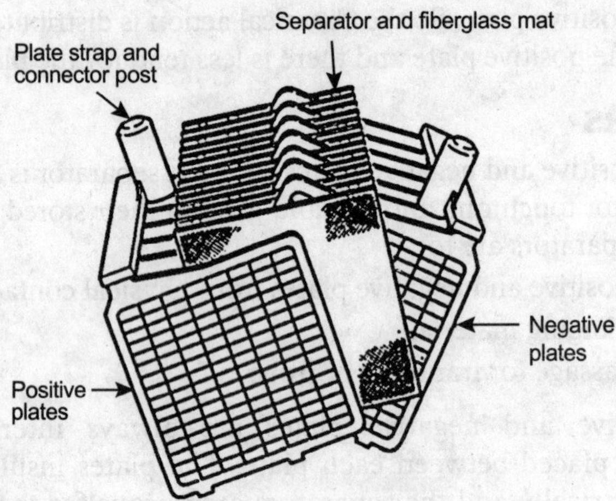


**Fig.3.8:** *Element*

An element is formed by alternately arranging a positive and a negative plate group. Separators are then inserted between the plates with the grooved faces of the separators next to the positive plates. Each plate has a separator to insulate the adjacent plates and each group of plates is held together by a connector at one side for holding the plates in place and for connecting the plates to the cells in making up a battery.

One element is used per cell. There may be any desired number or size of plates used in an element, depending upon how much energy is to be stored. The greater the number of plates used per element, the higher the voltage will be during discharge at high rates and low temperature.

## ELECTROLYTE

The spongy lead and lead peroxide, which fill the respective plates, are referred to as the active materials of the battery. But these materials cannot become active until they are covered by a water solution of sulphuric acid called the electrolyte.

Electrolyte is prepared by adding concentrated sulphuric acid to water. Electrolyte is both conductive and reactive. The purpose of the distilled water is to give freedom for the ions to move freely. Ordinary water is not used in an electrolyte because:

- the impurities in an ordinary water may short circuit the cells
- it may make the battery sulphated due to hardness

The electrolyte reacts chemically to combine with the active materials on the plates to form new compound and release electrons at the same time.

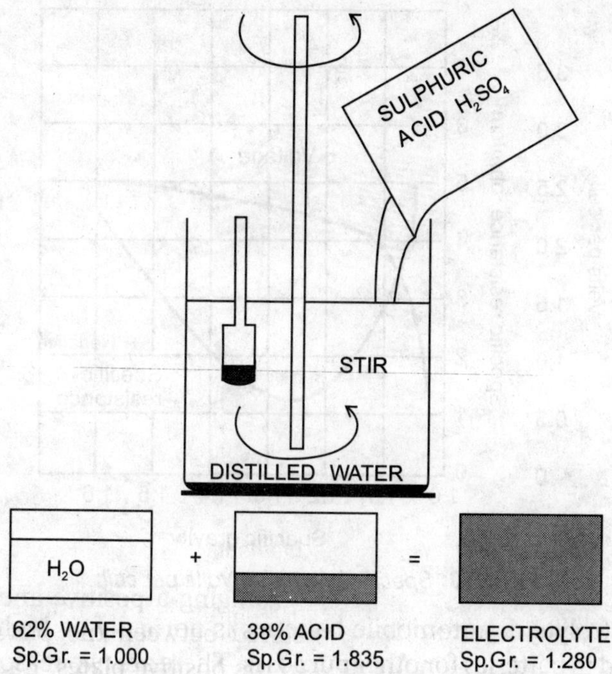


**Fig.3.9:** *Preparation of electrolyte*

· The amount of sulphate in the electrolyte is determined by its specific gravity. Specific gravity is the ratio of the weight of a given volume of a liquid divided by the weight of an equal volume of water. In other words, the more dense the material (liquid), the higher its specific gravity. Pure water is the basis for this measurement and is given a specific gravity of 1.000 at 80°F. Pure sulphuric acid has a specific gravity of 1.835. The correct concentration of water and sulphuric acid is usually contains about *38% sulphuric acid and 62% water by weight* (fully charged condition). This corresponds to a specific gravity of about 1.280 at 80°F. The following table shows the specific gravity of batteries used in different climates.

| State of charge | Specific gravity | | |
|---|---|---|---|
| | In cold climate | In temperate climate | In tropical climate |
| Fully charged | 1.280 | 1.260 | 1.225 |
| 75% charged | 1.260 | 1.215 | 1.180 |
| 50% charged | 1.220 | 1.170 | 1.135 |
| 25% charged | 1.190 | 1.120 | 1.090 |
| Discharged | 1.110 | 1.070 | 1.045 |

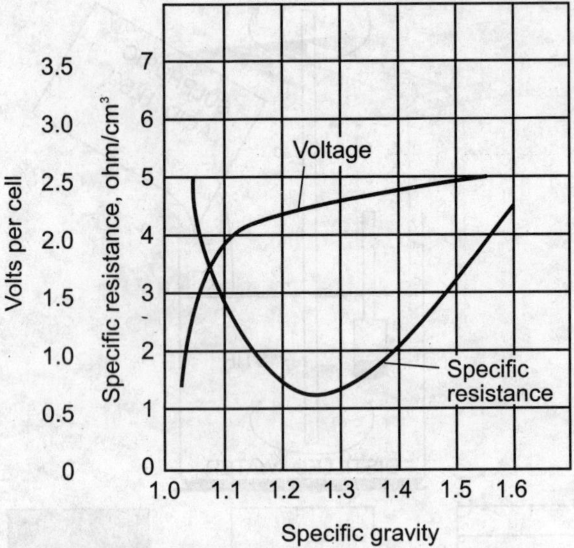


**Fig.3.10:** *Specific gravity Vs Volts per cell*

The electrolyte in automobile batteries is normally of higher density than that used in batteries for other purposes. There are two reasons for this, the first being that the volume of electrolyte is limited by consideration of battery size and weight and secondly starter batteries are frequently called upon to give very heavy discharge current. Normally, the specific gravity of the electrolyte in a cell is kept between 1.11 and 1.28 because under this range specific resistance (internal loss of current due to self-discharge) is minimum and the voltage is nominal. At high specific gravity (above 1.28) voltage is more and specific resistance is also more. At low specific gravity (below 1.11) voltage is low however specific resistance is again too high. That is why, specific gravity is maintained in between 1.11 and 1.28.

The electrolyte in a fully-charged battery does not freeze until very low temperatures are reached (about -65°C). A discharged battery has a weak electrolyte (mainly water) and so freezes more easily (-5°C). The electrolyte does not go entirely solid so permanent damage is unlikely if frozen and it will recover to work quite normally.

## CASE (CONTAINER)

The battery case is a one-piece moulded box with open at the top, made from hard rubber or plastic material, which is shock and acid resistant, and mechanically strong. They must also withstand extremes of heat and cold. Case has as many as the number of cell in the battery. In the bottom of each cell of a case, there are four-moulded narrow elements rest (or bridge) on which the elements sits and sediment chamber to collect the materials that slowly wear away from the plates and thus prevent short-circuiting.

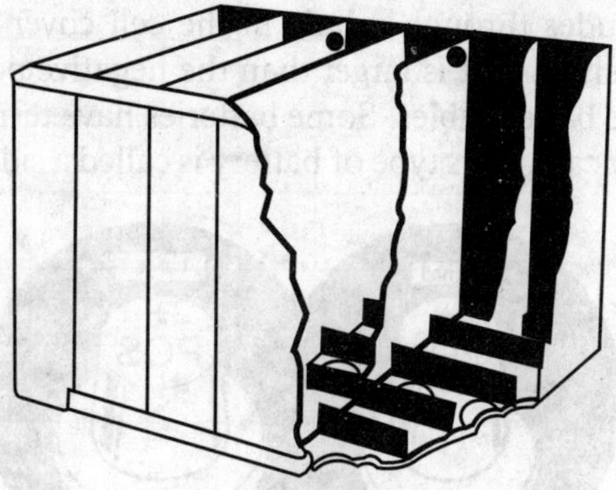

Fig.3.11: *Battery case*

## CELL COVER AND VENT PLUG

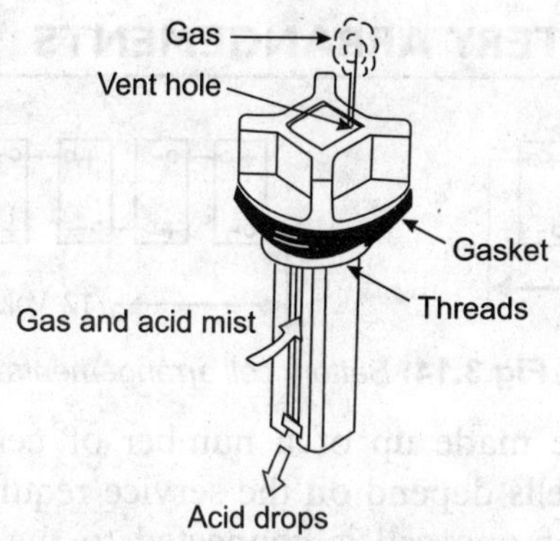


Fig.3.12: *Cell cover and vent plug*

Cell cover is also made of hard rubber. It has holes for terminal posts and for vent plug. Vent plugs provided in the cell cover facilitate the checking of electrolyte and pouring distilled water. Vent plugs either push into the plain holes or screw into threaded holes in the cell cover. The plugs can be removed so that water can be added to the cell when necessary.

Water is lost during battery charging. Some evaporates and some is converted into hydrogen and oxygen gases. The plug has a vent hole to permit these gases to escape and prevents battery explosion. Perforated baffle plates are usually included in the plug. When the battery is being charged, the outgoing gases will often contain droplets of electrolyte. The baffle plates tend to trap these droplets so that the electrolyte is returned to the cell instead of escaping with the gases.

## TERMINAL POSTS

The battery has top mounted terminal posts. The plate strap has a round post which protrudes through a hole in the cell cover and becomes cell terminal. The positive post is larger than the negative post to help correct installation of the batter cables. Some batteries have terminal posts on the side of the battery case. This type of battery is called a side terminal battery.

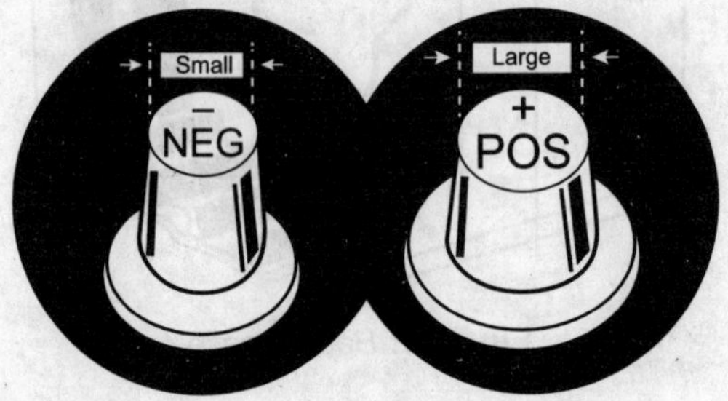


**Fig.3.13:** *Terminal post*

## CELL AND BATTERY ARRANGEMENTS

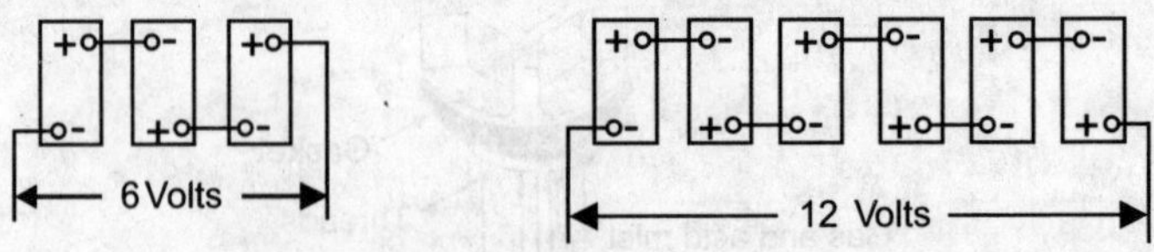


**Fig.3.14:** *Battery cell arrangements*

Storage batteries are made up of a number of cells. The number and dimensions of the cells depend on the service required from the battery. The negative post of one cell is connected to the positive post of the adjacent cell (in series). All the cells are connected in this way. A negative post and a positive post are left for connecting the battery cables.

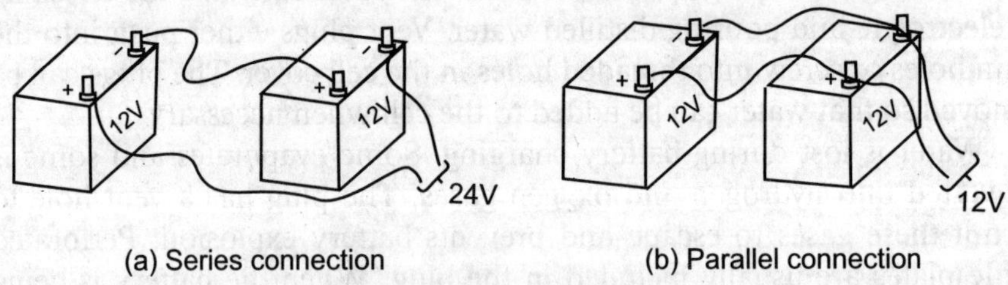


**Fig.3.15:** *Battery connections*

Voltage of the two batteries can be added if they are connected in series. For example, two 12V batteries connected in series provides a 24V supply. Some starting systems used with diesel engines have two batteries. The batteries are connected in parallel to provide the higher current required to crank a diesel engine. Two batteries connected in parallel provide the same voltage but can deliver larger currents. For example, two 12V batteries connected in parallel provides same 12V supply but this arrangement will double the Ah capacity.

## ELECTROCHEMICAL ACTION

## CHEMICAL REACTION

⇒ **Discharging**

| Lead Peroxide | Sulphuric Acid | Spongy Lead | Lead Sulphate | Water | Lead Sulphate |
|---|---|---|---|---|---|
| $PbO_2$ | $+ 2H_2SO_4$ | $+ Pb$ ⇔ | $PbSO_4$ | $+ 2H_2O$ | $+ PbSO_4$ |
| + plate | Electrolyte | – plate | + plate | Electrolyte | – plate |

⇐ **Charging**

### Discharge process

The term discharge means to use a battery to provide electric energy. If the two terminals of a lead storage cell are connected to each other through an electric load, electrons flow from the negative electrode through the load to the positive electrode because of the difference in potential. When a battery is being discharged, electrons (given the symbol $e^-$) leave the negative terminal. The battery produces electrons by means of chemical reactions. In the reactions the following changes take place:

- Sulphuric acid breaks down into irons: positive hydrogen ions and negative sulphate ions.
- Lead molecules change into positive lead ions and in doing so release electrons. The electrons build up on the spongy lead plates, which become the negative plates of the battery.
- Lead peroxide also breaks down into positive lead ions and oxygen. The oxygen reacts with the positive hydrogen irons to form water. When water is formed, the positive charges of the hydrogen ions draw electrons away from the lead peroxide plates. Because electrons are removed from the lead peroxide plates, they become positively charged and become the positive plates of the battery.
- The positive lead ions react with the sulphate ions to form lead sulphate.

As the battery discharges, the plate materials and the sulphuric acid are decreasing in quantity. The amount of water is increasing. If the discharge reactions were allowed to continue, the battery would become completely discharged. Much of the active material in the positive and negative plates would become lead sulphate. The electrolyte would become very weak in acid. The lead sulphate coats the plates and becomes a barrier to further chemical action. When the electrolyte is weak in acid, it is no longer effective as a supply of ions. Therefore, the battery no longer produces electric current.

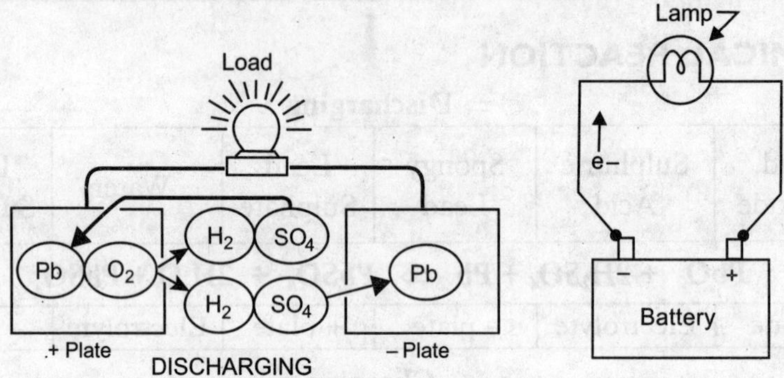


**Fig.3.16:** *Discharging process*

## Charging process

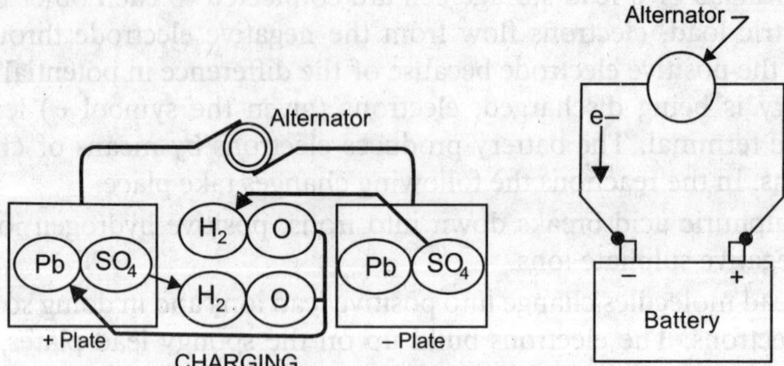


**Fig.3.17:** *Charging process*

To charge a battery, electrons must be made to flow into the negative terminal. This reverses the chemical reactions. In the automobile, the alternator charges the battery by forcing a flow of electrons into the negative terminal of the battery. In the charging reactions, the following changes occur:

- Lead sulphate on the plates reacts with the water in the electrolyte to form sulphuric acid.
- Lead sulphate on the plates is also converted into spongy lead on the negative plates and lead peroxide on the positive plates.
- Water is used up.

As the battery charges, the plate materials are recovered. At the same time, the electrolyte is increasing in acid content and decreasing in water content.

## ALKALINE BATTERY

Lead-acid batteries require a considerable amount of servicing to keep them in good condition, although this is not applicable with sealed and maintenance-free batteries. However, when a battery is required to withstand a high rate of charge and discharge on a regular basis, or is left in a state of discharge for long periods, the lead-acid cell is not ideal. Alkaline cells on the other hand require minimum maintenance and are able to withstand heavy discharge and over-charging. The disadvantages of alkaline batteries are that they are more bulky, have lower energy efficiency and are more expensive than a lead-acid equivalent. Bus and coach companies and some large goods vehicle operators have been using alkaline batteries.

Nickel-cadmium (NiCd) and nickel-metal-hydride (NiMH) batteries are alkaline batteries with which electrical energy is derived from the chemical reaction of a metal with oxygen in an alkaline electrolyte medium.

## NICKEL-CADMIUM BATTERY

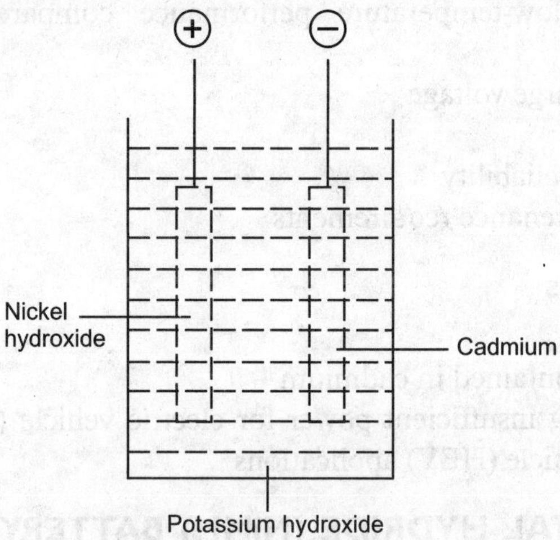


**Fig.3.18:** *NiCd alkaline battery cell*

The main components of the nickel-cadmium cell for vehicle use are as follows:

- Positive plate – Nickel oxide hydroxide (NiOOH)
- Negative plate – Cadmium (Cd)
- Electrolyte – Potassium hydroxide (KOH) and water ($H_2O$)

The process of charging involves the oxygen moving from the negative plate to the positive plate, and the reverse when discharging. When fully charged, the negative plate becomes pure cadmium and the positive plate becomes nickel oxide hydroxide. Fig.3.18 shows a simplified representation of a NiCd battery cell.

A simplified chemical equation to represent this reaction is given below.

$$2NiOOH + Cd + 2H_2O + KOH \longleftrightarrow 2Ni(OH)_2 + CdO_2 + KOH$$

The $2H_2O$ is actually given off as hydrogen (H) and oxygen ($O_2$) as gassing takes place all the time during charge. The electrolyte does not change during the reaction. This means that a relative density reading will not indicate the state of charge. These batteries do not suffer from over-charging because once the cadmium oxide has changed to cadmium, no further reaction can take place.

The cell voltage of a fully charged cell is 1.4V but this falls rapidly to 1.3V as soon as discharge starts. The cell is discharged at a cell voltage of 1.1V. The specific energy of NiCd batteries is 30 to 50 Wh/kg, which is similar to that of lead-acid batteries.

### Advantages

1. Superior low-temperature performance compared to lead-acid batteries
2. Flat discharge voltage
3. Long life
4. Excellent reliability
5. Low maintenance requirements

### Disadvantages

1. High cost
2. Toxicity contained in cadmium
3. Developing insufficient power for electric vehicle (EV) and hybrid electric vehicle (HEV) applications

## NICKEL METAL HYDRIDE (NiMH) BATTERY

The drawbacks of the NiCd batteries are the reasons for the development of NiMH batteries, which are deemed more suitable for EV and HEV applications.

The nickel-metal-hydride battery is a successor to the nickel-hydrogen battery and is already in use in production HEVs. In NiMH batteries, the positive electrode is a nickel hydroxide similar to that used in a NiCd battery, while the negative electrode is a metal hydride where hydrogen is stored. The concept of NiMH batteries is based on the fact that fine particles of certain metallic alloys, when exposed to hydrogen at certain pressures and temperatures, absorb large quantities of the gas to form the metalhydride compounds. Furthermore, the metal hydrides are able to absorb and release hydrogen many times without deterioration. The two electrode chemical reactions in a NiMH battery are:

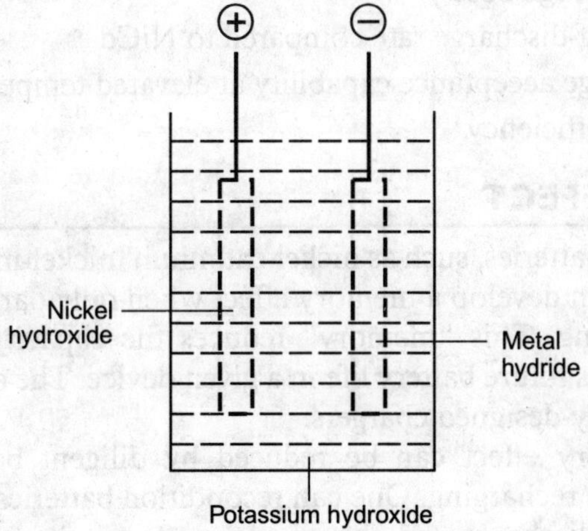


**Fig.3.19:** *Nickel metal hydride (NiMH) battery*

At the positive electrode,

$$NiOOH + H_2O + e^- \underset{\text{Charge}}{\overset{\text{Discharge}}{\rightleftarrows}} Ni(OH)_2 + OH^-$$

At the negative electrode,

$$MH_X + OH^- \underset{\text{Charge}}{\overset{\text{Discharge}}{\rightleftarrows}} MH_{X-1} + H_2O + e^-$$

"M" stands for metallic alloy, which takes up hydrogen at ambient temperature to form the metal hydride $MH_X$. The negative electrode consists of a compressed mass of fine metal particles. NiMH cells can last 40% longer than the same size NiCd cells and will have a life-span of up to 600 cycles. This makes them useful for high energy devices such as laptop computers, cellular phones, and camcorders.

## Advantages

1. Charge up quickly
2. High cycle stability

3. 20% higher specific energy and power, and in volumetric terms 40% higher
4. Longer life cycle than lead-acid batteries
5. Safe and abuse tolerant.

### Disadvantages

1. Relatively high cost
2. Higher self-discharge rate compared to NiCd
3. Poor charge acceptance capability at elevated temperatures
4. Low cell efficiency.

## MEMORY EFFECT

Some types of batteries, such as nickel-cadmium, nickel-metal hydride and Lithium-ion, can develop a memory effect when only partially discharged before recharging. This "memory" reduces the capacity of subsequent charges and thus future battery life in a given device. The effect can also be caused by poorly-designed chargers.

The memory effect can be reduced by diligent battery use: fully discharging and recharging. One can recondition batteries with a memory effect by deep discharge and charge cycles. Some chargers have specific battery reconditioning modes as well.

The other terms used for memory effect are voltage depression of a battery, battery effect, lazy battery effect or battery memory.

## SODIUM-SULPHUR BATTERY

Fig.3.20 shows a representation of a sodium-sulphur battery cell. The typical sodium sulphur battery consists of a cathode (negative) molten sodium electrode and also a molten sulphur anode (positive) electrode. The two are separated by a layer of beta-alumina ceramic solid electrolyte that primarily only allows sodium ions through. The charge and discharge process can be described by the chemical equation,

$$2Na + 4S \leftrightarrow Na2S4$$

In the discharge process, the two elements combine to form sodium polysulphides but in the charging process, the sodium ion is released back through the electrolyte. The discharge process produces roughly 2 Volts.

The major problem with this system is that the running temperature needs to be 300-350°C. A heater rated at a few hundred watts forms part of the charging circuit. This maintains the battery temperature when the vehicle is not running. Battery temperature is maintained when in use due to internal resistance ($I^2R$ losses) which produces sufficient heat for the electrode in the battery.

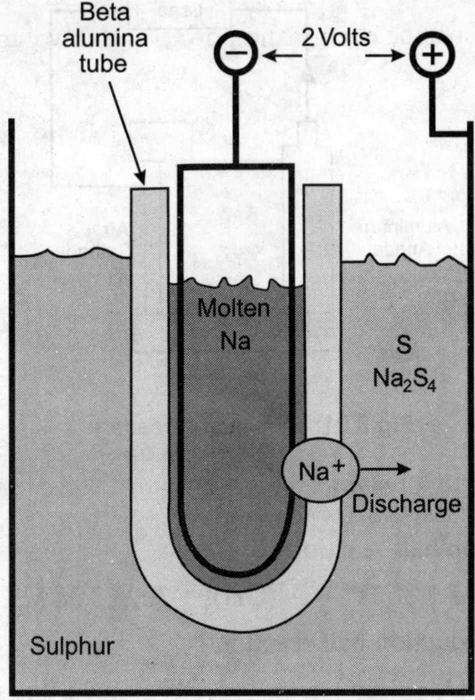


**Fig.3.20:** *Sodium-sulphur battery*

### Advantages

1. High cycle life
2. Potential low cost
3. Flexible operation
4. Good power density
5. High energy efficiency
6. State-of-charge identification
7. Insensitivity to ambient conditions

### Disadvantage

1. Requires a heat source for operational conditions

## ALUMINUM-AIR BATTERY

The aluminum-air battery has been developed as a power source for electric vehicles. It consists of an aluminum alloy anode and an air cathode. The air cathode comprises a semi-porous mixture of carbon, wet proofing agents, and catalyst on a metal screen current collector. An aqueous sodium or potassium hydroxide solution serves as the electrolyte.

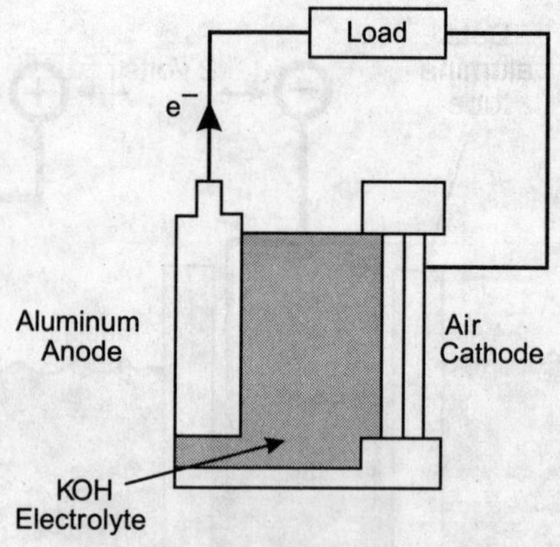


**Fig.3.21:** *Aluminum-air battery*

### Chemical Equation

The anode oxidation half-reaction is

$$Al + 3OH^- \rightarrow Al(OH)_3 + 3e^- - 2.31\ V$$

The cathode reduction half-reaction is

$$O_2 + 2H_2O + 4e^- \rightarrow 4OH^- + 0.40\ V$$

The balanced equation is

$$4Al + 3O_2 + 6H_2O \rightarrow 4Al(OH)_3 + 2.71\ V$$

Aluminum plates are added every 300 km to replace the used aluminum. The aluminum plates react with oxygen in a sodium hydroxide electrolyte solution to form sodium aluminate. The sodium aluminate produces a by-product of aluminum, aluminum trihydroxide, which is removed and replaced with fresh aluminum. Aluminum-air batteries will probably be used by larger vehicles because of their size.

## LITHIUM-ION BATTERY

The use of Lithium batteries in cars, motorcycles, heavy vehicles and off-road equipment is increasing rapidly. Lead Acid batteries are being replaced by Lithium batteries to improve charge/discharge performance, save weight, save space, increase lifetime and avoid Pb (lead) in the environment.

This battery is now used primarily for electric vehicles, as they are much lighter which provides a more fuel efficient vehicle. The Porsche 911 2010 model uses the lithium ion battery for starting. These are very expensive batteries, and starting costs for cars are about Rs.60,000.

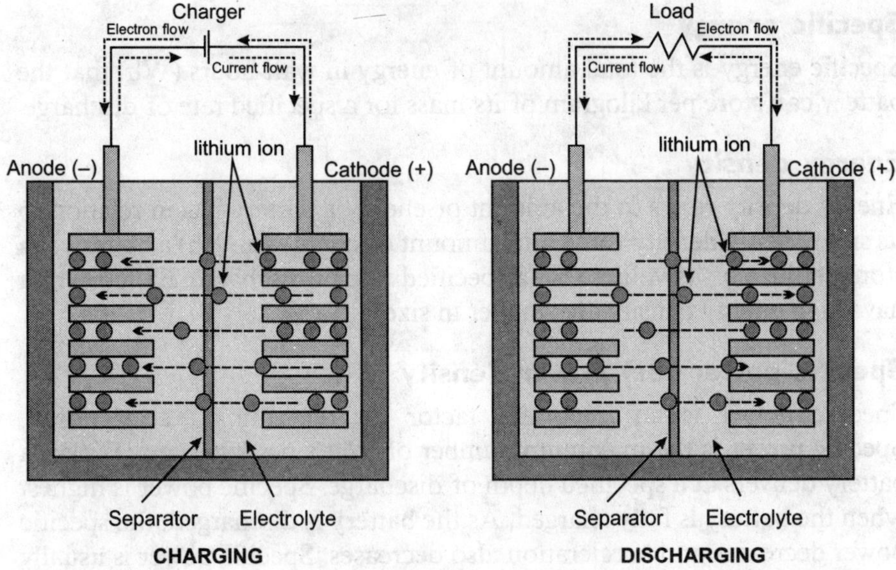


**Fig.3.22:** *Lithium-ion battery*

Lithium-ion cells, in their most common form, consist of a graphite anode, a lithium metal oxide cathode and an electrolyte of a lithium salt and an organic solvent. Lithium is a good choice for an electrochemical cell due to its large standard electrode potential (3.04 V) resulting in a high operating voltage (which helps both power and energy) and the fact that it is the metal with the lowest density (which reduces weight). In an automotive application a lithium-ion battery consists of tens to thousands of individual cells packaged together to provide the required voltage, power and energy.

### Advantages over conventional batteries

1. Up to 80% weight saving
2. Charge and discharge at high currents
3. Offer phenomenal starting power and massive deep cycle reserve power. They are capable of providing engine cranking pulses of 2250 amps for 5 seconds.
4. Longer life span with a minimum of 5 years' service life, up to 10 years depending on use.

## PERFORMANCE PARAMETERS OF A BATTERY

Below is a short description of performance parameters which are important in determining which battery should be used for a particular application.

### Specific energy

Specific energy is the total amount of energy in watt-hours (Wh) that the battery can store per kilogram of its mass for a specified rate of discharge.

### Energy density

Energy density refers to the amount of energy a battery has in relation to its size. Energy density is the total amount of energy (in Wh) a battery can store per litre of its volume for a specified rate of discharge. Batteries that have high energy density are smaller in size.

### Specific power (or) Power density

Specific power is an important factor for determining acceleration. Specific power is the maximum number of Watts per kilogram (W/kg) a battery delivers at a specified depth of discharge. Specific power is highest when the battery is fully charged. As the battery is discharged, the specific power decreases and acceleration also decreases. Specific power is usually measured at 80% depth of discharge.

### Cycle life

Cycle life is the measure of the total number of times a battery can be discharged and charged during its life. When the battery can no longer hold a charge over 80%, its cycle life is considered to be finished. This is typically between 500 and 1200 cycles.

### Shelf life

The battery shelf life is the time a battery can be stored inactive before its capacity falls to 80%. The reduction in capacity with time is caused by the depletion of the active materials by undesired reactions within the cell.

## CHARACTERISTICS OF A BATTERY

The following battery characteristics must be taken into consideration when selecting a battery.

## DISCHARGE CHARACTERISTICS

The discharge curve is a plot of voltage against percentage of capacity discharged. A flat discharge curve is desirable. This means that the perfect battery during discharge has constant voltage at the terminals. When depth of discharge (DOD) is 100%, the voltage drops suddenly to zero. Fig.3.23 has a perfect battery curve. The curves A, B, and C give terminal voltage as the battery is being discharged. Curve A has a modest current with curve B at higher value. Curve C is for a very large current and very rapid discharge. As the voltage curves bend downwards, the useful state of charge (SOC) is decreased. The curves decline mainly because of battery internal resistance (impedance).

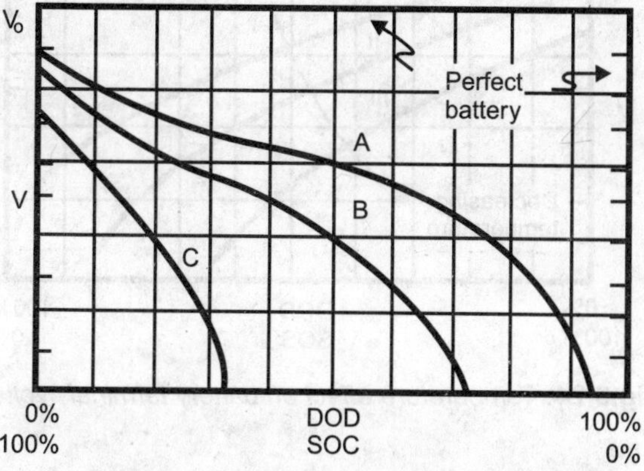


**Fig.3.23:** *Battery terminal voltage on the SOC*

If a battery is discharged at a constant rate to the final discharged voltage, the capacity will be higher when the electrolyte temperature is higher and lower when the temperature is lower.

## TEMPERATURE CHARACTERISTICS

The process of charging and discharging generates heat within the battery. As a result, during charging and discharging, temperature has a direct effect on the performance of a lead-acid battery. The concentration of sulphuric acid inside the battery increases and decreases with temperature. A battery being used in 32°F (0°C) weather will only operate at 70% of its capacity. The most efficient temperature that battery manufacturers recommend is 80°F (26.7°C).

### Effect of high temperature

Some batteries operate best at higher temperatures but conditions that are too hot can shorten battery life. A very hot environment may cause rapid deterioration of the electrolytes. This is because some of the chemicals in the electrolyte may decompose at high temperature.

Battery voltage for different temperatures is shown in Fig.3.24. Each curve is for the same discharge current. As the temperature of a metal conductor, for example copper wire, increases, the electrical resistance increases. In contrast, most electrolytes have decreasing resistance with increasing temperature. This results because the electrolyte can diffuse more easily when its temperature is high, allowing more active materials to react chemically with it. As a result, the battery voltage increases. Curve I has a higher voltage than curve III. For curve I, the ohmic loss (current-resistance loss) is less since resistance is less at elevated temperature.

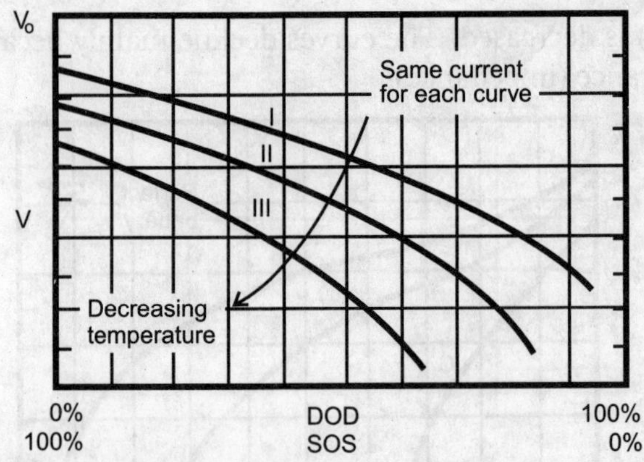


**Fig.3.24:** *Temperature effect on battery terminals voltage*

### Effect of low temperature

At low temperatures, the electrolyte will be more concentrated and will have a higher specific gravity, but against this, the chemical reactions within the battery are slower and the overall result is a lowering of the battery capacity with temperature as shown in Fig.3.25.

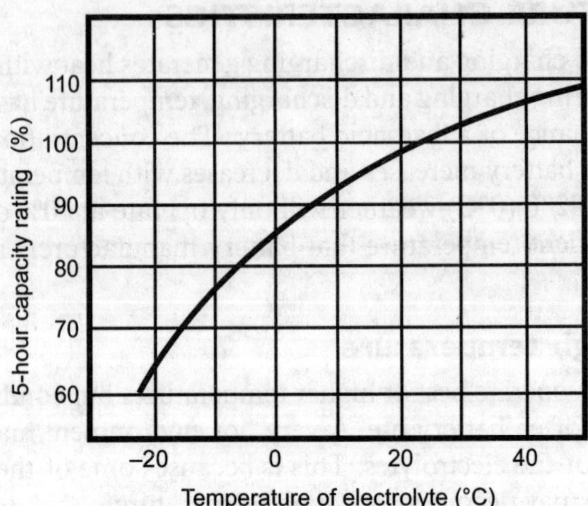


**Fig.3.25:** *Battery capacity with electrolyte temperature*

## FREEZING CHARACTERISTICS

An environment that is too cold may cause freezing of the electrolyte. The battery should be protected from freezing, particularly when in a low charge state, because then the proportion of water in the electrolyte is higher, giving a greater tendency to freeze. A good state of charge is a good protection against freezing.

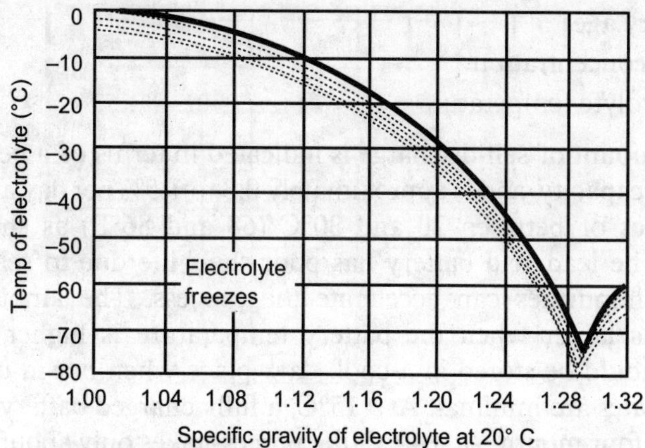


**Fig.3.26:** *State of charge of a battery against freezing*

A frozen battery will give a very small current only, but does not normally suffer permanent damage. This is because frozen acid electrolyte does not expand appreciably and remains in a gelatin-like state. It should be noted that recharging in a frozen state is difficult since only a small current can be passed through. Topping up should not be carried out if the outside conditions are below the freezing point of water.

## SELF-DISCHARGE CHARACTERISTICS

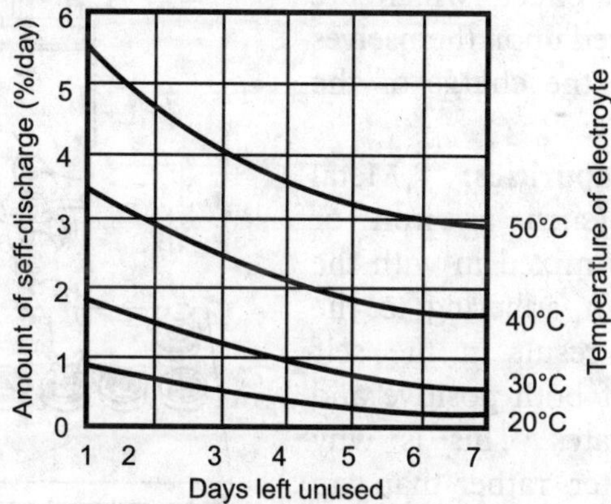


**Fig.3.27:** *Self-discharge curves*

A battery discharges by itself even if it not used. This phenomenon is called self-discharge. Self-discharge takes place inside the battery as a result of chemical activity even if no external circuit is connected. Instead of appearing as output power in form of the product amperes and voltage, chemical energy is dissipated as heat. The self-discharge of batteries depends on the following factors. These are:

- Battery age
- Acid concentration
- Electrolyte temperature

The amount of self-discharge is indicated in terms of a percentage of the battery capacity, and is approximately 0.3 to 1.5% per day at electrolyte temperatures of between 20 and 30°C (68 and 86°F) as shown in the Fig.3.27. The lead-acid battery has poor shelf life due to self-discharge. Heat and impurities can accelerate the process. The amount of self-discharge is larger when the battery temperature is higher. Therefore, batteries should be stored in a cool, dark place when not in use. At 0°C, discharge rates are minimal. At +15°C, a fully charged battery will go flat after about four month. At +40°C, the process takes only about two weeks. Batteries of quality use lead of highest purity in the active plate area with the antimony proportion kept small.

Self-discharge can be caused by any of the following:

- **Internal chemical processes:** Batteries which have been in service for some time suffers from the effect of antimony deposits on the negative plates. These set up miniature batteries, in effect, which are short circuited upon themselves and use up the charge of the negative plate.
- **Metal impurities:** Metal impurities (such as iron or manganese) mixed in with the electrolyte or adhering to the plates will result in the self discharge of both positive and negative plates. This is why distilled water, rather than tap water, is used in batteries since it contains minimal impurities.

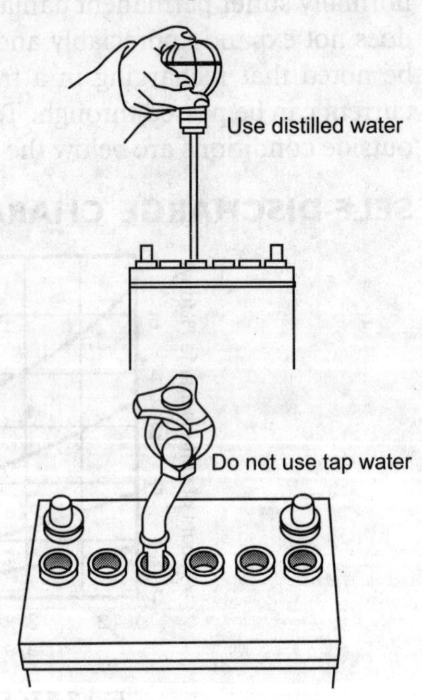


**Fig.3.28:** *Adding distilled water*

- **Leakage currents:** Dirt and the effects of acid fumes on the top surface of the battery can result in conducting film paths between the positive and negative battery lugs. This trouble can be minimised by periodic cleaning of the battery top. Sediments at the bottom of the battery case can also lead to leakage paths. This problem is much reduced with sealed batteries.

The remedy for self-discharge is to charge the battery periodically or to maintain a trickle charge approximately equal to the self discharge rate. This is being found by trail and error or by a rule of thumb which gives the trickle charge rate as:

$$\frac{1}{1000} \times \text{ampere – hour rating, amps}$$

For example, a 50 Ah battery might be trickle charged at:

$$\frac{50}{1000} = 0.05 \text{ amp}$$

## INTERNAL RESISTANCE

All batteries have an internal resistance, but this is very low in the case of a lead-acid battery in good condition. This is why this type of battery is favoured for vehicle where starting current demand can be very high.

A little arithmetic will show that if, for example, the internal resistance is to be 0.05 Ohms and the open circuit voltage 12.0 Volts then, with a lighting load of 10 ampere:

$$\begin{aligned} \text{Battery terminal voltage} &= 12 - \text{internal volt drop} \\ &= 12 - (10 \times 0.05) \\ &= 11.5 \text{ Volts} \end{aligned}$$

The table shows terminal voltages for this battery under various loads.

| Open circuit voltage | Load current | Internal drop | Terminal voltage |
|---|---|---|---|
| 12 V | 10 amps | 0.5 Volts | 11.5 Volts |
| 12 V | 20 amps | 1.0 Volts | 11.0 Volts |
| 12 V | 50 amps | 2.5 Volts | 9.5 Volts |
| 12 V | 100 amps | 5.0 Volts | 7.0 Volts |

The internal resistance is made up of various individual resistances, namely between electrodes and the electrolyte, the plate resistances, internal connectors and resistance of the electrolyte to ion flow (ions are particles moving through the electrolyte and carrying a positive or negative charge).

Additionally, the internal resistance depends upon the state of discharge and cell temperature, the value being higher as the battery is discharged. The internal resistance can be controlled by the surface area of the plates. Batteries with a larger number of plates (and therefore ampere-hour capacity) will have a lower internal resistance.

As batteries age, one of the effects is a rise in the internal resistance. For a cold morning start, extra torque will be required to free the crankshaft and the minimum engine turnover speed for firing will be about 100 rpm. It is under these conditions that the end of the battery life is determined.

## BATTERY RATING AND CAPACITY

Energy stored in a battery expressed in terms of period during which it gives the rated current before dropping to the specified final voltage (1.75 Volts per cell) is called its capacity. In simple terms, the characteristics or rating of a particular battery are determined by how much current it can produce and how long it can sustain this current. The rate at which a battery can produce current is determined by the speed of the chemical reaction. This in turn is determined by a number of factors:

- Surface area of the plates and volume of active plate materials
- Temperature
- Strength and amount of electrolyte
- Current demanded

The actual current supplied therefore determines the overall capacity of a battery. The rating of a battery has to specify the current output and the time. The common ratings are (i) Ampere-hour rate (ii) Watt-hour capacity, (iii) Reserve capacity rate and (iii) Cold rate.

## AMPERE-HOUR CAPACITY

$$\text{Ah capacity} = \text{Rate of discharge in amperes} \times \text{Number of hours required for discharge}$$

This is now seldom used but describes how much current the battery is able to supply for either 10 or 20 hours. A battery rated at 40 Ah, based on a 10 hour rate, should supply a steady current of 4 A for 10 hours before the final voltage is reached. The 20-hour rate is the most common. This means the amount of current that a battery can deliver for 20 hours before the final voltage is reached. For example, if a battery is capable of delivering 5 amperes for 20 hours before the cell voltage falls to 1.75 Volts, it is rated as 100 Ah.

## WATT-HOUR CAPACITY

A battery may also be rated by Watts, which represent the available cranking power. Wattage ratings for most 12-Volt batteries range from 2000 to 4000 Watts.

Watt-hour rating is determined by calculating the ampere-hour rating of the battery times the battery voltage at 0°C (-17.8°C).

$$\text{Wh capacity} = \text{Ampere-hour capacity} \times \text{Average voltage during discharge period}$$

## RESERVE CAPACITY

The reserve capacity rating is the time in minutes a vehicle can be driven after the charging system fails. This is roughly equivalent to the conditions after the alternator fails while the vehicle is being driven at night with the head lights on. The battery alone must supply current to the headlights and the ignition system. The assumed battery load is a constant discharge current of 25A. It is the length of time in minute that a fully charged battery at 80°F can deliver 25 amperes without the cell voltage dropping below 1.75 Volts. This shows for how long the battery can deliver the electrical operating load, if the alternator fails to supply current.

## COLD RATING

Large engines require that a starter motor deliver a large amount of cranking torque. Smaller engines require less torque. The greater the torque supplied by a starter motor, the more current it draws from the battery. As a rough guide, a starter motor requires approximately 1A of current for each cubic inch ($in^3$) of engine displacement. Therefore, a 350 $in^3$ engine would require a battery with a cold cranking rating of approximately 350 amperes. This rating indicates the power a battery can supply for engine cranking under low temperature conditions. It is the number of amperes that a battery delivers for 30 seconds at 0°F without the cell voltage falling below 1.2 Volts. The ampere values may be listed as 305, 315, 350 or 440.

Cranking power available from a fully charged battery based on temperature is:

| Temperature | Cranking power |
|---|---|
| 80°F (26.7°C) | 100% |
| 32°F (0°C) | 65% |
| 0°F (–17.8°C) | 40% |

## BATTERY EFFICIENCY

This is the ability of battery to supply current. It depends upon temperature and rate of discharge. At low temperature, sulphuric acid cannot work so actively on plates and thus chemical action is highly reduced. So the battery cannot supply high current for long time. At high discharge rates, chemical action takes place only on the surface of the plates due to short time. But during slower rates of discharge, there is sufficient time for the chemical activities to penetrate deep below the plate's surfaces and act in the active material.

When the battery is discharged slowly, it is far more efficient than when it is discharged rapidly. High rates of discharge do not produce as many ampere-hours as low rates of discharge. Hence lower the discharge rate, higher the battery efficiency and higher the discharge rate, lower the efficiency.

The efficiency of a battery can be determined either as the quantity (ampere-hour efficiency) efficiency or the energy efficiency (power efficiency).

$$\text{Quantity efficiency} = \frac{\text{Ampere – hour output}}{\text{Ampere – hour input}} \times 100\%$$

$$\text{Energy efficiency} = \frac{\text{Watt – hour output}}{\text{Watt – hour input}} \times 100\%$$

Battery output is dependent upon the rate of discharge. The quantity efficiency may vary from 75 to 95% and as low as 50% if the discharge rate is very high. For finding energy efficiency, average voltage during charge and discharge is taken into consideration. Owing to difference in cell voltage on charge and discharge, energy efficiency is much lower than quantity efficiency. For a 75 Ah battery, energy efficiency is 85% for normal discharge and 66% for higher discharge rate.

Battery efficiency is also affected by temperature. At 0°C, efficiency falls by 65%. Batteries in warm, tropical climates are more efficient, but may have reduced life spans. Batteries in cold climates have increased operating lives, but are less efficient.

The following table shows the approximate battery efficiencies in relation to battery temperatures:

| Battery temperature, °C | Battery efficiency, % |
|---|---|
| 26.7 | 100 |
| 0 | 65 |
| –17.8 | 50 |

## HYDROMETER

A hydrometer is an instrument for measuring the specific gravity of liquids.

It generally consists of a small cylindrical glass tube with a lead-weight at its bottom. The hydrometer floats in an upright position. The top part of the hydrometer extends above the liquid surface and the divisions on it indicate directly the specific gravity of the liquid. The glass float is a large hollow bulb which increases the buoyancy so that the hydrometer floats. The narrow stem increases the sensitivity of the hydrometer. The bottom of the hydrometer is made heavier by loading it with lead shots so that it floats vertically.

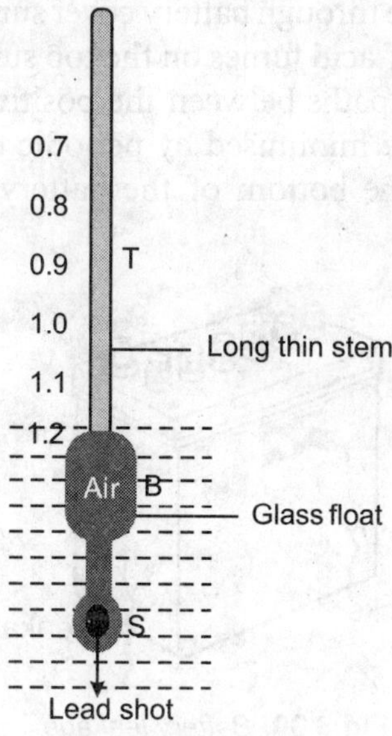


**Fig.3.29:** *Hydrometer*

## Principle

It works on the principle of flotation. Float sinks more in a lighter liquid so as to displace more volume of the lighter liquid whose weight is equal to the weight of the float. Hence, it will sink less in a denser liquid so that it has to displace less volume of the denser liquid whose weight will be equal to the float. Thus, the lowest density is marked on the top of the stem of a hydrometer and the maximum density is marked at the bottom of the graduated stem.

## VARIOUS TESTS ON BATTERY

The purpose of testing a battery is to determine whether it:

- is in good condition
- needs recharging
- is defective and must be discarded

Battery tests are made to determine the state of charge and battery capacity. Battery state of charge is tested with a hydrometer or voltmeter and battery capacity is determined by a high rate discharge test. The following tests have been used to test the batteries:

## LEAKAGE TEST

The outer surface of the battery is covered with dirt, moisture or electrolyte (contaminated with adverse impurities, especially with iron or copper), which result in a discharge through battery cover surface (external leakage).

Dirt and the effects of acid fumes on the top surface of the battery can result in conducting film paths between the positive and negative battery posts. This trouble can be minimised by periodic cleaning of the battery top. Sediment paths at the bottom of the battery case can also lead to leakage paths.

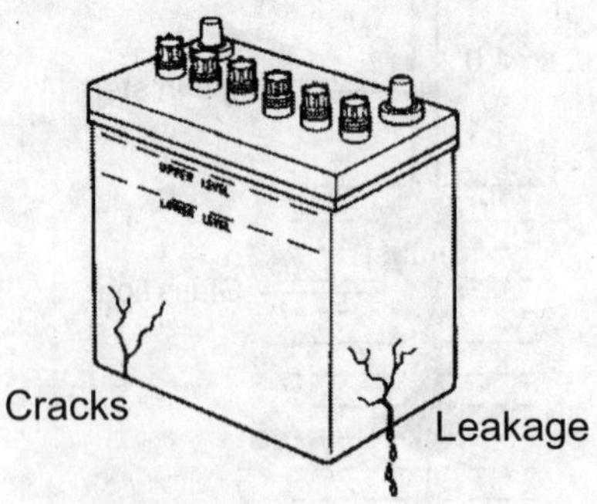


**Fig.3.30:** *Battery leakage*

Battery external leakage is measured with the help of voltmeter. This is executed by connecting the voltmeter positive probe to the positive terminal post of the battery, and slides the negative probe against the battery cover towards the negative terminal post side. Leakage voltage should be V £ 0.1V. If not, clean the top cover and make a test again.

## HYDROMETER TEST (SPECIFIC GRAVITY TEST)

During discharge, chemical reaction converts the active material of both plates to lead sulphate, water being liberated in the process. This results in the lowering of the specific gravity of the electrolyte. When charging, this chemical action is reversed, with a consequent rise in the specific gravity of the electrolyte. The specific gravity is therefore a good indication of the state of charge of the battery. The specific gravity reflects the percentage of sulphuric acid present in the electrolyte. This indicates the amount of charge still left in the battery.

The specific gravity of the electrolyte is determined with the help of hydrometer. The specific gravity of batteries is usually measured at a standard temperature or else temperature correction has to be made because the specific gravity is affected by temperature. The amount the stem sticks out above the electrolyte gives battery specific gravity or state of charge.

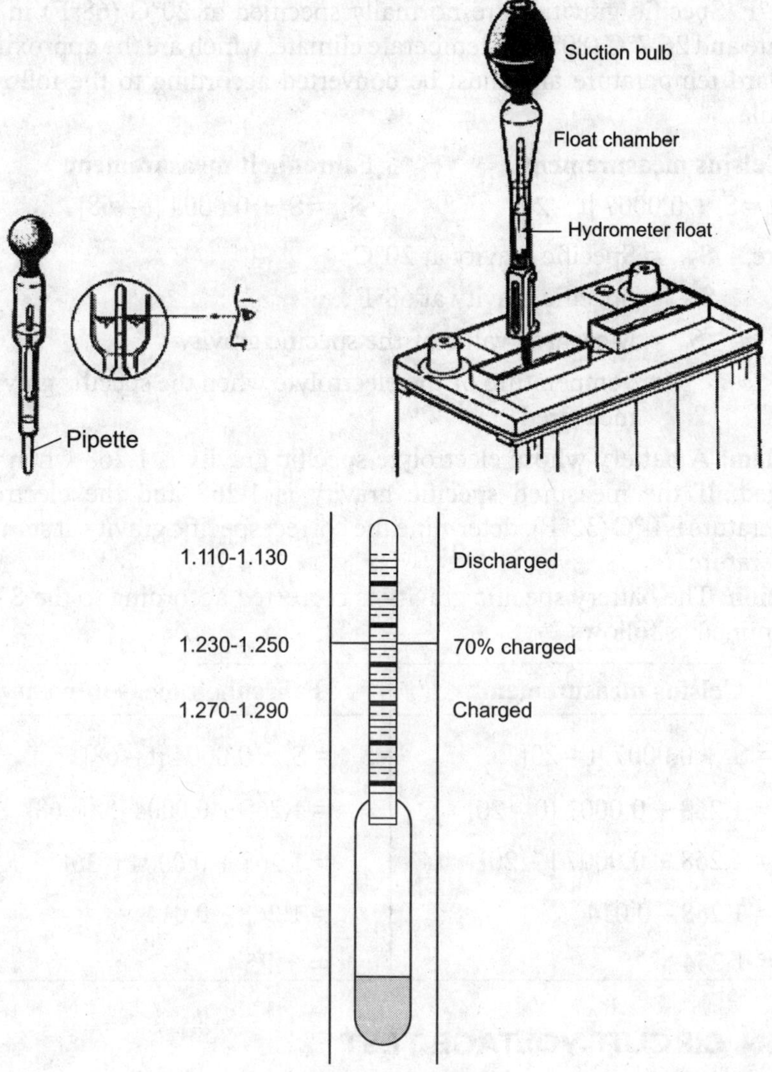


**Fig.3.31:** *Hydrometer*

## Procedure

- Squeeze and release the rubber bulb to draw electrolyte into the glass tube.

- To take true reading, the eye should be level with the surface of the electrolyte in the hydrometer glass tube.

## Temperature correction on specific gravity

The specific gravity of an electrolyte varies by 0.0007 per 1°C or 0.0004 per 1°F. Specific gravities are normally specified at 20°C (68°F) in cold climate and 26.7°C (80°F) in temperate climate, which are the approximate standard temperature and must be converted according to the following formula.

**Celsius measurement**

$$S_{20} = S_t + 0.0007\,[t - 20]$$

**Fahrenheit measurement**

$$S_{68} = S_t + 0.0004\,[t - 68]$$

Where, $S_{20}$ = Specific gravity at 20°C

$S_{68}$ = Specific gravity at 68°F

$S_t$ = Measured value of the specific gravity

$t$ = Temperature of the electrolyte when the specific gravity is measured

**Problem:** A battery whose electrolyte specific gravity is 1.268 when fully charged. If the measured specific gravity is 1.268 and the electrolyte temperature is 0°C (32°F), determine the correct specific gravity at standard temperature.

**Solution**: The battery specific gravity is corrected according to the STP is determined as follows.

| Celsius measurement | Fahrenheit measurement |
|---|---|
| $S_{20} = S_t + 0.0007\,[t - 20]$ | $S_{68} = S_t + 0.0004\,[t - 68]$ |
| $= 1.268 + 0.0007\,[0 - 20]$ | $= 1.268 + 0.0004\,[32 - 68]$ |
| $= 1.268 + 0.0007\,[-20]$ | $= 1.268 + 0.0004\,[-36]$ |
| $= 1.268 - 0.014$ | $= 1.268 - 0.014$ |
| $= 1.254$ | $= 1.254$ |

## OPEN CIRCUIT VOLTAGE TEST

Open circuit voltage test is a simple way of testing the state of charge of a battery. It is measured between the terminals of the battery using a voltmeter. During this test, no load is applied to the battery.

The voltmeter reading can be compared with state of charge table below in order to determine the condition of the battery. The voltages shown are for batteries at or near room temperature (20°C – 26.7°C).

| Battery voltage | State of battery charge |
|---|---|
| 12.6 V (or) Higher | 100% Charged |
| 12.4 V | 75% Charged |
| 12.2 V | 50% Charged |
| 12.0 V | 25% Charged |
| 11.9 V (or) Below | Discharged |

## PEAK LOAD TEST

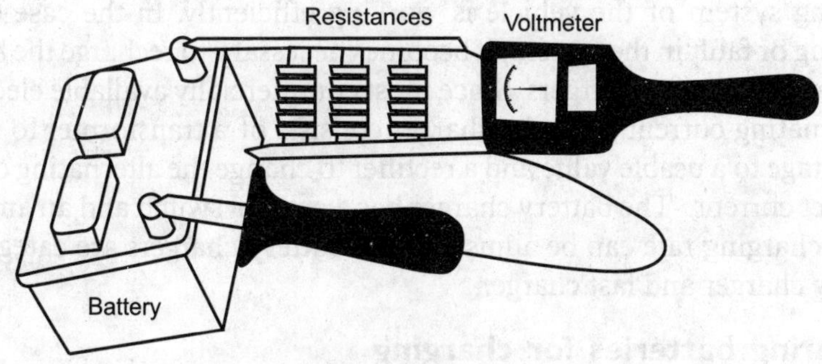


**Fig.3.32:** *Peak load tester*

This unit is provided with a voltmeter and has a measuring scale of 12-0-12 Volts and 6-0-6 Volts and coloured dial. It is used for measuring the total battery voltage with load applied to the battery.

- Green colour indicates good battery condition
- Red colour indicates weak (or) poor battery condition

During peak load test, the battery should be in a good state of charge to three-fourth (3/4) or better charged, preferably fully charged. This is because a rundown battery does not accurately reflect its condition. The test should be done within 5 to 10 seconds. This load placed on the battery during this test is approximately the same as that of the starter motor.

During this test, the voltmeter reading should not drop below 9.6 Volt.

$V \geq 9.6 \rightarrow$ battery is in good condition

$V < 9.6 \rightarrow$ battery is defective

The load chosen to be placed on the battery is shown in the table.

| Battery capacity | Battery voltage | Resistors connected | Load applied |
|---|---|---|---|
| | 6 V | 1 | 50 amps |
| Up to 40 Ah | 12 V | 1 | 100 amps |

| | | | |
|---|---|---|---|
| 40 to 80 Ah | 6 V | 2 | 100 amps |
| | 12 V | 2 | 200 amps |
| 80 Ah and above | 6 V | 3 | 150 amps |
| | 12 V | 3 | 300 amps |

## BATTERY CHARGING

Charging a battery means causing electrons to flow into the negative terminal. On the car, this job is performed by the alternator. Usually the batteries do not require charge from the external source provided the charging system of the vehicle is working efficiently. In the case of less charging or fault in the system, it becomes necessary to recharge the battery by means of battery chargers. Since most commercially available electricity is alternating current (AC) the charger consists of a transformer to reduce the voltage to a usable value and a rectifier to change the alternating current to direct current. The battery charger has a control switch and an ammeter so the charging rate can be adjusted. The battery chargers are categorized as slow charger and fast charger.

### Preparing batteries for charging

- Clean and wash dirt from the battery terminals.
- Inspect the case, cell covers, sealing compounds and vent plugs and make repairs if damaged.
- Check the electrolyte level and fill it if necessary.

### Safety precautions

- Do not overfill as it may bubble out of the vent openings during charging.
- Because the battery emits explosive hydrogen gas, be sure not to allow fire or sparks near the battery.
- Never disconnect the battery charger's cables from the battery terminals when the battery is being charged. Turn the main switch of the charger off before disconnecting them.
- The battery electrolyte temperature must not exceed 45°C (113°F). If it goes above this, reduce the charging amperage or stop charging.

## SLOW CHARGING

In this method, battery is removed from the vehicle and connected to a slow charger to recharge it. Slow chargers are usually constant current machines which use a rectifier to convert AC to DC and a rheostat to adjust current flow to the battery. Slow battery chargers operate at low rate and take considerably longer time to charge a battery. Slow charging permits the battery to be completely charged. The charging rate is one ampere for each positive plate in a cell. For example, a battery having four positive plates in a cell should be charged at a rate of 4 amps. Batteries of different size can be charged at the same time. But the charging rate is determined by the smallest battery. Batteries are safely charged at a high rate if the temperature is below 45°C (113°F). The length of time required to recharge a good battery to fully charged condition is 12 to 16 hours at normal rates.

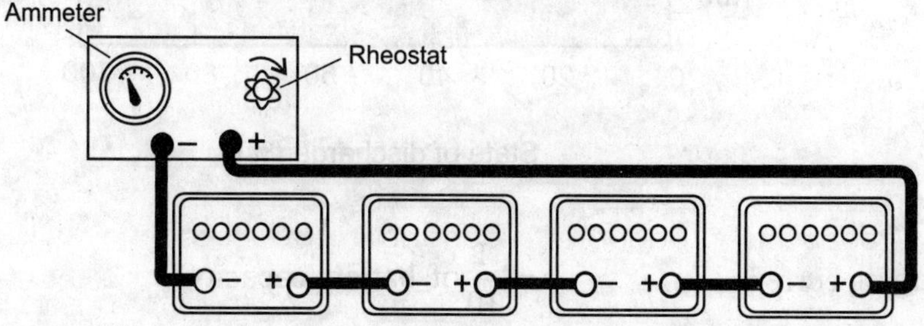


**Fig.3.33:** *Slow charger*

During slow charging it is also important to ensure that the maximum charging current should be less than 1/10th of the battery capacity. For example, the battery capacity is 40 Ah, and then the charging current is 1/10th of 40 Ah.

i.e. 40/10 = 4 amps or less

Slow charging time can be calculated by the following formula:

$$\text{Charging time (h)} = \frac{\text{State of discharge capacity (Ah)}}{\text{Charging current (A)}} \times (1.2 \text{ to } 1.5)$$

> ***Note:*** *If more than one battery are to be charged, connect the batteries in series up to the capacity of the charger.*

**Problem**: The battery capacity is 60 Ah and specific gravity is 1.16, find the charging current and charging time.

**Solution**: From the graph for specific gravity 1.16, state of discharge is about 50%.

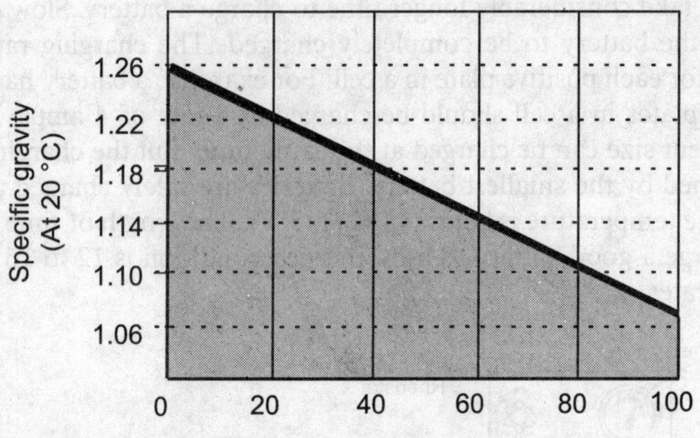


$$\text{Charging current} = \frac{1}{10} \text{ of battery capacity}$$

$$= \frac{1}{10} \times 60$$

$$= 6\text{A}$$

$$\therefore \text{ State of discharge capacity} = 50\% \text{ of battery capacity}$$

$$= 50\% \text{ of } 60 \text{ Ah}$$

$$= 30 \text{ Ah}$$

$$\text{Charging time} = \frac{30}{6} \times (1.2 \text{ to } 1.5)$$

$$= 6 \text{ to } 7.5 \text{ hours}$$

## FAST CHARGING

This method does not require battery to be removed from the vehicle, as most fast chargers are portable. Its purpose is to recharge the battery sufficiently to allow the starting motor to crank the engine. This is done in a short period of time, usually about 30 to 45 minutes. The battery is brought to a fair state of charge before it reaches at harmful high temperature. Fast charger is usually constant voltage machines with a high capacity so that common charging rates of 80-100 amperes for 6 Volt battery and 50-60 amperes for 12 Volt battery. 30 to 45 minutes of charging time will normally charge the battery so that the starting motor will crank the engine and the battery will not be over heated. Fast chargers do not fully charge a battery in the time allotted for charging. The remaining recharge is done by the alternator when the engine is operating. Some fast chargers have controls for finishing the charging cycle at low rate so the battery can be brought up to a full charge.

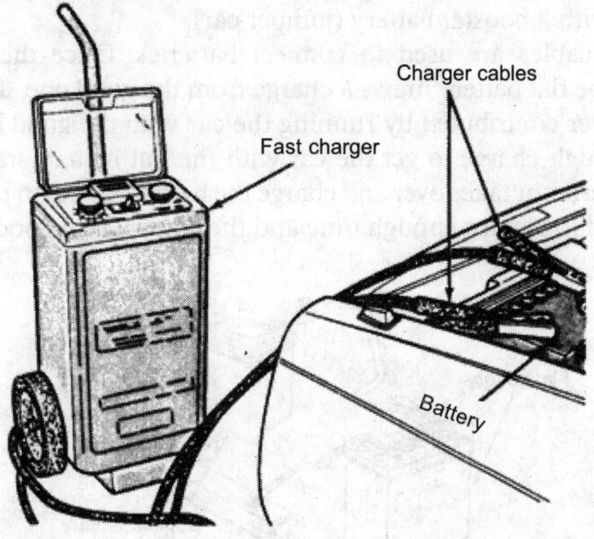


**Fig.3.34:** *Fast charger*

Determination of the allowable charging amperage and time is very essential if the battery chargers have no test device. Determine the state of the battery from the specific gravity, using the graph, then calculate the correct charging amperage using the following formula. The charging time for quick charging is usually from half to one hour.

$$\text{Correct charging amperage (A)} = \frac{\text{State of discharge capacity (Ah)}}{1 + \text{Charging time}}$$

> ***Note:*** *If more than one battery are to be charged, connect the batteries in parallel up to the capacity of the charger.*

***Problem:*** The battery capacity is 40 Ah specific gravity at 20°C (or 68°F) is 1.18. What would be the charging current for half an hour?
From the chart, the specific gravity 1.18 tells us it is 40% discharged,

$\therefore$ State of discharge capacity = 40% of battery capacity
= 40% of 40 Ah
= 16 Ah

To charge the battery for half an hour, the correct charging amperage would be:

$$\text{Charging currect } \frac{16}{1 + 0.5} \cong 11 \text{ amps}$$

## JUMP STARTING

A jump start is a process of starting a car with a flat battery (dead car) using a car with a booster battery (jumper car).

Booster cables are used to connect batteries. Once the cables are connected, the flat battery draws a charge from the good one, and together with the power contributed by running the car with the good battery, this provides enough charge to get the car with the flat battery started. From there, the alternator takes over and charge the battery up again provided the car is allowed to run for enough time and the battery is in good condition.

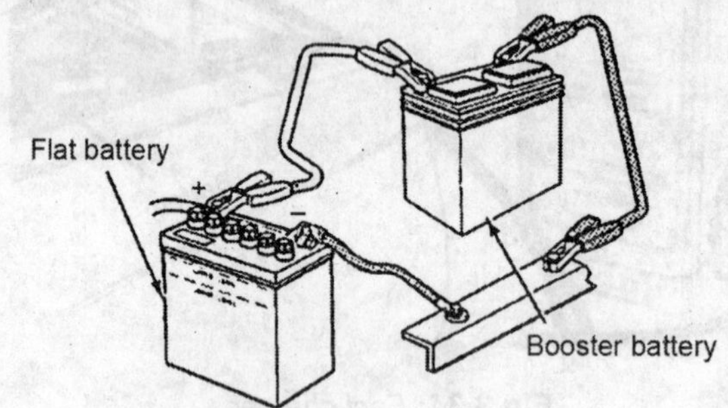


**Fig.3.35:** *Jump start*

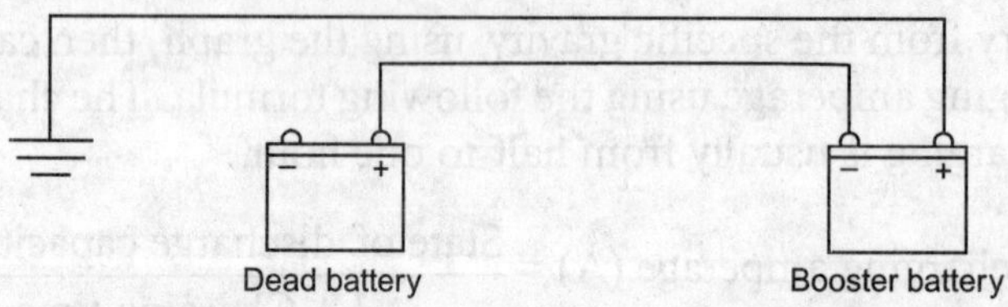


**Fig.3.36:** *Jump starting cable connection*

In the cold winter weather, a flat battery is one of the most common break-down faults. Jump-starting is one of the most basic break-down remedies. It is a safe and quick way to get a car with a flat battery moving again. And it can save the time and the cost of calling out a breakdown recovery vehicle.

### Jump start procedure

**Fig.3.37:** *Procedure for jump starting*

1. Pull jumper car near to dead car. Make sure both cars are turned off and their bonnets (hoods) are opened.
2. Connect one end of the red (positive) jumper cable to the positive terminal on the flat battery. Connect the other end to the positive terminal of the booster battery.
3. Connect one end of the black (negative) jumper cable to the negative terminal of the booster battery. Connect the other end to a clean, unpainted metal surface under the bonnet of a dead car. Do not connect the negative cable to the negative terminal of the flat battery.
4. Now start the jumper car and allow it to run for 2-3 minutes before starting the dead car.
5. Remove cables in reverse order of how they are connected. Keep the jumped car running for at least 30 minutes to give the battery sufficient time to recharge itself.

### Problems with jump starting

1. If not done properly, jump-starting a car can be dangerous.
2. If the battery is connected the wrong way, it could overheat and catch fire, or explode.
3. If there are sparks when terminals are connected, that could cause an explosion of the hydrogen gas within or around the battery.
4. There is the risk of electrocution. It is important to make sure hands and other body parts are protected from metal parts of the clips and the battery throughout the process.

## BATTERY GASSING

As part of the chemical reactions during charging, hydrogen gas is liberated at the negative plates and oxygen gas is liberated at the positive plates. This is known as gassing and it causes water loss in the battery. The higher the rate of charge, the greater the gassing. Vent holes are provided to prevent pressure build up in the battery and hence prevent battery explosion. Gassing causes the battery to release the above mentioned gases and sulphuric acid fumes,

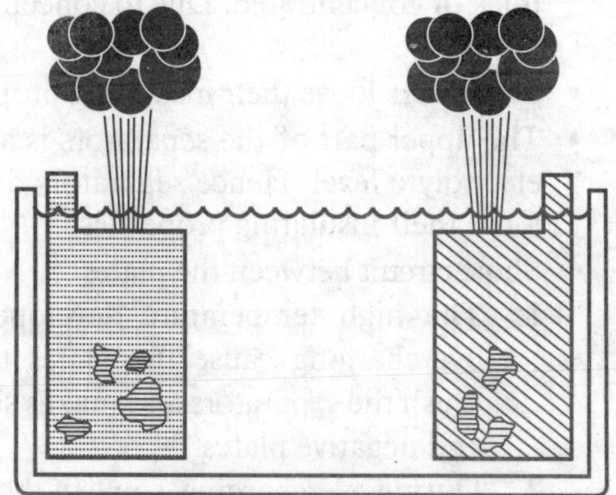

**Fig.3.38:** *Battery gassing*

which can cause corrosion to the battery terminals, battery tray and other areas surrounding the battery.

## BOOST CHARGING

The fast charger can also be used to give a battery an emergency boost when a customer is in a hurry or in some other emergency requiring quick service. The battery is charged at the normal fast charging rate for one-third to one-half the normal charging time. This puts back sufficient current for starting the engine. The vehicle should be operated for a long enough time so the car's alternator can complete recharging the battery.

## OVERCHARGING

Batteries that are completely charged need not be charged furthermore. Because the current that is applied to charge the battery makes electrolysis of water and the water will be broken into oxygen (O) and hydrogen ($H_2$) gases. Oxygen find no Pb and hydrogen find no $SO_4$ to combine with and they bubble out of the electrolyte through vent holes (battery gassing). Since oxygen and hydrogen are constituent parts of water, water is rapidly lost from the battery cells.

The effects of overcharging are as follows:

- Plates loose their ability to take charge
- As a result of water loss, the electrolyte level falls down which makes plates exposed to air. The dry plates get hard and loose their ability to take charge.
- Plates deteriorate rapidly
- Due to water loss, the acid at the lower parts of plates becomes more thick or concentrated. Due to concentrated acid the plates deteriorate rapidly.
- Separators loose their insulating properties
- The upper part of the separators is also exposed to air due to fall in electrolyte level. Hence separators dry out and crack at the top and loose their insulating properties.
- Short circuit between the plates
  1. The high temperature developed in the battery cells during overcharging cause the plates to buckle. The buckling plates crush the separators and makes short circuit between the positive and negative plates.
  2. During overcharging, some of the oxygen gathering at the positive plates attacks the grid structure of the plates causing them to oxidize. This result in swelling of the plates resulting in pushing up of cell covers. Too much swelling causes the upper edges of

positive plates to be pushed against the negative plate straps. This results in direct short circuit and hence battery failure.

## SULPHATION

During discharging process the active materials of the plate is converted into lead sulphate and restored back into the active material by the charging process. This is a normal cycle which is continually repeated in the battery. However, if the battery stands for long periods in a discharged condition, the lead sulphate changes into a hard crystalline substance. As it forms, the crystalline sulphate tends to break the plate grids. It is very difficult to reconvert hard crystalline lead sulphate into active materials by normal charging process. Such a battery should be charged at half the normal rate for 60 to 100 hours. Batteries that are heavily sulphated generally cannot be restored to a serviceable condition.

## COMPENSATION CHARGING

When batteries are kept in store or not used for a long period of time with their electrolyte, they constantly loose some of their charge due to the chemical reaction (self-discharge). Their daily self-discharge rate can make up to 1% of their capacity.

To overcome this self discharging process, compensation charging is executed on a very low charging current which is about 0.1% of the value of the normal capacity of the battery. If compensation charging is not possible, a normal charging of the batteries for every one or two months can be done.

## MAINTENANCE FREE BATTERY

Maintenance free batteries are sealed and never need water. They contain a lifetime supply of water, which is put in during manufacturing. The reason is that the plate grids are made of a special lead alloy with calcium, strontium or cadmium, which greatly reduces gassing. Because there is less of this, there is less water loss. The maintenance free batteries are also called “low water loss” batteries.

The grid structure of the lead acid battery is made from a lead alloy. Pure lead is too soft and would not support itself, so small quantities of other metals are added to get the mechanical strength and improve electrical properties.

The two most common alloys used today to harden the grid are antimony and calcium. A normal battery uses up to 5% antimony in the construction of the plate grids to add strength. However, the greater the amount of antimony, the greater the amount of gassing and therefore the more water the battery will use. Maintenance free batteries use calcium instead of antimony because 0.2% calcium has the same strength as 6% antimony. The calcium or strontium provides the needed strength in the grids and in addition greatly reduces gassing and self-discharge.

Sealed batteries have a much longer storage life than non-sealed batteries. Because the gassing rate is so low, these batteries can be sealed. However, one or more small vent openings are provided to release any pressure formed by gassing.

## Charge indicator

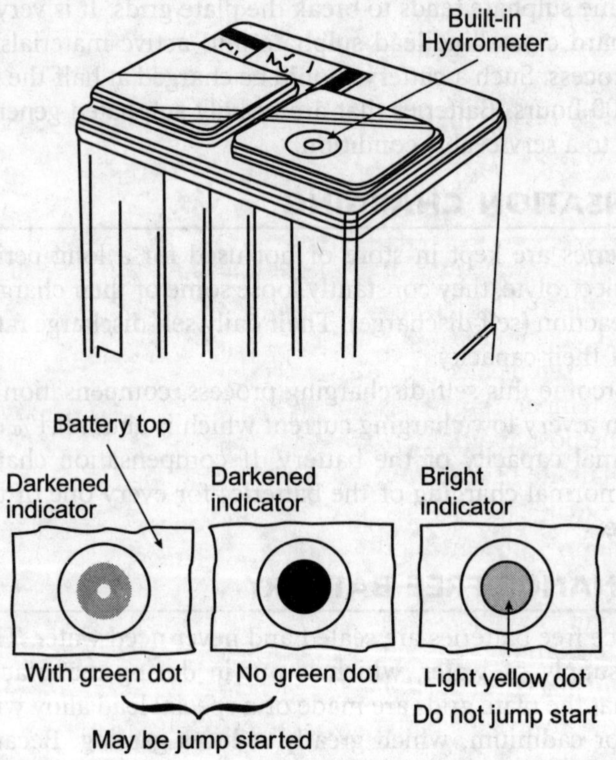


**Fig.3.39:** *Maintenance free battery*

Many maintenance free batteries have a charge indicator (built-in-hydrometer) built into the cover. The indicator shows the state of charge of the battery (actually only one cell). If the dot that is at the bottom of the indicator is green, then the battery is charged (more than 65% charged). If the dot is black, the battery state of charge is low. A yellow or clear eye means the electrolyte is low and the battery must be replaced.

### Advantages

1. Generally do not use water during normal usage.

2. Reduce the corroding of battery cables, terminals, and battery trays. Lead-calcium cells do not "gas" until the cell voltage exceeds 2.15 V (12.9 battery voltage), whereas non-maintenance free (lead-antimony) batteries start to gas at 2.0 V per cell (12.0 battery voltage). Gassing causes the battery to release gases and sulphuric acid fumes, which can cause corrosion to the battery terminals, battery tray and other areas surrounding the battery.
3. A lead-calcium battery has a longer shelf life than most other types of batteries (three times longer than a lead-antimony battery), "Shelf life" means how long a battery can remain in storage (on a shelf) at a store before being sold and still be able to provide new-battery performance.
4. Because of the higher internal resistance of lead-calcium plates, a maintenance free battery resists overcharging.

## MAINTENANCE OF BATTERIES

*Battery Maintenance* is an important issue. The battery should be cleaned using a baking soda and water solution (a couple of table spoons to a pint of water). Cable connections need to be cleaned and tightened as battery problems are often caused by dirty and loose connections. Coat the washer with high temperature grease or petroleum jelly (Vaseline), then place cable on the post and tighten. Coat the exposed cable end with the grease.

The main items of battery maintenance are as follows:

- Top up the electrolyte level periodically with distilled water. Use only mineral free water, Distilled is best as all impurities have been removed, and there is nothing left that could contaminate cells. The frequency of topping up depends on how much the vehicle is used, on the charging rate, and on the ambient temperature. The electrolyte should just cover the top of the plates (10 to 15 mm). Avoid overfilling.
- Charge always the batteries to their full rated capacity.
- Do not permit discharge of the cells below a terminal voltage of 1.75 Volts.
- Keep the terminals clean and dry. A coating of petroleum jelly retards corrosion.
- Batteries should never be allowed to remain in a semi-discharged or discharged condition.
- Mount the battery securely, to avoid damage.
- Don't overfill battery cells especially in warmer weather because the natural fluid expansion in hot weather can push excess electrolyte from the battery.

## REVIEW QUESTIONS

1. Is electricity generated by a battery?
2. What is a cell?
3. How are the cells of a battery grouped?
4. What is the difference between series and parallel cell connection?
5. Name the materials for separators.
6. Why is battery called storage battery?
7. What is a primary cell?
8. What is a secondary cell?
9. On what factors the power of a battery depends?
10. What does the capacity of a battery indicate?
11. How many cells are there in a 12V storage battery?
12. How are positive and negative plates identified in a battery?
13. What type of battery is used in automobiles?
14. What is an electrolyte? What substance is used as an electrolyte in automobile batteries?
15. Why ordinary water is not used in an electrolyte?
16. How is the electrolyte prepared for a battery?
17. Explain why a relative density reading of the electrolyte gives an indication of a battery's state of charge.
18. What will happen when the specific gravity of the electrolyte is more than 1.31?
19. What is meant by a grid in a battery?
20. Why does the terminal of a battery get corroded? How these are cleaned?
21. What is the effect of neglecting to top up the battery?
22. If the specific gravity of the acid in a battery is low, is it desirable to add some more acid?
23. If the specific gravity reading of a storage battery is 1.29 at 90°F, what is the correct specific gravity?
24. What is the specific gravity for (1) a fully charged (2) a half-charged (3) a fully discharged battery and (4) a dead battery?
25. What is ampere-hour rating?
26. What is known as cold cranking amperes of a battery?
27. What is reserve capacity of a battery?
28. What do you understand by battery sulphation?

29. How often is battery service required?
30. What instrument is used to measure the specific gravity of a liquid?
31. What is a hydrometer used for?
32. What precautions to be taken before the battery is put in charge?
33. Explain the chemical reaction that takes place when a battery is being discharged.
34. Discuss about the characteristics of lead acid battery.
35. Explain how batteries are rated?
36. What do you mean by efficiency of batteries?
37. What are the various tests done on batteries?
38. Write short note on battery maintenance.
39. Write short note on battery charging.
40. With a neat sketch, explain the construction of a lead acid battery.
41. What is the advantage of applying an inhibitor to the battery terminal connection?
42. What causes batteries to gas?
43. What care must be taken to dissipate battery acid gas?
44. What causes batteries to overheat while recharging?
45. How are the terminals of a battery charger identified?
46. What is a high rate discharge tester used for?

UNIT

4

# Charging System

The storage battery is a necessary source of power for the ignition of the air-fuel mixture, starting of engine, lighting, and other duties, but the amount of energy which it can store is limited so that it cannot supply the power required by the automobile continuously.

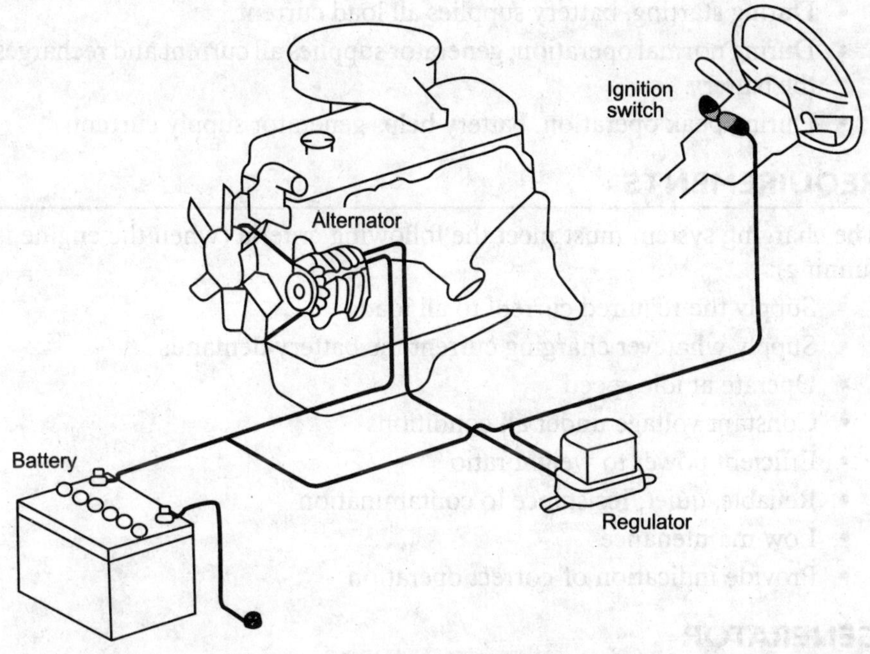


**Fig.4.1:** *Charging system*

Therefore it is necessary to have a charging system that serves to supply power to the various electric parts while the automobile is in operation and with the reserve power, charges the battery so that the battery will always be in the state to supply power when required. The charging system is comprised essentially of the generator, regulator, battery, and wiring harness.

## OPERATION

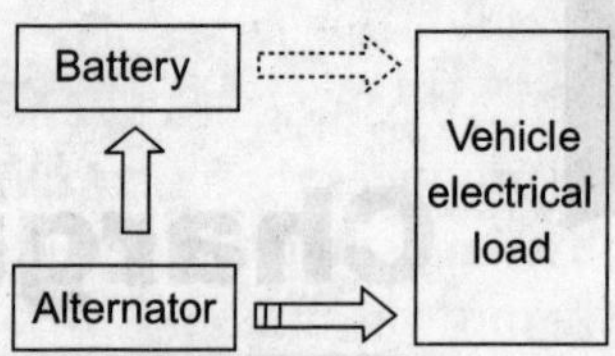


**Fig.4.2:** *Charging flow diagram*

When the engine is running, the charging system generates electricity for the vehicle's electrical systems, and supplies all the current needed by the vehicle. The charging system also sends current into the battery to restore its charge.

All charging circuits operate in three stages:

- During starting, battery supplies all load current.
- During normal operation, generator supplies all current and recharges the battery.
- During peak operation, battery helps generator supply current.

## REQUIREMENTS

The charging system must meet the following criteria (when the engine is running):

- Supply the required current to all loads
- Supply whatever charging current the battery demands
- Operate at idle speed
- Constant voltage under all conditions
- Efficient power to weight ratio
- Reliable, quiet, resistance to contamination
- Low maintenance
- Provide indication of correct operation

## GENERATOR

The generator is driven by the V-belt which gets drive from the crankshaft. It converts mechanical energy obtained from the engine into electrical energy. Generators can be DC (dynamo) or AC (alternator).

All power required while the automobile is in operation is supplied by the generator, but when this is not enough, the battery also serves as the power source. In the reverse case, when the generator is producing more power than is required, the surplus power goes to charge the battery. The generator is, therefore, the power source when the vehicle is in operation and at the same time, it serves to charge the battery.

## DIRECT-CURRENT (DC) GENERATOR

A generator which produces direct current is called a dynamo. This book will not describe dynamos in great detail as few cars now have them. But in the past the dynamo was the only convenient way to generate electrical energy.

## PRINCIPLE OF ELECTRIC CURRENT GENERATION

If a conductor cuts magnetic lines of force, electromotive force is generated in the conductor, and that the electromagnetic force produced will vary with the number of magnetic lines of force, number of conductors (number of coil windings), and the speed at which the conductor cuts the lines of magnetic force.

As shown in the Fig.4.3, if a conductor is turned clockwise between the N and S poles of a magnet, electromotive force will be generated in the arrow direction in accordance with Fleming's right hand rule.

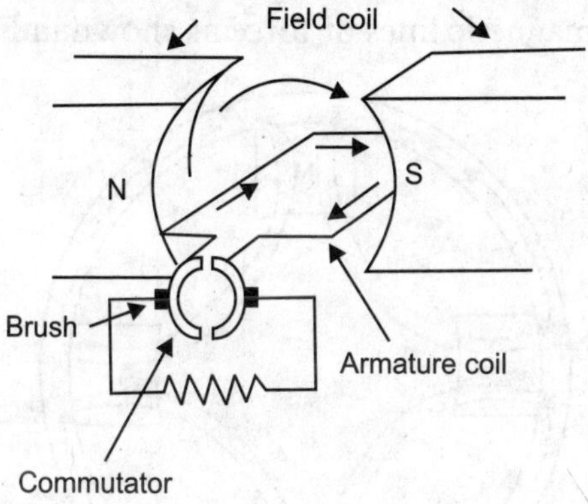


**Fig.4.3:** *Principle of generator*

The generator can be roughly divided up into the field coils that make the magnets to produce the magnetic lines of force, the armature coils that cut the magnetic lines of force, and the commutator that draws out the armature electromotive force.

When the armature coil rotates, current flows in the arrow direction through the armature coil. The induced current flows through the commutator and brush at one side to the load such as the battery and lights, and returns through the brush and commutator at the other side to the armature coil. A part of the current induced in the armature coil flows through the field coils to strengthen the flux.

As the armature coil turns, the position of the conductors reverse, so that current now flows in the opposite direction through the coil; but through the commutator sections and brushes, the current flows in one direction to the outside (direct current).

Generators used in automobiles have a large number of armature coils so that the voltage is generated continuously.

## CONSTRUCTION

The generator is comprised of the pole cores, field coils, armature, brushes, pulley, and frame.

### Pole cores

Pole cores become the N and S poles of magnet when the current flows through the field coils. Most of the generators have four pole cores, bolted to the inner side of the generator frame.

### Field coils

Field coils comprise insulated wire wound in coil form and installed around the pole cores. The current flows through the field coils and magnetises the pole cores to form magnetic lines of force as shown in Fig.4.4.

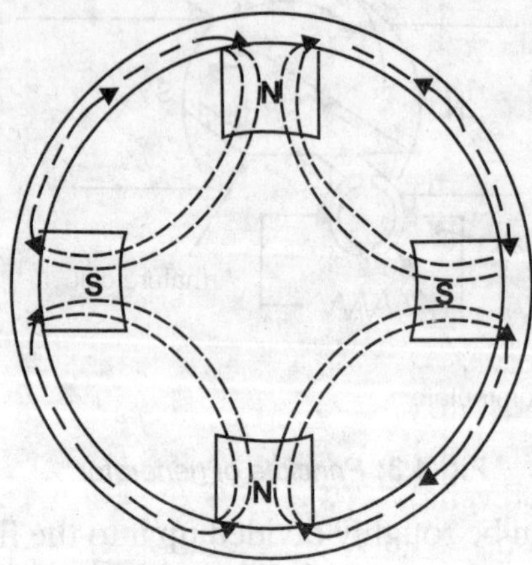


**Fig.4.4:** *Magnetic lines of force*

Even if there is no current flowing through the field coils, the pole cores have a certain amount of residual magnetism. Thus, the generator can start producing power as soon as the armature begins to turn (when the engine starts up).

### Armature

The armature rotates inside the pole cores and produces the electromotive force. It is comprised of a rotating shaft with the armature coil at the center and the commutator at one end. The coil windings are wound in the slots of the core built of laminated silicon steel sheets, and are connected to the commutator segments. The commutator serves to convert the alternating current produced to direct current.

### Brushes

The brushes contact against the commutator and serve to pass the current generated in the armature to the field coils and to the outside of the generator. The brushes are generally made of carbon with a pair of positive and negative brushes held in brush holders, and pressed against the commutator by springs.

### Pulley

The pulley is provided with a fan to cool the generator which produces heat when generating power.

## REGULATOR

Continued high speed operation will cause large currents to flow through the battery, generator, and other circuits, and shorten their service life. On the other hand, during low speed operation, or when the engine is stopped, the voltage produced by the generator will become lower than that of the battery, so that the current tends to flow from the battery to the generator. To prevent such adverse conditions, a regulator has been provided in the generating system. The regulator is comprised of three relays: a voltage regulator to control the generated voltage, a current regulator to control the output current, and a cut-out relay to prevent reverse flow of current from the battery.

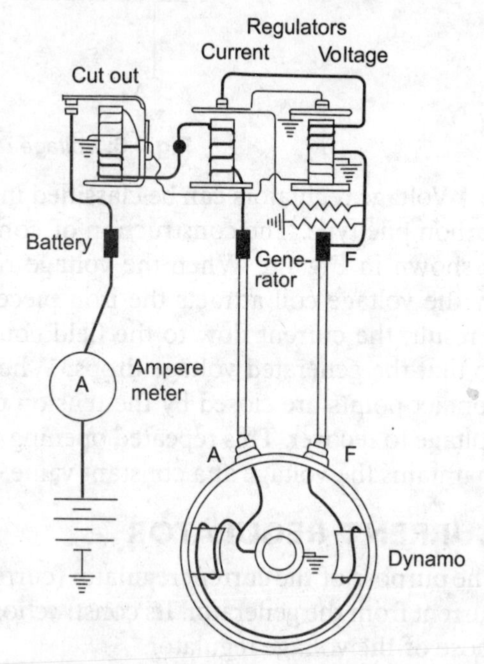


**Fig.4.5:** *Cut-out and regulators for dynamo*

## VOLTAGE REGULATOR

The purpose of this relay is to maintain the generated voltage at a constant value. When the voltage rises, a resistance is inserted in series with the field coil circuit to limit current flow through the field coils and lower the generated voltage. When the voltage drops, the resistance is removed from the field coil circuit to allow the generated voltage to build up again. The repetition of this action maintains the voltage at a constant value.

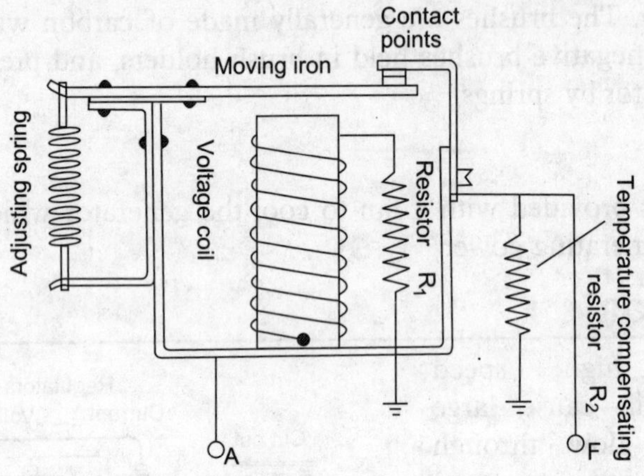


**Fig.4.6:** *Voltage regulator*

Voltage regulators can be classified into the contact point type and the carbon pile type. The construction of contact point type voltage regulator is shown in Fig.4.6. When the voltage rises too high, the magnetisation by the voltage coil attracts the iron piece to open the contact points. As a result, the current flow to the field coils is limited by the resistance $R_1$ so that the generated voltage drops. When the voltage drops too low, the contact points are closed by the tension of regulating spring to allow the voltage to recover. This repeated opening and closing of the contact points maintains the voltage at a constant value.

## CURRENT REGULATOR

The purpose of the current regulator (current limiter) is to prevent excessive current from the generator. Its construction and operation is very similar to those of the voltage regulator.

The contact points are normally closed, but when the generated current rises over a specified value, the points P open so that the current flows through the resistance R to the field coil F. This reduces the current flow to the field coil so that the generated current becomes less. When the current becomes too small, the points close to restore normal state.

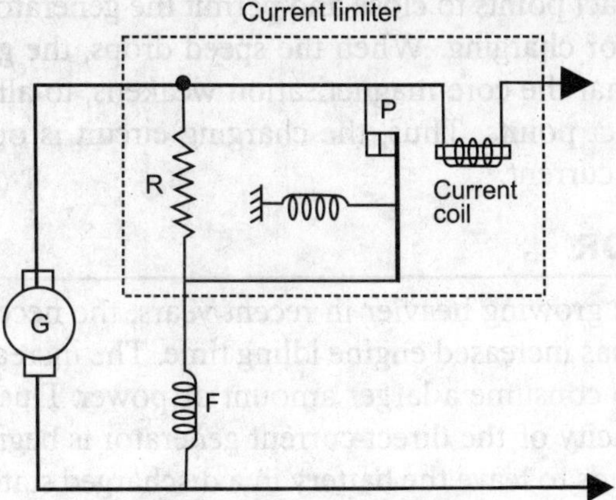


**Fig.4.7:** *Current regulator*

## CUT-OUT RELAY

A voltage regulator has an additional component known as a cut-out relay, which prevents the battery from discharging back into the generator's field coils when the key is turned to OFF or the engine is idling. Without the cut-out relay, the generator would always be on and the battery would go dead, even with the engine not running.

When the generator speed drops so that the generated voltage becomes lower than the battery voltage, the cut-out relay contacts open to prevent the battery from discharging back to the generator. The cut-out relay is merely a form of automatic switch.

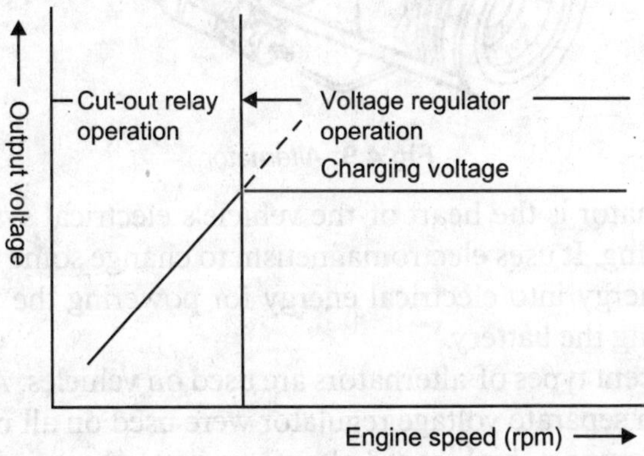


**Fig.4.8:** *Cut-out relay*

The construction of the cut-out relay is as shown in Fig.4.8. When the generator is not rotating, the contact points are held open by the tension of the spring, but when the generator starts rotating and its generated voltage reaches the charging voltage, the relay core becomes strongly magnetized through the voltage and current coils, and attracts the iron piece. This causes the contact points to close and permit the generator current to flow to the battery for charging. When the speed drops, the generated voltage also drops so that the core magnetisation weakens, to allow the spring to open the contact points. Thus, the charging circuit is opened to prevent reverse flow of current.

## ALTERNATOR

With the traffic growing heavier in recent years, the necessity for making frequent stops has increased engine idling time. The increasing numbers of accessories also consume a larger amount of power. Due to these causes, the output capacity of the direct-current generator is beginning to be over strained and tends to leave the battery in a discharged state. The alternator can overcome this problem. This alternator is compact and lightweight. It also provides better serviceability.

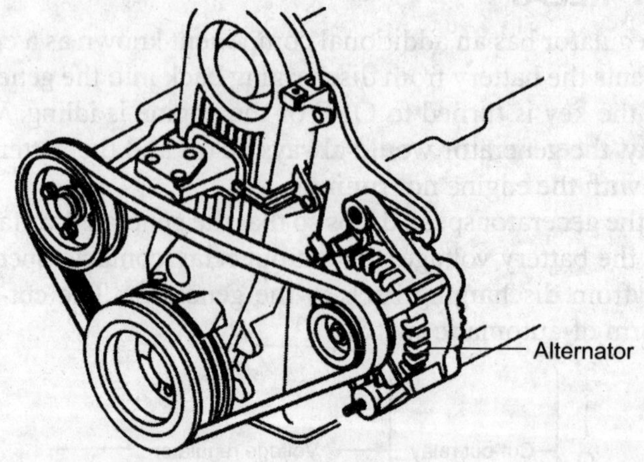


**Fig.4.9:** *Alternator*

The alternator is the heart of the vehicle's electrical system when the engine is running. It uses electromagnetism to change some of the engine's mechanical energy into electrical energy for powering the vehicle's loads and for charging the battery.

Two different types of alternators are used on vehicles. A conventional alternator with separate voltage regulator were used on all models prior to 1986. A new compact, high-speed alternator with a built-in IC regulator is now used on most models. Both types of alternators are rated according to current output. Typical ratings range from 40 to 80 amps.

## CONSTRUCTION

The alternator is comprised primarily of the stator, rotor, bracket, rectifier (silicon diodes) and brushes.

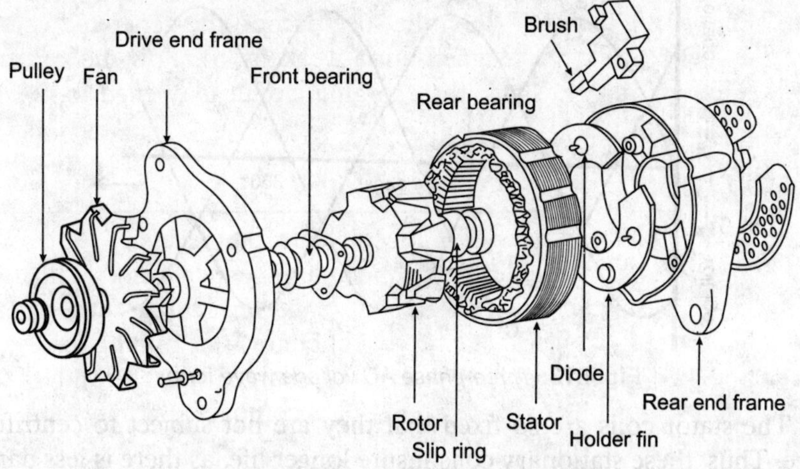


**Fig.4.10:** *Exploded view of alternator*

## STATOR

The stator has three sets of armature coils wound on it to induce a three-phase current.

The coils in the stator can be connected in either star or delta form. The star connection is generally used in automobile alternators. The generated wave forms, phased 120º apart, are shown in Fig.4.13.

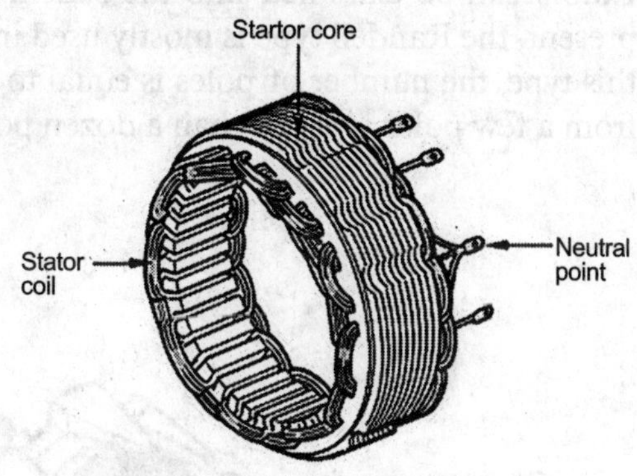


**Fig.4.11:** *Stator*

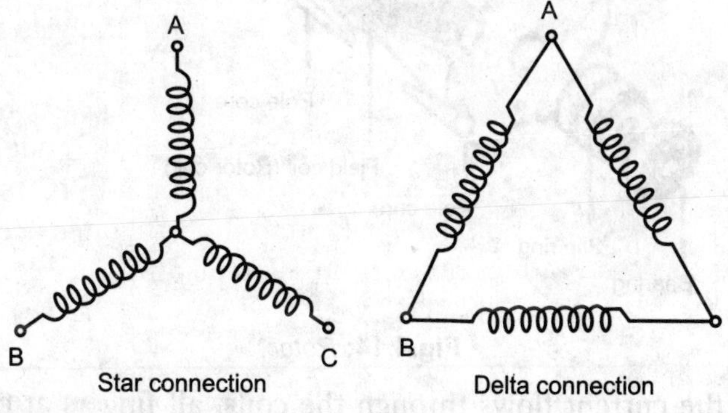


**Fig.4.12:** *Stator coil connection method*

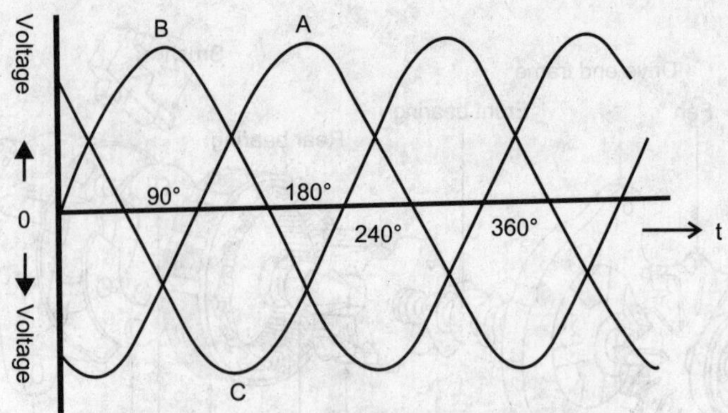


**Fig.4.13:** *Three-phase AC voltage wave forms*

The stator coils are so fixed that they are not subject to centrifugal force. Thus, these stationary coils insure longer life, as there is less danger of the insulation becoming damaged and developing short circuits.

## ROTOR

Rotors can be classified into the salient pole type and Randell type. At present, the Randell type is mostly used in automotive alternator rotors. In this type, the number of poles is equal to the number of fingers and range from a few poles to more than a dozen poles.

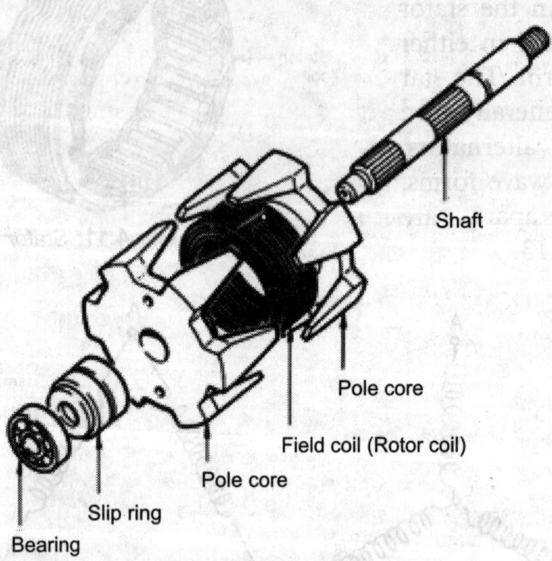


**Fig.4.14:** *Rotor*

When the current flows through the coils, all fingers at one side of the disk become N poles and all those at the other side become S poles, resulting in the N and S pole to be arranged alternately. The features of this type are simplicity of construction and high strength against centrifugal force.

## RECTIFIER (SILICON DIODES)

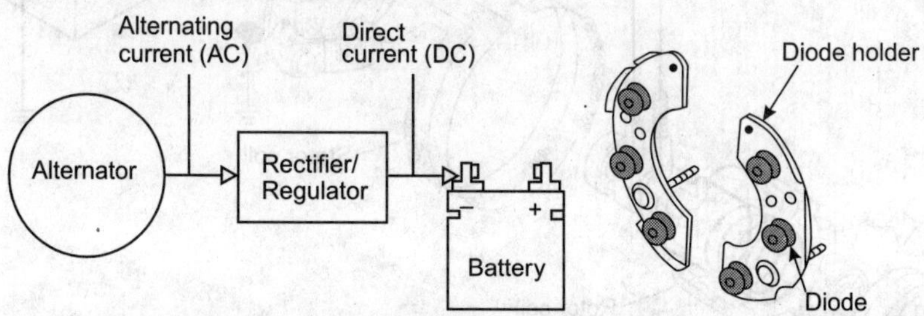


**Fig.4.15:** *Diode*

A diode is the simplest semiconductor device. It is an electrical one-way check valve that allows current to flow in one direction only. When the diode has the positive voltage on the anode, the diode is considered forward bias and current will flow through it. When a positive voltage is present on the cathode, the diode is reverse bias and current flow is prevented.

Three positive diodes and three negative diodes are contained in their respective diode holders. The current generated by the alternator is supplied from the positive side diode holder so it is insulated from the end frames.

During rectification, the diodes become hot, so the diode holders act to radiate this heat away and prevent the diodes from overheating. When the diodes heat up to above 150°C, they will no longer be able to rectify. Therefore, they are under forced cooling by the fan on the pulley to prevent them from heating over the specified temperature.

### Rectification

The electrical components of an automobile require direct current for operation and the battery needs direct current for charging.

The alternator produces three-phase alternating current but the charging system of the automobile cannot use this electricity unless it is converted to direct current. Converting alternating current to direct current is called rectification. Rectification can be accomplished in several ways but the automobile alternator uses a simple but effective diode.

As shown in the illustration 4.17, when six diodes are used, the three-phase alternating current is converted to direct current by full wave rectification. As the automobile alternator has built-in diodes, the electricity output is direct current.

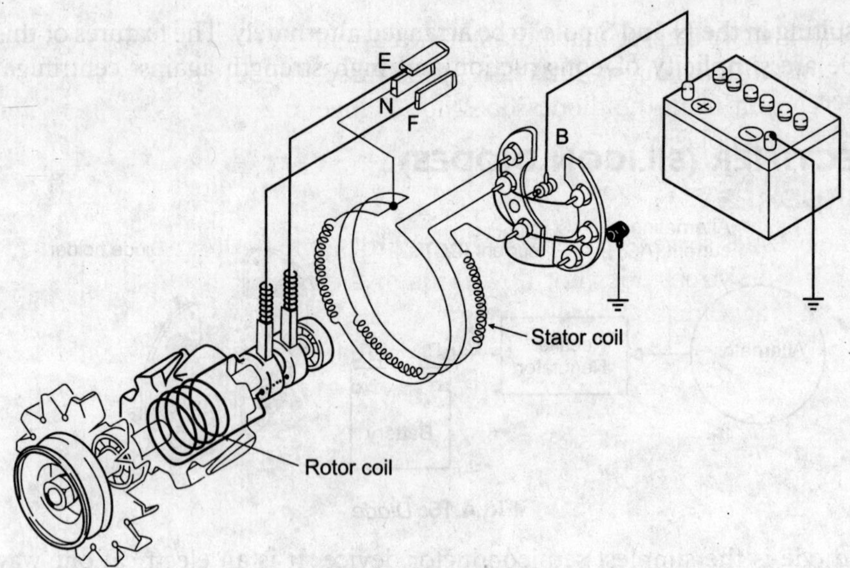


**Fig.4.16:** *Rectification circuit*

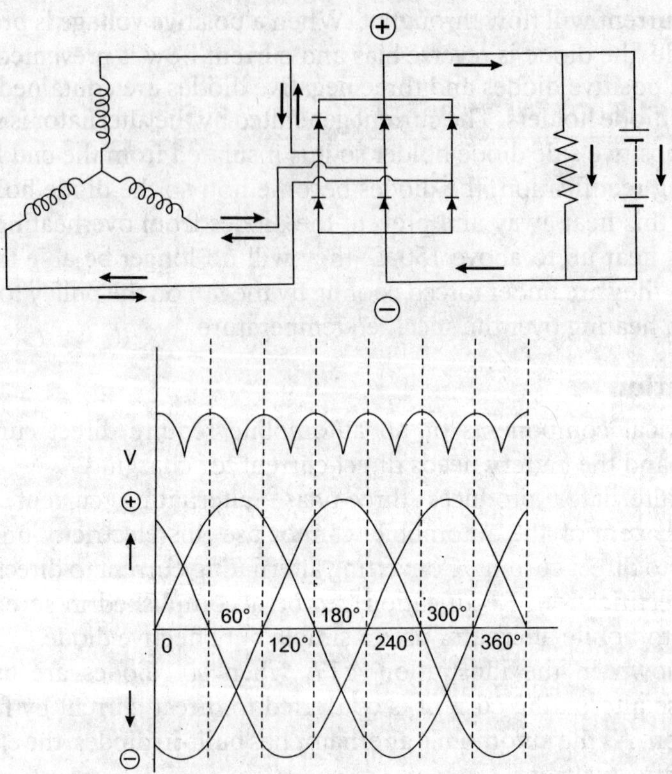


**Fig.4.17:** *Current rectification*

Thus, we can see that although the current from each coil to the diodes is constantly changing direction in the three lead wires, the direction of the current from the diodes does not change and it forms a circuit with unvarying polarity.

## BRUSHES

The brushes serve to pass a current of 2 to 3 amperes to the rotor coils. As the main current does not flow through these brushes, the brushes have much longer life and hardly require replacement.

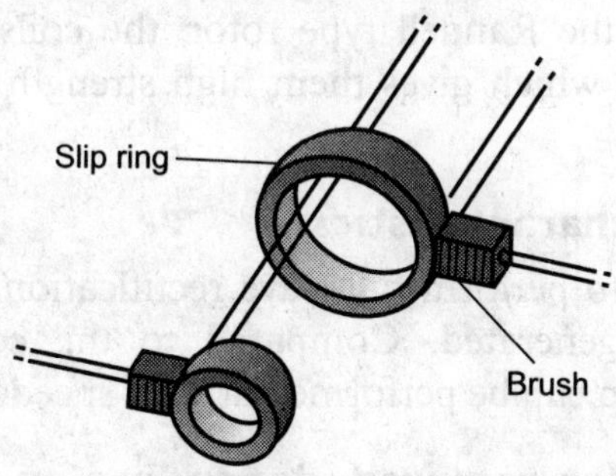


**Fig.4.18:** *Slip ring and brush*

## PRINCIPLE OF ALTERNATOR

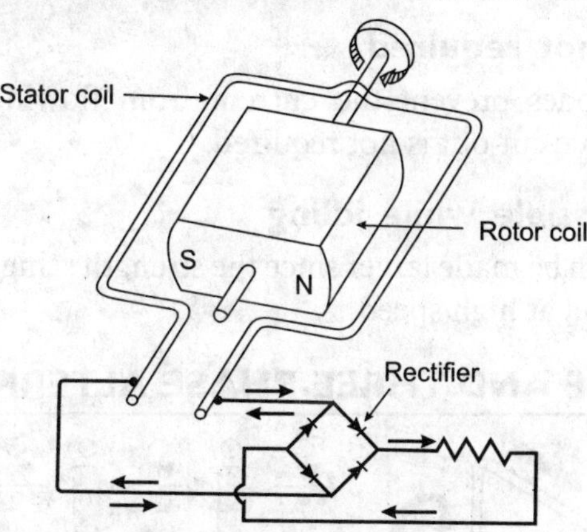


**Fig.4.19:** *Principle of alternator*

An alternator is designed to change mechanical energy into electrical energy. The working principle of an alternator is the electromagnetic induction (EMI). When the rotor spins, its strong magnetic field cuts across the stator windings. This induces current in the stator windings. If stator windings are connected to a load (for example Bulb), the load would operate.

## ADVANTAGES OF ALTERNATOR

The brushes contact on slip rings so that there is no sparking to shorten the brush life. In the alternator, the field coils (rotor coils) rotate so that the brushes are required to pass only a small current, and moreover there are no commutator grooves to wear out the brush.

### Rotor has high strength against speed

Only the field coils are wound on the rotor so that there are a small number of windings, which makes the rotor lighter and also stronger against high speed operation. In the Randell type rotor, the coils are wound in the direction of rotation which gives them high strength against centrifugal forces.

### Good rectifying characteristics

Six diodes are used to perform full-wave rectification of the three-phase alternating current generated. Compared to the commutator, higher efficiency rectification can be performed at high speeds without trouble.

### Current limiter not required, depending on the design

Due to the inductance resistance of the stator coil, the current produced by the alternator will not increase even if the output speed rises above the specified value.

### Cut-out relay not required

The rectifying diodes prevent the current from flowing in the reverse direction, and hence cut-out is not required.

### Charging is possible while idling

The pulley ratio can be made larger since the strength is high and generating characteristics good at high speed.

## SINGLE-PHASE AND THREE-PHASE ALTERNATORS

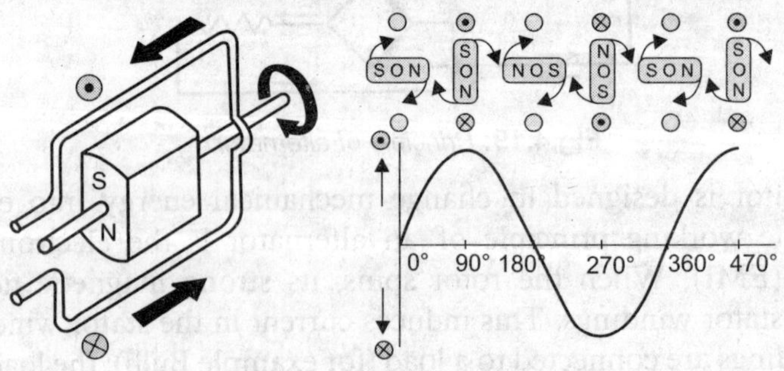


**Fig.4.20:** *Single phase alternating current*

When a magnet rotates within a coil, a voltage will be created between each end of the coil. This will give rise to an alternating current.

The relationship between the current generated in the coil and the position of the magnet is shown in the Fig 4.18. The largest amount of current is generated when the N and S poles of the magnet are closest to the coil. However, the current flows in opposite direction with each half turn of the magnet. Current which forms a sine wave in this manner is called "*single phase alternating current*".

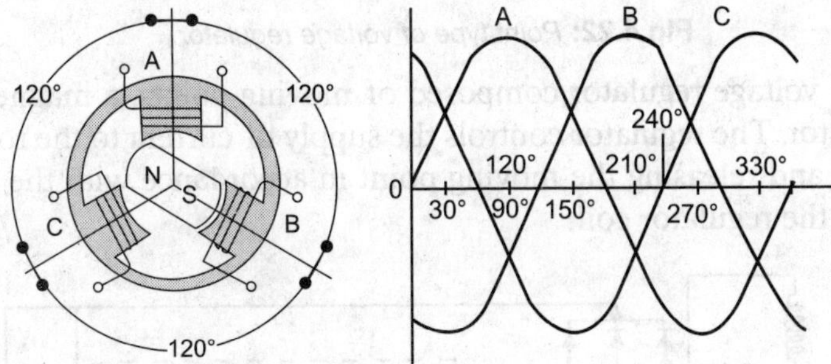


**Fig.4.21:** *Three phase alternating current*

To generate electricity more efficiently, the automobile alternator uses three coils arranged as shown in Fig.4.21. Each coil A, B and C is spaced 120° apart. When the magnet is rotated among them, alternating current is generated in each coil. The illustration shows the relationship between the three alternating currents like this is referred to as "three-phase alternating current". All modern automobile alternators generate three-phase alternating current.

## VOLTAGE REGULATOR

Under most driving conditions the load on the vehicle's electrical system is somewhat constant. An unregulated generator will have a variable output as the engine speed changes. To keep the head lamps at the same brightness, at both idle and high speed, a voltage regulator is needed.

The voltage regulator controls the alternator's output current to prevent over-charging and under-charging of the battery. The regulator increases and decreases the amount of field current to the rotor in order to control the amount of voltage produced by the alternator. When the voltage rises, a resistance is inserted in series with the field coil circuit to limit current flow through the field coils and lower the generated voltage. When the voltage drops, the resistance is removed from the field coil circuit to allow the generated voltage to build up again. The repetition of this action maintains the voltage at a constant value. Voltage regulators can be classified into the contact point type and the IC type.

## POINT TYPE VOLTAGE REGULATOR

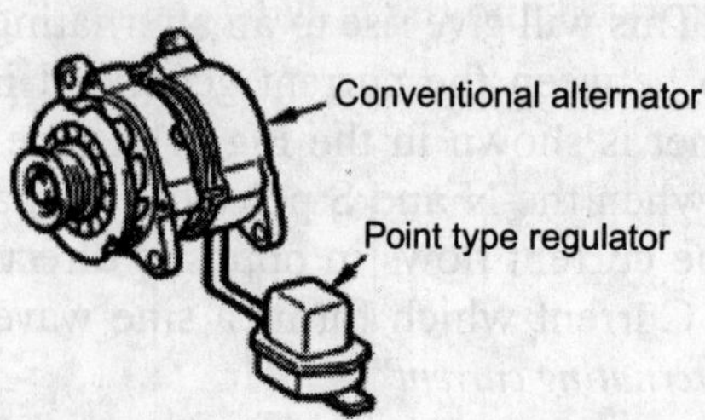


**Fig.4.22:** *Point type of voltage regulator*

Point type voltage regulator composed of moving points, a magnetic coil and a resistor. The regulator controls the supply of current to the rotor coil by pulling and releasing the moving point in accordance with the voltage applied to the regulator coil.

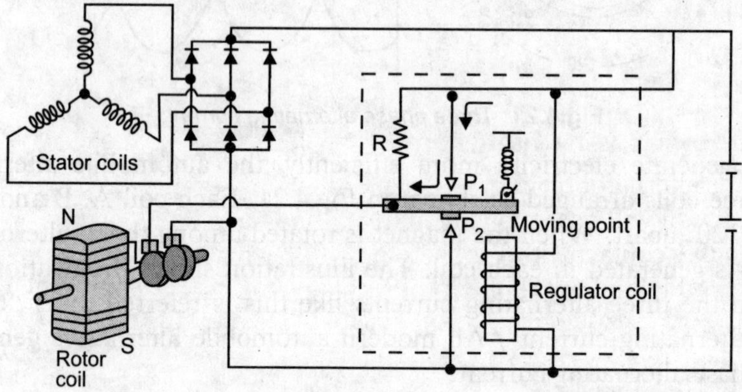


**Fig.4.23:** *Circuit of point type voltage regulator*

When the alternator is turning at a low rpm and a heavy load is applied, the voltage at the stator coil will become lower than the regulating voltage (approximately 14V) at the regulator. Accordingly, the moving point will make contact with $P_1$, thus allowing the battery current to flow to the rotor coil through $P_1$.

### Operation at high rpm and light load applied

When the alternator is turning at a high rpm and the voltage of the stator coil rises above that of the regulating voltage (approximately 14V) at the regulator, this voltage is applied to the regulator coil so it is affected by a greater pulling force which separates it from $P_1$.

When the moving point moves away from $P_1$, the current flowing to the rotor coil passes through resistor R and therefore is reduced in intensity. Less current to the rotor coil will cause the stator coil to generate less voltage and there will be reduced pulling force on the coil, so the moving point will return and make contact with $P_1$. This, in turn, causes a rise in the current to the rotor coil and the moving point again moves away from $P_1$.

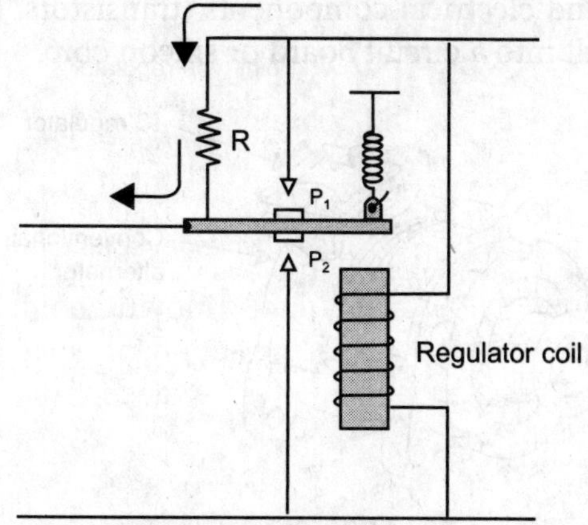


**Fig.4.24:** *Operation at high rpm and light load applied*

## Operation at very high rpm

When the alternator turns at an even faster speed, the voltage generated by the stator coil will rise, applying more pulling force to the regulator coil so that it makes contact with $P_2$. Therefore, the current through resistor R will then flow to $P_2$ and not to the rotor coil.

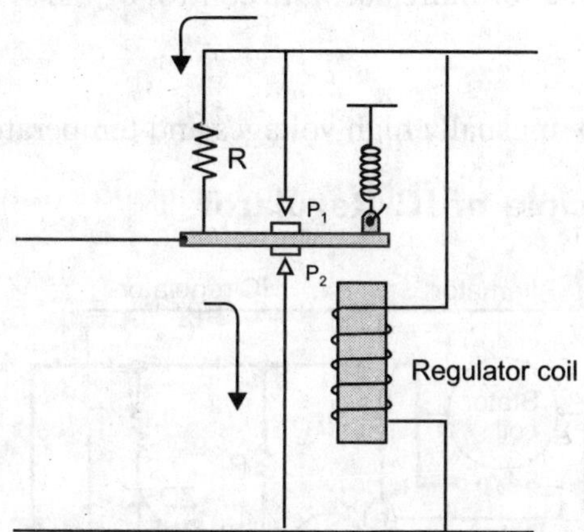


**Fig.4.25:** *Operation at very high rpm*

When there is no longer any current to the rotor coil, the stator cannot generate electromotive force, so the voltage of the alternator drops and the moving point moves away from $P_2$. Thus, once again the alternator voltage will rise and the moving point will be pulled.

## IC VOLTAGE REGULATOR (POINTLESS TYPE)

In IC regulator, Integrated Circuit (IC) is used instead of relay as in the point type regulator. An IC is a miniaturized circuit composed of many small electronics and electrical components (transistors, diodes, resistors, capacitors, etc) built into a circuit board or silicon chip.

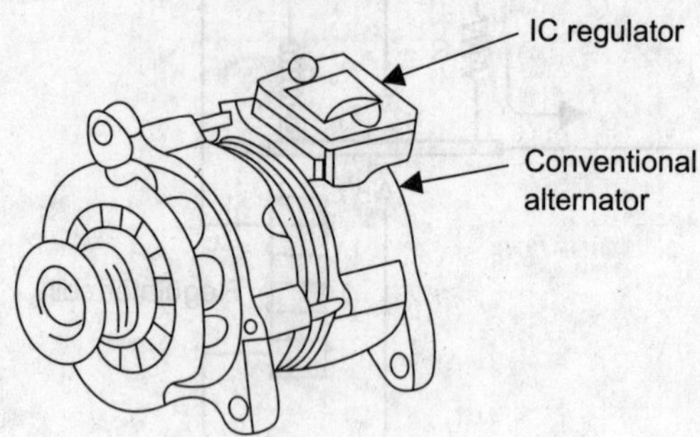


**Fig.4.26:** *IC regulator*

### Advantages

- Compact and light weight
- No moving parts to wear
- Excellent reliability due to its lack of mechanical points
- Good resistance to vibration and high durability due to lack of moving parts
- Faster switching for more accurate control of the system voltage

### Disadvantages

- Susceptible to unusually high voltages and temperatures

### Operating Principle of IC Regulator

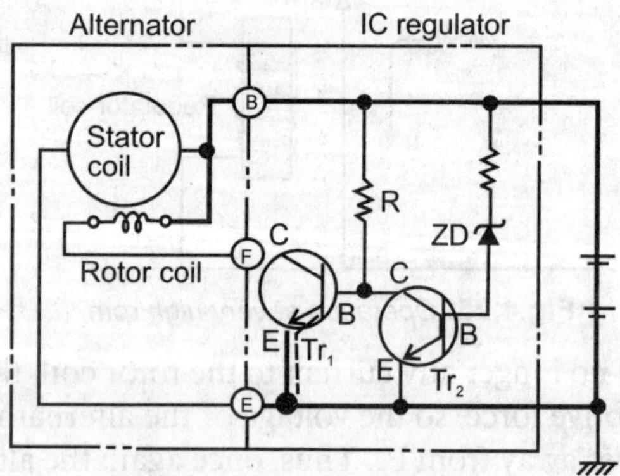


**Fig.4.27:** *Output voltage is low*

When the output voltage at terminal B is low, the battery voltage is applied to the base of $Tr_1$ through resistor R, and $Tr_1$ turns on; at the same time, field current to the rotor coil flows from B Þ rotor coil Þ F Þ $Tr_1$ Þ E.

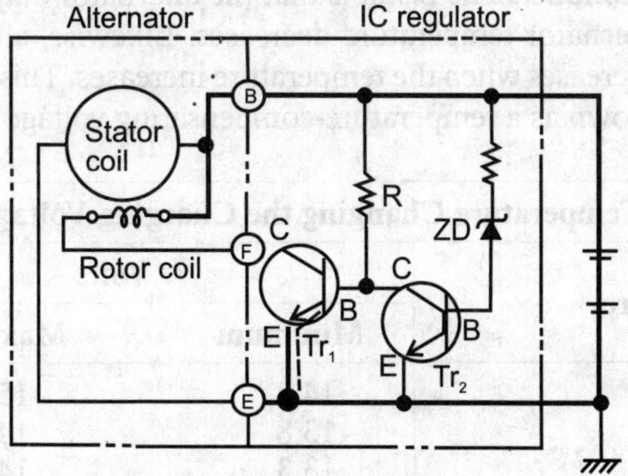


**Fig.4.28:** *Output voltage is high*

When the output voltage at terminal B is high, a higher voltage is applied to the Zener diode (ZD) and when this voltage reaches the Zener voltage, the ZD becomes conductive. Accordingly, as $Tr_2$ turns on, $Tr_1$ turns off. This interrupts the field current, regulating the output voltage.

> ***Note:*** *The difference between the Zener diode and a normal diode is that when any voltage more than a certain level is applied in the reverse direction, the Zener diode will become conductive and will allow a current to flow.*

## COMPENSATED VOLTAGE REGULATOR

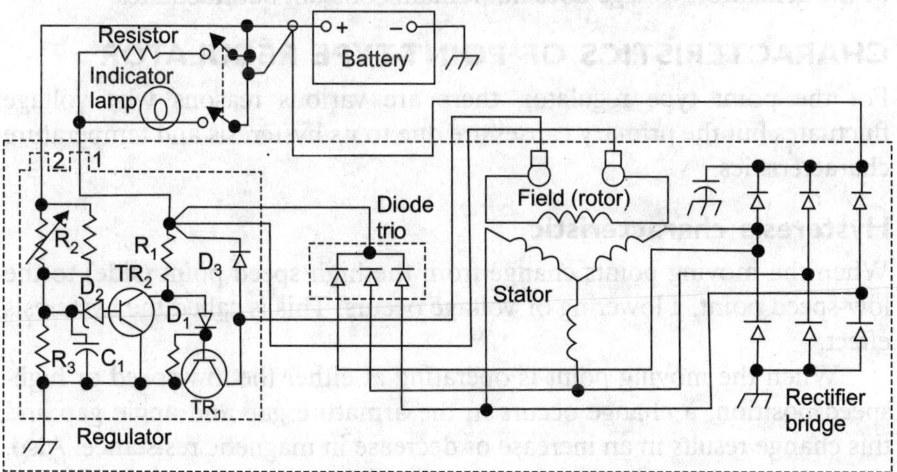


**Fig.4.29:** *Compensated voltage regulator*

Notice that the resistor $R_2$ in Fig.4.29 is shown as a variable resistor. The resistance of $R_2$ increases as the temperature decreases. This change in resistance requires the alternator output voltage to increase before the Zener diode $D_2$ will conduct. The result is that the alternator output voltage will increase as alternator temperature decreases. Likewise, alternator output voltage also decreases when the temperature increases. This type of voltage regulator is known as a temperature-compensating voltage regulator.

| Temperature Changing the Charging Voltage | | |
|---|---|---|
| **Temperature** | **Volts** | |
| | **Minimum** | **Maximum** |
| 20°F | 14.3 | 15.3 |
| 80°F | 13.8 | 14.4 |
| 140°F | 13.3 | 14.0 |
| Over 140°F | < 13.1 | 13.0 |

The increase in alternator output voltage in cold temperatures recharges the battery more efficiently. The decrease in alternator output voltage in hot weather prevents overcharging the battery. This temperature-compensated voltage regulator is utilized in some trucks, although many modern trucks have a non-temperature compensated voltage regulator.

## REGULATOR CHARACTERISTICS

The function of the regulator is to maintain the voltage generated by the alternator at a constant level. Actually, however, due to the characteristics of the generator, voltage does not remain constant but fluctuates.

## CHARACTERISTICS OF POINT-TYPE REGULATOR

For the point type regulator, there are various reasons why voltage fluctuates but the primary causes are due to its hysteresis and temperature characteristics.

### Hysteresis characteristic

When the moving points change from the high speed point (side) to the low-speed point, a lowering of voltage occurs. This is called the hysteresis effect.

When the moving point is operating at either the low-speed or high-speed position, a change occurs in the armature gap and angle gap and this change results in an increase or decrease in magnetic resistance. Also, when the moving point changes from the high-speed side to the low-speed side, residual magnetism from high-speed operation remains in the coil core and continues to pull on the moving point for a short time. These two phenomena cause the alternator output voltage to decline.

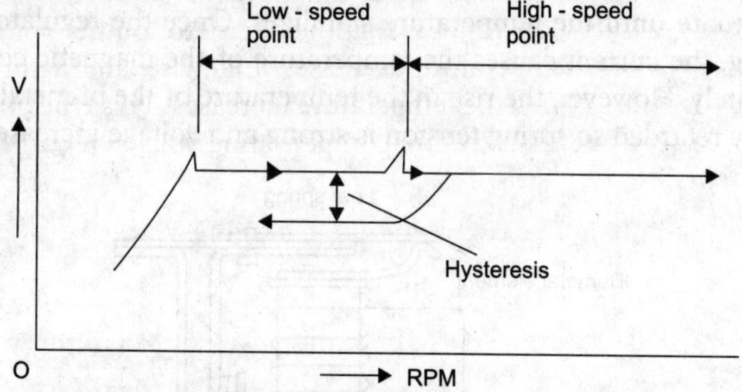


**Fig.4.30:** *Hysteresis characteristic*

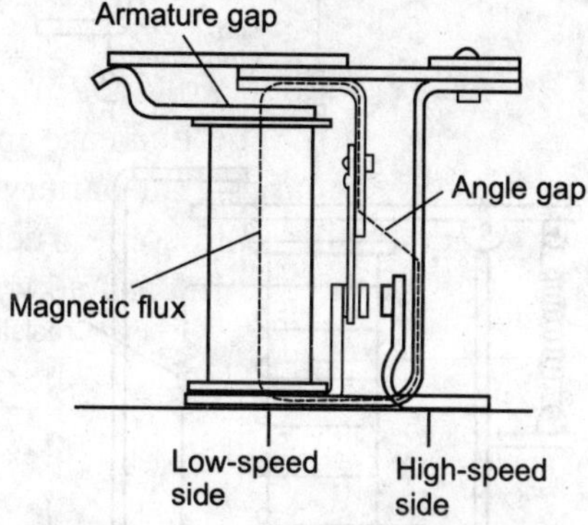


**Fig.4.31:** *Hysteresis characteristic*

## Temperature characteristic

The magnetic coil of the voltage regulator employs copper wiring, and if the temperature of this wire rises, resistance will be increased and there will be a reduction in the pulling force (electromagnetic force) of the magnetic coil, resulting in a higher alternator output voltage.

To prevent such a rise in voltage, the regulator utilizes either a *resistor* or a *bi-metal element* for temperature compensation, but some regulators use both.

The resistor has either a Nichrome wire or a carbon element, with a low temperature resistance coefficient, and is connected in series to the coil. It reduces the ratio of the overall resistance in accordance with the temperature fluctuation.

The bi-metal element is used together with a spring which supports the moving point. The bi-metal element reduces the spring tension as the temperature rises. After the regulator begins operating, the voltage will fluctuate until the temperature stabilizes. Once the regulator begins operating, the current causes the temperature of the magnetic coil to rise immediately. However, the rise in the temperature of the bi-metal element is slightly retarded so spring tension is strong and voltage increases.

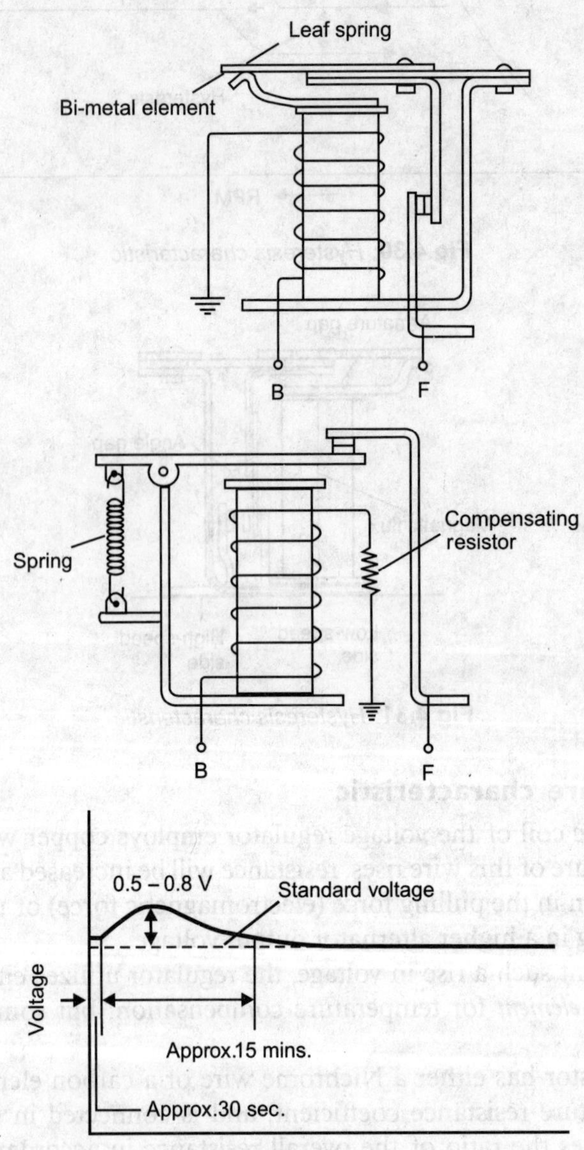


**Fig.4.32:** *Temperature characteristics*

# CHARACTERISTICS OF IC REGULATOR

## Battery load characteristics

There is little or no variation in output voltage (not more than 0.1 to 0.2 V) with changes in alternator speed, and no hysteresis characteristics as with the point type.

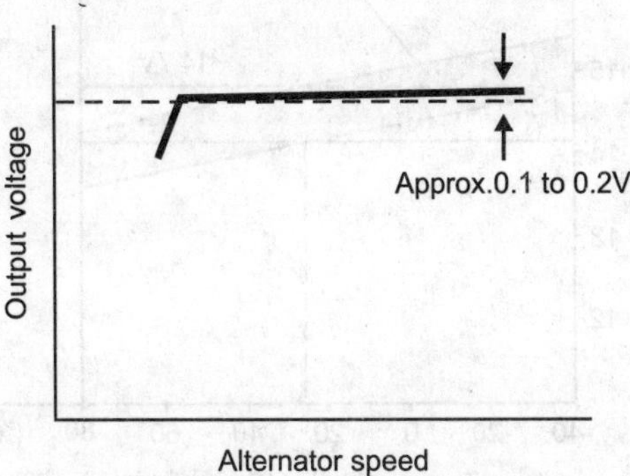


**Fig.4.33:** *Battery load characteristics*

## External load characteristics

Output voltage becomes lower as the load current increases. There are no hysteresis characteristics as with the point type regulator. Voltage variation, even at the rated load, or maximum output current of the alternator, is between 0.5 V and 1 V.

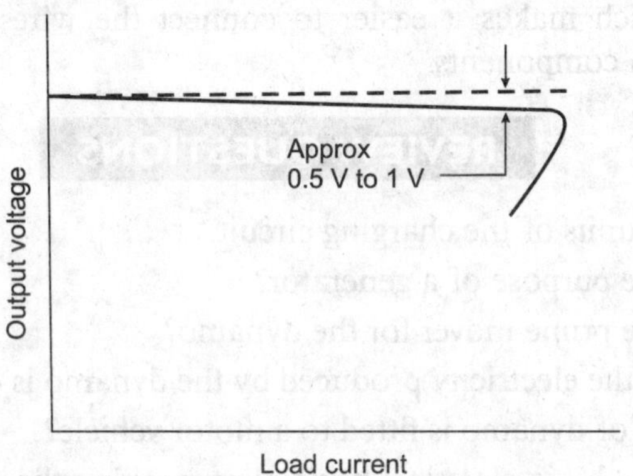


**Fig.4.34:** *External load characteristics*

## Temperature characteristics

Since the Zener diode used for regulating the output voltage tends to become more conductive as the ambient temperature rises, the output voltage generally becomes lower as the temperature rises. Since the output voltage drops at high temperatures (eg, during summer) and rises at low temperatures (e.g., during winter), proper charging suited to the battery's characteristics is performed at all times.

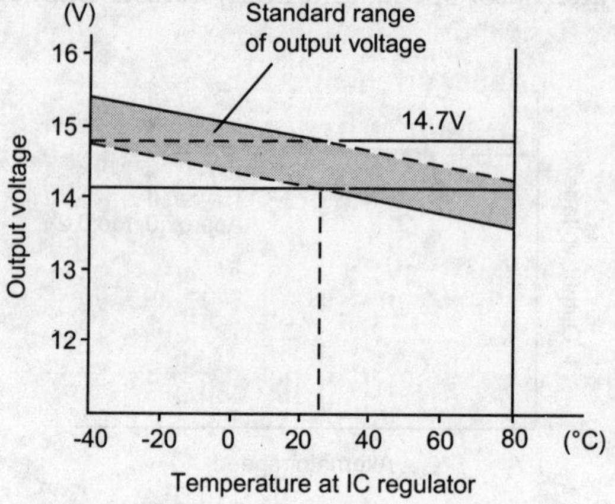


**Fig.4.35:** *External load characteristics*

## Wiring harness

A wire harness is a group of individually insulated wires and cables, connecting parts, circuit protecting parts, etc., all grouped together for easy connection between electrical components of a vehicle. The wires are colour coded for tracing ease. Wire ends are equipped with lugs or connectors, which makes it easier to connect the wires to the various electrical system components.

## REVIEW QUESTIONS

1. Name the units of the charging circuit.
2. What is the purpose of a generator?
3. What is the prime mover for the dynamo?
4. How does the electricity produced by the dynamo is controlled?
5. What type of dynamo is fitted to a motor vehicle?
6. What is the basic electrical principle governing the operation of an alternator?
7. Name and state the purpose of the components of an alternator.
8. How are the field windings of a generator energised?
9. What is a cut out in an automobile? What will happen if a cut out is not provided in a DC generator?
10. State the three purposes of a generator regulator.

11. What is the purpose of the voltage regulator?
12. What is the purpose of the current regulator?
13. What is the purpose of alternator?
14. Explain the construction of alternator.
15. Describe the constant current and voltage regulator with a sketch.
16. Draw the circuit diagram of Bridge wave rectifier.
17. On what principle dynamo or generator works?
18. How the EMF is produced in the dynamo and alternator?
19. What is the difference between dynamo and alternator?
20. What do you mean by three brush control?
21. What may be the reasons when a generator fails to produce the voltage?
22. What are the advantages for using the alternator instead of DC generator?
23. How the AC voltage produced is rectified in alternators?

# UNIT 5

# Starting System

Automotive engines are not capable of self-starting. They must be rotated or started before it will run under its own power. In the early days of automobiles, engines were cranked by hand. But this was troublesome, unreliable and often dangerous. Also engines became larger. So starting system replaced the hand crank. In 1912, Charles kettering invented the first practical self-starter. The main elements in a starting system are battery, start switch, transmission neutral safety switch, starter motor control switch, starter motor and cables and wires to connect these components.

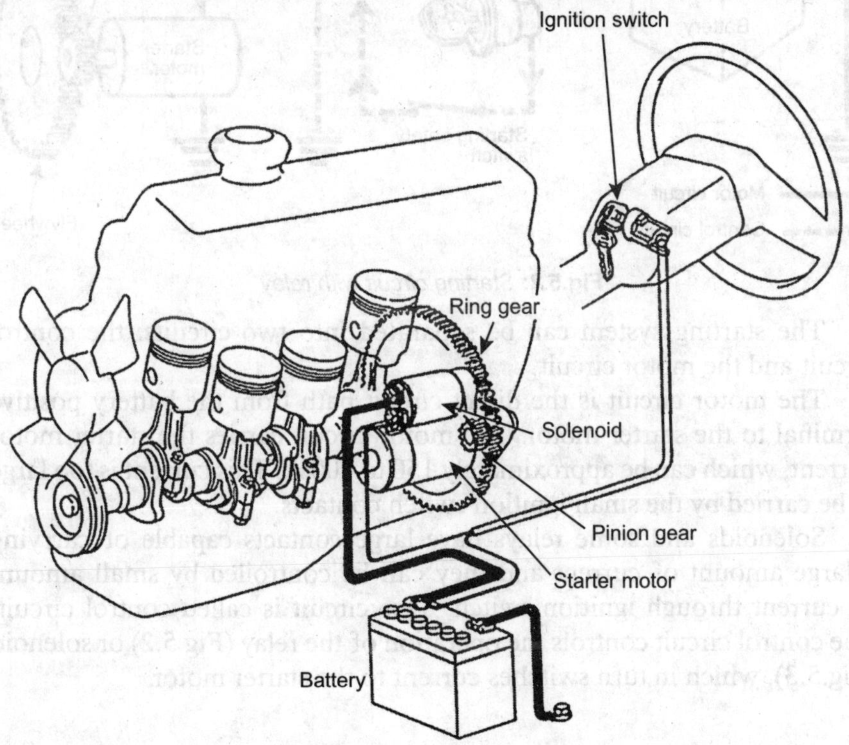


**Fig.5.1:** *Starting system*

The battery sends current to the starter motor when the driver turns the ignition switch to start. This causes a pinion gear in the starter motor mesh with teeth on the ring gear around the engine flywheel. Then the starter motor rotates the crankshaft for starting.

When starting, starter motor gets a heavy dose of current from the battery. Normally high current creates more heat. If the starter motor cranks the engine for long time, more heat will be generated. This results in damage of the motor. For this reason, the starter motor must never be used for more than 30 seconds continuously when starting. Also cranking should not be repeated without a pause of at least 2 minutes. During the first start, heat will be induced and it should be dissipated. Two minutes gap should be given between one start and the subsequent start to allow the heat dissipation.

## Starting circuits

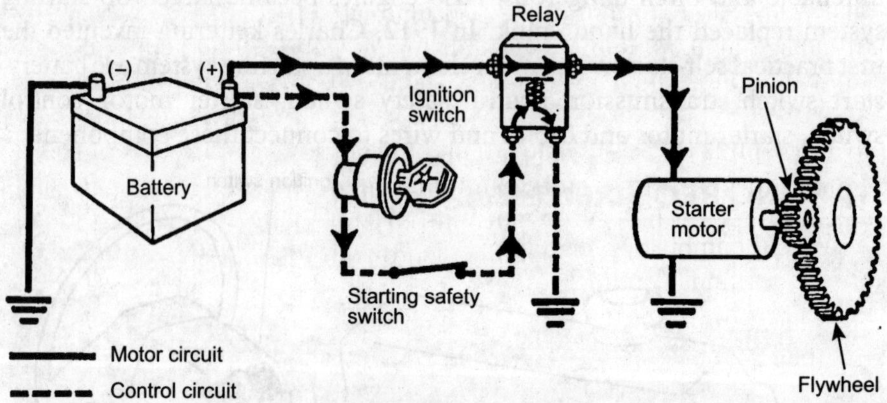


**Fig.5.2:** *Starting circuit with relay*

The starting system can be separated into two circuits: the control circuit and the motor circuit.

The motor circuit is the direct circuit path from the battery positive terminal to the starter motor. The motor circuit carries the starter motor current, which can be approximately 150 to 400 A. This current is too large to be carried by the small ignition switch contacts.

Solenoids and some relays have large contacts capable of carrying a large amount of current and they can be controlled by small amount of current through ignition switch. This circuit is called control circuit. The control circuit controls the operation of the relay (Fig.5.2) or solenoid (Fig.5.3), which in turn switches current to the starter motor.

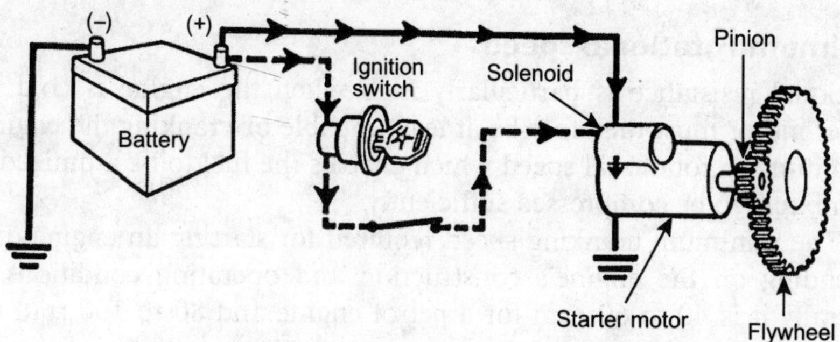


**Fig.5.3:** *Starting circuit with solenoid*

## FUNCTIONS

The starting system:

1. Uses a powerful electric motor to drive the engine at about 200 rpm (fast enough to allow the fuel and ignition systems to operate).
2. Drives the engine through a pinion gear engaged with a ring gear on the flywheel.
3. Disengages as soon as the engine starts.

## STARTING REQUIREMENTS

An internal combustion engine requires the following criteria in order to start and continue running.

### Combustible mixture

An engine to be started has to have proper air-fuel mixture so that the combustion can easily happen to make the engine to start. The quick start will reduce the strain on the starter motor. The fuel system of an engine has to ensure to provide the proper combustible mixture so that the starter motor can easily start an engine.

### Compression stroke

The effective compression stroke helps for easy starting of an engine. During the compression stroke, the air/air-fuel mixture is brought to the ignition temperature for easy combustion.

### A form of ignition

The combustible compressed air-fuel mixture or compressed air is once ready, then a suitable source has to be provided to ignite the air-fuel mixture or air in order to initiate combustion so that the engine gets started. An effective form of ignition should be powerful and should occur at the correct moment. This will make the starter motor to start the engine easily.

### Minimum rotational speed

Frictional resistance is particularly high when the engine is cold. The starter motor must therefore be at least capable of cranking the engine at the minimum rotational speed which enables the fuel to be atomized and the mixture to be compressed sufficiently.

The minimum cranking speed required for starting an engine differs depending on the engine's construction and operating conditions, but generally it is 40 to 60 rpm for a petrol engine and 80 to 100 rpm for a diesel engine.

The following table gives empirical values for minimum cranking speeds at the minimum starting temperature.

| Required cranking speed at -20°C | Cranking speed, rpm |
|---|---|
| Reciprocating piston SI engine | 60 – 90 |
| Direct Injection diesel engine | |
| • Without starting aid | 80 – 200 |
| • With starting aid (Glow plug) | 60 – 140 |
| Pre-chamber and swirl chamber diesel engine (IDI engine) | |
| • Without starting aid | 100 – 200 |
| • With starting aid (Glow plug) | 60 – 100 |

### Sufficient cranking torque

The compression ratio of a diesel engine and a modern petrol engine are high, which demand high cranking torque. To obtain sufficient cranking torque from the starter motor, a large current (amperage) is required. On a cold morning, this may be as high as 500A although between 100 and 150A are more normal values for light vehicles. To carry such high amperages, suitable thick, short supply cables and a good battery capable of delivering this high current are required.

### Minimum starting temperature

This is the lowest engine and battery temperature at which the system must ensure that the engine starts. 2.2 KW starter and a 12V, 90 Ah, 450A battery is required for a minimum starting temperature of -23°C.

| Engines for | Starting limit temperatures |
|---|---|
| Passenger cars | -18 to -25°C |
| Trucks and Buses | -15 to -20°C |
| Tractors | -12 to -15°C |
| Diesel locomotives | +5°C |

At low temperature, the internal resistance of the battery is low. Starter speed decreases due to increased internal resistance of the battery. Similarly minimum required initial engine speed increases due to increased cranking resistance i.e. increased cranking resistance needs more starting speed. The intersection of both curves (Fig.5.4) gives the starting limit temperature.

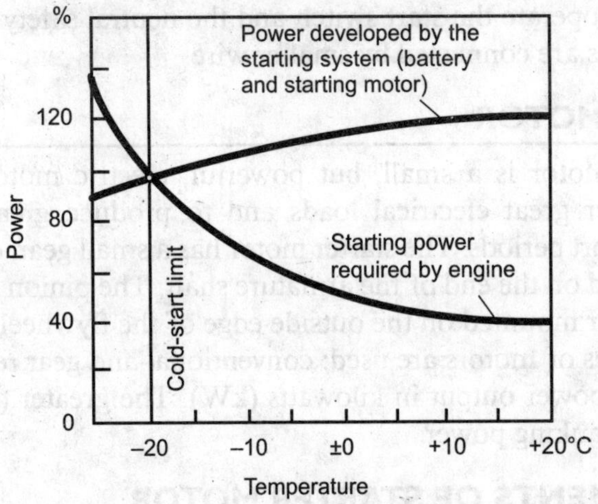


**Fig.5.4:** *Supply and demand power curves*

It also shows a typical characteristic giving a crossover at -20°C between supply and demand of power for starting a given vehicle.

### Starter pinion/ring gear ratio

The gear ratio of ring gear teeth to pinion gear teeth is usually between 8:1 to 20:1. A typical ratio is 20 to 1. It means starter motor rotates 20 times faster than the flywheel. There are 20 teeth on the flywheel for every tooth on the pinion. To turn the engine over at 100 rpm, the starter motor must rotate at 2000 rpm (100 ′ 20).

### Rated voltage of the starting system

Higher output can be obtained from high voltage compact starters.

| Engines for | Rated voltage |
|---|---|
| Passenger cars | 12 V |
| Tractors & Marines | 12 V |
| Trucks & Buses | 12 V & 24 V |
| Large commercial vehicles | 24 V |

### Length of the cable

In a starting circuit, there are two current flow paths. One path by short heavy block line and other path by long smaller wire. The heavy block line running from the battery to starter motor which has low resistance, so that starter motor gets a heavy dose of current from the battery. The other path by long smaller wire running from the start switch to the starter motor control switch. Smaller wire gives high resistance so that current flow is reduced. Long cable will have voltage drop also. Less amount of current is enough to operate the start switch and the neutral safety switch so that these switches are connected by smaller wire.

## STARTER MOTOR

The starter motor is a small, but powerful, electric motor designed to operate under great electrical loads and to produce great amounts of torque for short periods. The starter motor has a small gear, called a pinion gear, mounted on the end of the armature shaft. The pinion gear engages a large ring gear mounted on the outside edge of the flywheel.

Two types of motors are used: conventional and gear reduction. Both are rated by power output in kilowatts (kW). The greater the output, the greater the cranking power.

## REQUIREMENTS OF STARTER MOTOR

The requirements of starter motor are:

- Continuous readiness for starting
- Sufficient starter power at different temperatures
- Long service life for a high number of starts
- Robust design (to withstand meshing, cranking, high temperature, etc)
- Compact and light weight, helps mounting easier
- Longest possible maintenance free service life

## PRINCIPLE

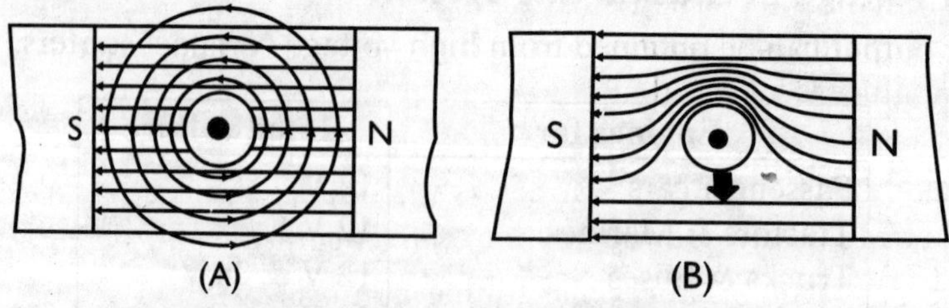


**Fig.5.5:** *Principle of a starter motor*

A current passing through a conductor produces concentric lines of force around the conductor, the strength of which is proportional to the current.

The direction of the lines of force can be determined by the right-hand screw rule. Assume the conductor to be placed between magnetic N and S poles as shown in Fig.5.5A. In this case, the lines of force in opposed direction will counteract each other, while those in the same direction will add up to make up bent lines of force as shown in Fig.5.5B.

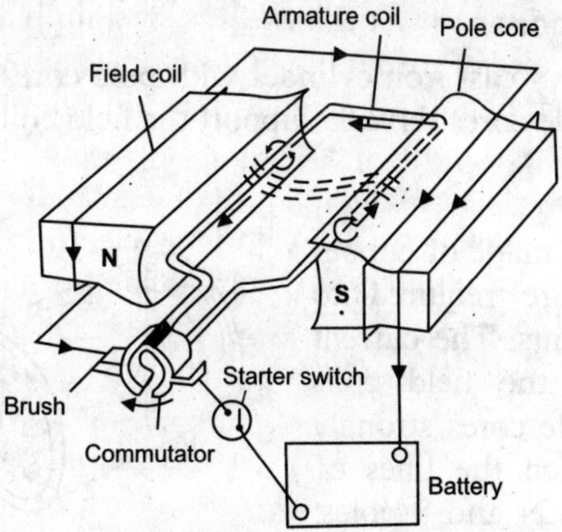


**Fig.5.6:** *Starter motor*

This pattern develops because the lines of force below the conductor run in opposite direction and counteract each other, while those above the conductor add up and become twice as strong. The dense and bent lines of force act like rubber bands and produce a force tending to push the conductor downward. The starting motor utilises this force to produce torque.

The current from the battery flows through the field coils and armature coils and returns to the battery. The S pole side of the coil receives the magnetic force in a downward direction while the N pole side of the coil receives the magnetic force in a upward direction so that torque is produced in the armature coil in a clockwise direction to cause the armature to rotate in the arrow direction. Since the armature is provided with a commutator and brushes, the direction of current flow through the armature is reversed when the armature coil rotates 180° so that the direction of torque produced in the armature is always the same.

## STARTER CONSTRUCTION

The starter can be divided roughly into three operating parts, as follows:

1. The part that produces the torque comprising the armature, field coils, brushes, etc.
2. The power transmitting mechanism including the overrunning clutch (starter clutch), and

3. A starter drive mechanism that advances the starter clutch to engage with the flywheel ring gear.

The starters used in automobiles are the same in construction and operation, but differ in working voltage, size and number of pole cores.

### Yoke and pole cores

The yoke is a heavy cast iron cylinder with pole cores screwed on to the inner side. The pole cores serve to support the field coils.

### Field coils

The field coils are made of copper bands as they are required to carry heavy currents. The current flowing through the field coils magnetize the pole cores strongly so as to strengthen the lines of force between the N and S poles. The yoke serves as the return path for the lines of force. The field coils are connected in series with the armature coil so that the current flowing through the field coils also flows through the armature coil. Four field coils are generally used.

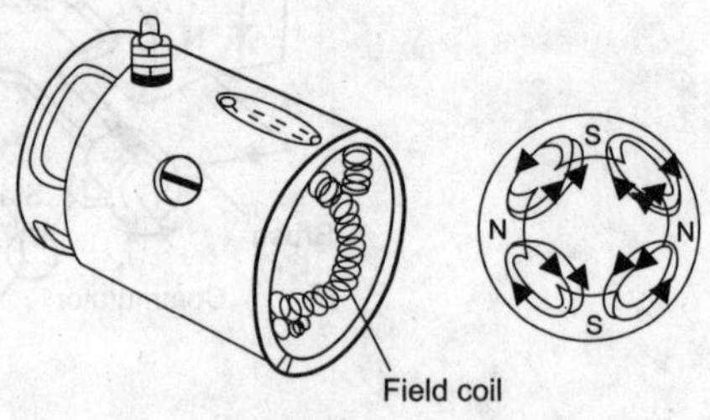


**Fig.5.7:** *Field coil*

### Armature

The armature is comprised of the slotted iron cylinder, shaft, commutator, and armature coils, and rotate between the pole core supported at both ends by bearings. The coils are made of copper bands to allow carrying heavy currents. They are insulated and wound in the slots of the iron cylinder. A large number of coils are used, each connected to the commutator. As discussed in the principle of motor, magnetic force acts on each coil, causing the armature to rotate.

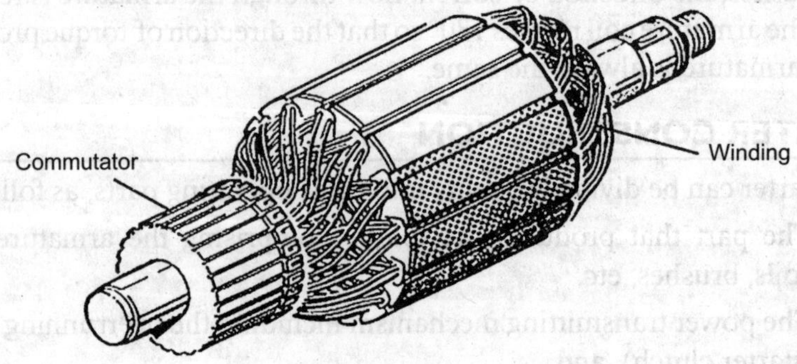


**Fig.5.8:** *Armature*

### Brushes

There are normally four brushes. Two of these are held in insulated holders and connected to the armature coil through the commutator, and also to the field coils. The other two brushes are held in grounded holders and connected to the ground.

### Power transmission mechanism

The power transmission mechanism transmits the torque developed by the motor to the engine flywheel to crank the engine.

The pinion that transmits the power to the flywheel ring gear has a small number of teeth when compared to the ring gear, usually in the reduction ratio of 8 ~ 20:1 (reducing the speed raises the torque).

When the engine starts, it begins to turn at a very fast speed. If the ring gear and pinion were to remain in mesh at this time, the armature would be spun at very high speed, and be damaged. For example, when the engine starts, the flywheel speed will be 1500 rpm. Then the starter pinion speed will be very high i.e. 30000 rpm (1500×20). At this high speed, starter motor shafts, armature slots and the segments will be thrown outward and motor will be damaged. Therefore, a provision must be made to connect the pinion to the ring gear when the engine is being cranked and to disengage the gears after the engine has started. This is done by the starter drive mechanism.

To prevent such a condition, the pinion is made to turn free and separate from the ring gear as soon as the engine starts.

## CONVENTIONAL STARTER MOTOR

The conventional starter motor contains the components shown in Fig.5.9. The pinion gear is on the same shaft as the motor armature and rotates at the same speed. A plunger in the magnetic switch (solenoid) is connected to a shift lever. When activated by the plunger, the shift lever pushes the pinion gear and causes it to mesh with the flywheel ring gear. When the engine starts, an over-running clutch disengages the pinion gear to prevent engine torque from ruining the starter motor.

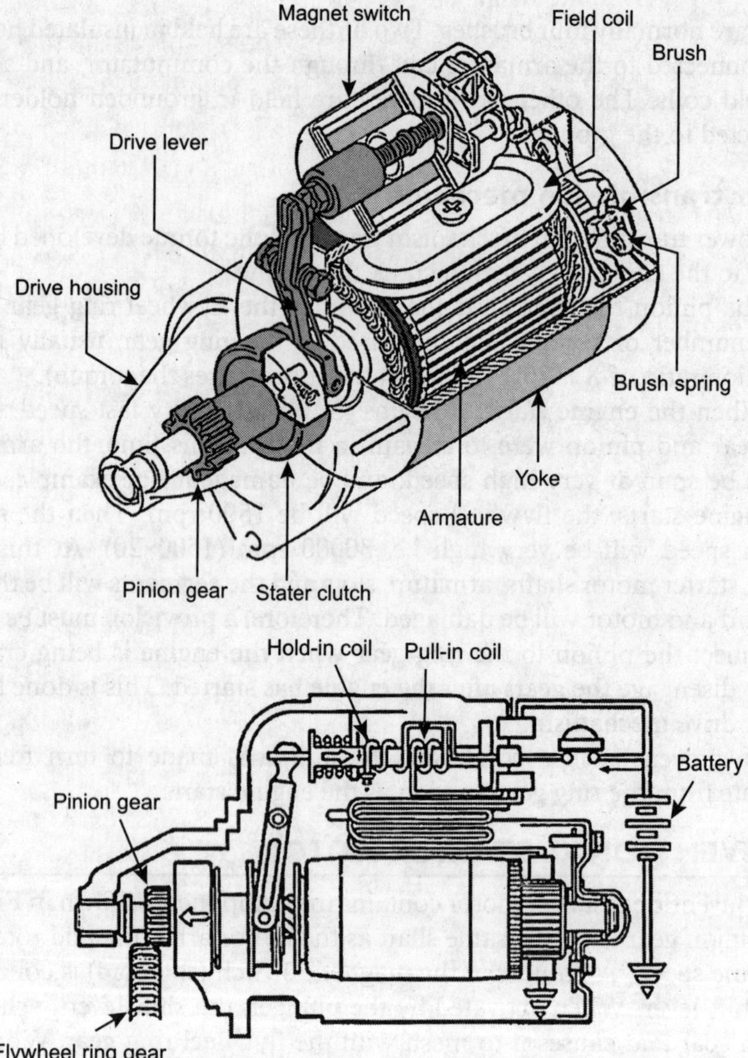


**Fig.5.9:** *Conventional starter motor*

## REDUCTION TYPE STARTER MOTOR

Many petrol engines are today fitted with a very compact, light-weight, reduction-type starter motor. The reduction-type starter motor is fitted with a high speed motor, a set of reduction gears together with other conventional pre-engaged starter motor components.

The motor rotates at a much higher speed than the conventional type of pre-engaged starter motor. The use of the reduction gears reduces the speed of the pinion by approximately three times that of the motor. The use of a high-speed motor increases the torque of the motor, allowing the engine to rotate at a greater rate from rest, thus improving the engine's ability to start. The solenoid acts directly on the pinion, pushing the pinion into the flywheel ring gear.

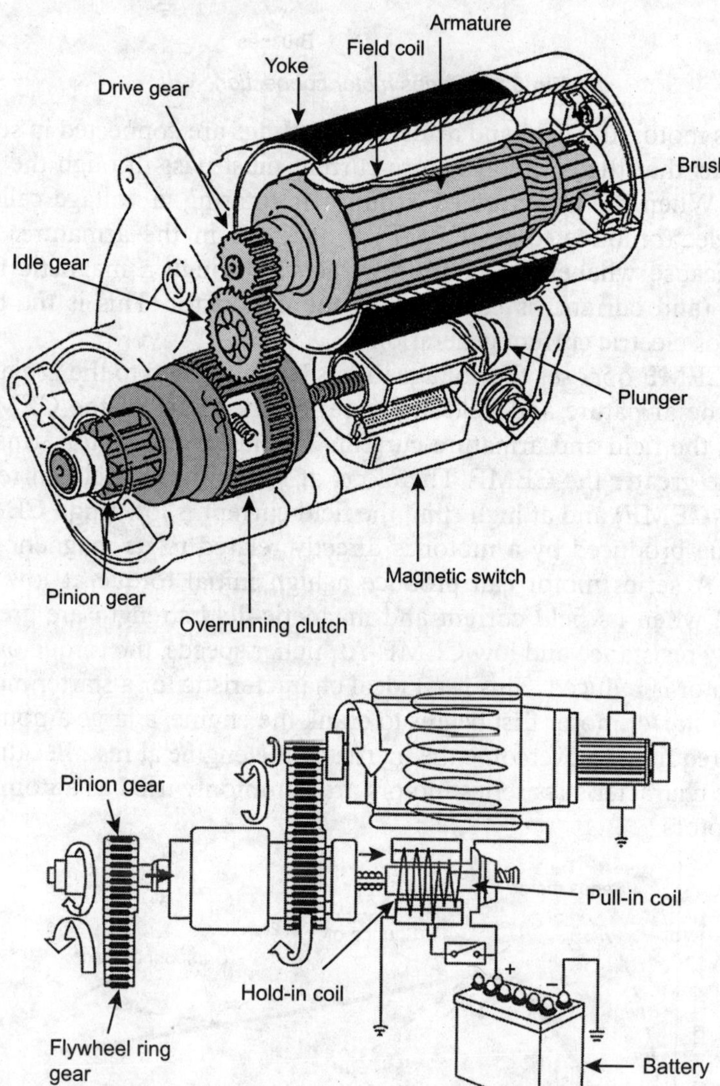


**Fig.5.10:** *Reduction type starter motor*

## TYPES OF STARTER MOTOR

The four main types of starter motor are referred to as series, shunt, compound, and permanent magnet excitation.

## SERIES MOTOR

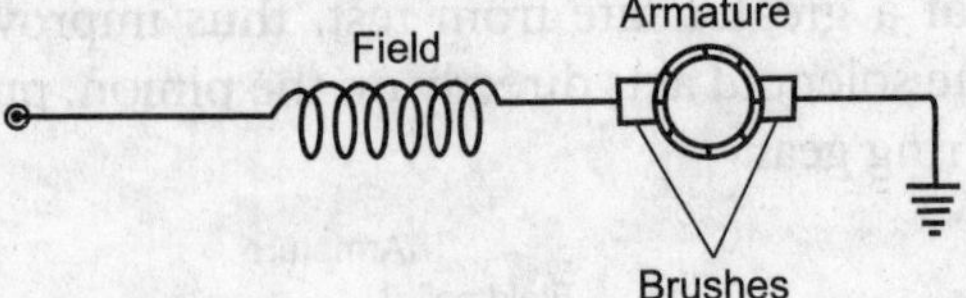


**Fig.5.11:** *Series motor connection*

In a series motor, the field and armature windings are connected in series. This means that the entire armature current must pass through the field winding. When the armature of a motor is rotating, a voltage called a counter electromotive force (CEMF) is induced in the armature. This occurs because whenever a conductor moves through a magnetic field, a voltage (and current) is generated in the conductor. This is the basic principle of electric current generation.

The CEMF opposes the voltage that is being applied to the armature. Because the armature and field are in series, the effect of the CEMF is to reduce the field and armature current. The faster the motor armature rotates the greater the CEMF. Therefore, at low rpm, the field current is high (low CEMF) and at high rpm, the field current is low (high CEMF). The torque produced by a motor is directly related to its magnetic-field strength. A series motor can produce a high initial torque at low rpm (Fig.5.12) when its field current and magnetic-field strength are greatest due to low resistance and low CEMF. At higher speeds, the torque output of the motor is reduced. This is an ideal characteristic for a starter motor. When the starter motor first begins to crank the engine, a large amount of torque is required to overcome the inertia of the engine at rest. Because of the above characteristics series motors are commonly used as automotive starter motors.

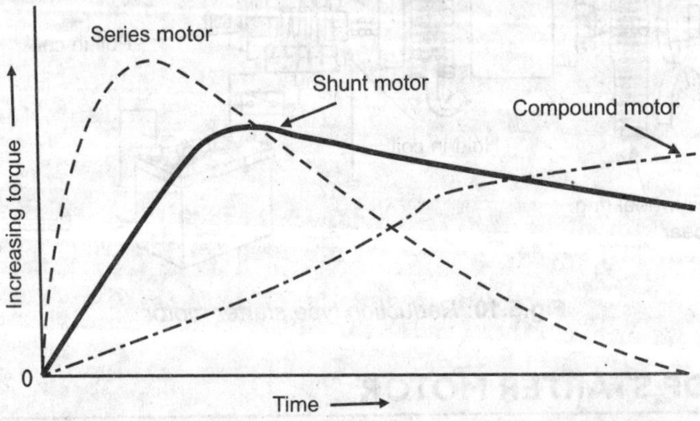


**Fig.5.12:** *Torque and speed relationships of starter motors*

## SHUNT MOTOR

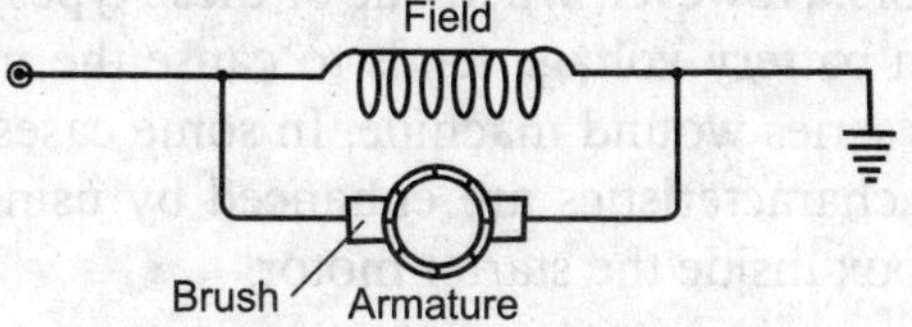


**Fig.5.13:** *Shunt motor connection*

In a shunt motor, the armature and field are connected in parallel. This means the CEMF which is generated in the armature does not affect the field current. The magnetic field strength is constant as motor speed changes. A shunt motor tends to regulate its own speed. If the load on the motor increases, the motor slows down and the CEMF is reduced. A reduced CEMF causes an increase in armature current which, in turn, increases the motor speed. These motors are used if constant speed is more important than starting torque.

## COMPOUND MOTOR

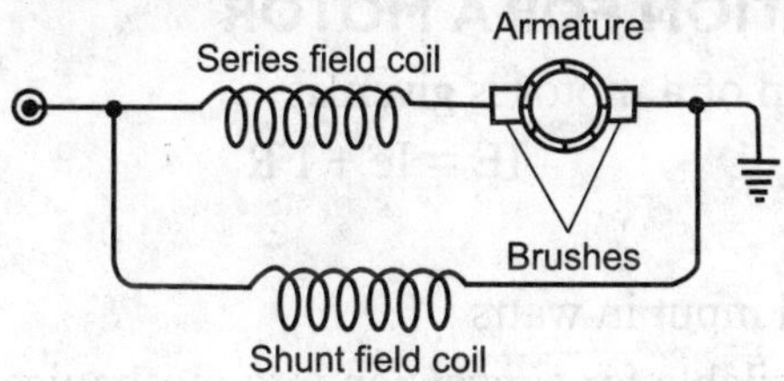


**Fig.5.14:** *Compound motor connection*

In a compound motor, some of the field windings are in series, and some of the windings are in parallel or shunt. This type of motor combines the characteristics of the series motor and the shunt motor. The compound motor has a high starting torque and a fairly constant speed. Compound motors are also used as automotive starter motors.

## PERMANENT MAGNET MOTOR

Permanent magnet motors are smaller and simpler compared with the other three discussed. Field excitation, as the name suggests, is by permanent magnet. This excitation will remain constant under all operating conditions.

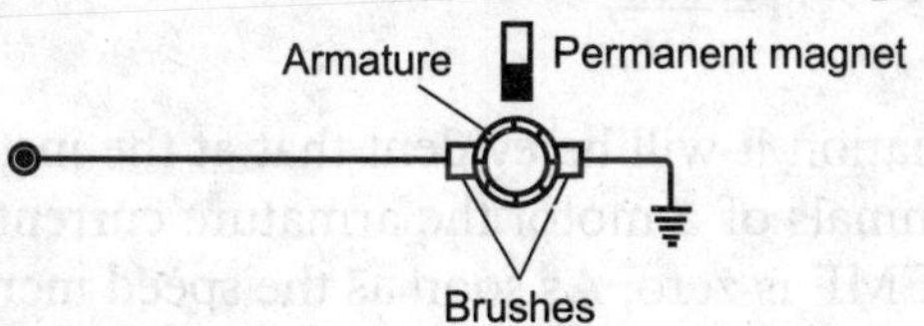


**Fig.5.15:** *Permanent magnet motor connection*

The characteristics of this type of motor are broadly similar to the shunt wound motors. However, when one of these types is used as a starter motor, the drop in battery voltage tends to cause the motor to behave in a similar way to a series wound machine. In some cases, the higher speed and lower torque characteristics are enhanced by using an intermediate transmission gearbox inside the starter motor.

## CHARACTERISTICS OF A STARTER MOTOR (SERIES)

Let,

I = Armature current in amperes
E = Applied EMF in volts
e = Total back EMF (CEMF) in volts
R = Resistance of armature in ohms
Z = Total number of active armature conductors
P = Number of poles
X = Number of parallel paths through armature winding
$\phi$ = Flux per pole

## POWER EQUATION FOR A MOTOR

The power equation of a motor is given as

$$IE = Ie + I^2R \qquad ......(1)$$

Where,

IE – Electrical input in watts
Ie – Watts available for conversion into mechanical power
$I^2R$ – Watts loss due to the heating of the armature windings

A small percentage of the ' Ie ' is expended in overcoming frictional resistance and iron losses, and the remainder in producing the necessary driving torque demanded by the load conditions.

## EQUATION FOR ARMATURE CURRENT

When conductors rotate in the magnetic field, CEMF will be induced in them. As CEMF opposes the applied EMF, it will tend to diminish the current flowing in the conductors. The armature current is given by the equation

$$I = \frac{E - e}{R} \qquad ......(2)$$

From this equation it will be evident that at the instant of applying a voltage to the terminals of a motor the armature current will be relatively great, since the CEMF is zero. As soon as the speed increases, the current will necessarily diminish to satisfy the equation.

With increase in load, the armature current will increase, and it will be seen from armature current equation that this will necessarily involve a reduction in the CEMF. Therefore any increase in load will be followed by a reduction in speed.

## EQUATION FOR MOTOR SPEED

The motor speed is given by the equation

$$N \alpha \frac{E}{\phi} \qquad \text{......(3)}$$

It will be seen that the speed of the motor is proportional to the applied voltage and inversely proportional to the flux per pole.

## TORQUE EQUATION

If the radius of a motor armature is r, the developed torque may be derived from equation as follows.

$$T \alpha \phi I \qquad \text{......(4)}$$

From the above relationship it is seen that the torque is directly proportional to the product of the two variables, namely flux and current.

## CURRENT Vs TORQUE

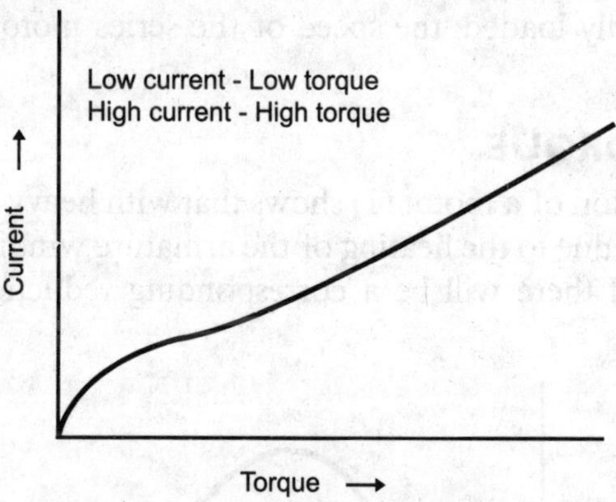


**Fig.5.16:** *Current-torque curve*

At starting the current taken by any motor is great due to the absence of any induced CEMF. As the field winding is in series with the armature winding, the full current taken by the motor also serves to excite the main field. Torque equation (4) shows that the torque of a motor is directly proportional to the product of the flux and current. Since, with a series motor, these two factors are a maximum at starting, it follows that the torque is then also a maximum.

## SPEED Vs TORQUE

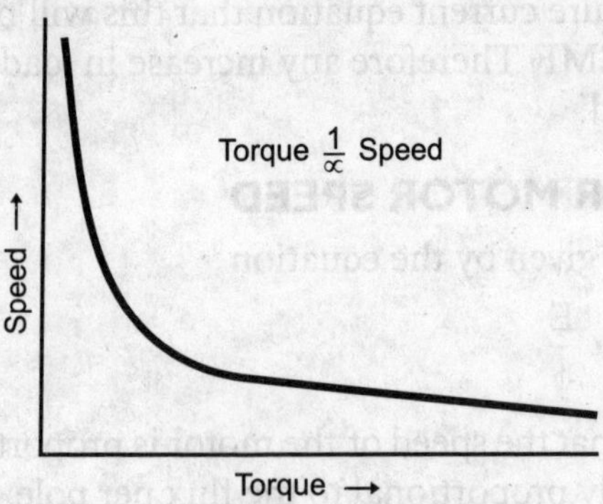


**Fig.5.17:** *\Speed-torque curve*

Motor speed equation (3) shows that the speed of a motor varies inversely as the field strength. With a series wound motor, the effect of additional load is to increase the field strength, which in turn causes a reduction in speed. This type of motor is essentially a variable speed motor and the speed varies appreciably with load variation. The speed variation is not so great at the higher torque values because the flux density in the field system is approaching the saturation point of the iron. A considerable increase in current is then necessary to produce any marked increase in the magnet flux. When heavily loaded, the speed of the series motor is substantially constant.

## POWER Vs TORQUE

The power equation of a motor (1) shows that with heavy load and current values, the losses due to the heating of the armature windings will increase very rapidly, and there will be a corresponding reduction in the power developed.

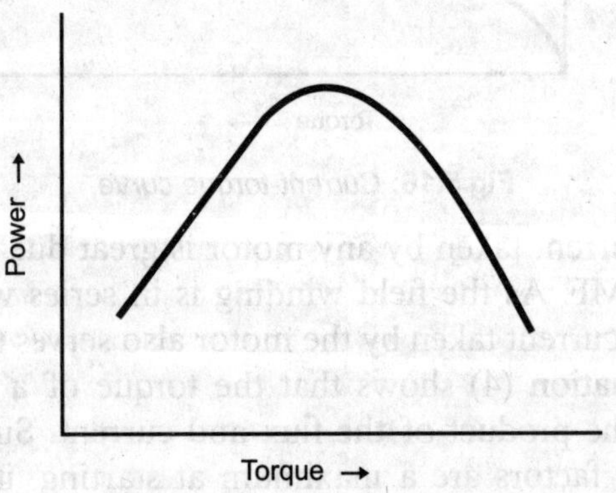


**Fig.5.18:** *Power-torque curve*

## TORQUE Vs APPLIED VOLTAGE

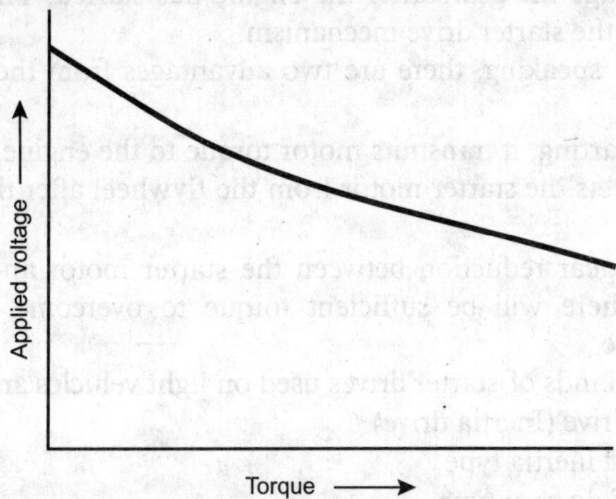


**Fig.5.19:** *Applied voltage-torque curve*

Speed $\alpha$ Applied voltage (from speed formula)

Speed $\frac{1}{\alpha}$ Torque

Applied voltage is proportional to speed, and torque is inversely proportional to the speed. From these relationships it is seen that the torque is inversely proportional to the applied voltage.

The CEMF opposes the voltage that is being applied to the armature. Because the armature and field are in series, the effect of the CEMF is to reduce the field and armature current. The faster the motor armature rotates the greater the CEMF. Therefore, at high rpm, the armature current is low due to high CEMF. This results in low torque.

## STARTER DRIVE MECHANISMS

A typical ratio of ring-gear teeth to pinion-gear teeth is 20 to 1. This gear ratio means that the starter motor rotates 20 times faster than the flywheel (and engine crankshaft). This ratio also means that there is a mechanical advantage of 20. It is 20 times easier for the motor to turn the crankshaft by driving the flywheel ring gear than for the motor to turn the crankshaft directly. This feature permits a relatively small starter motor to crank the engine.

Because the starter motor turns 20 times faster than the engine, the pinion and ring gears cannot be meshed together permanently. At high engine speeds, the starter motor shaft would be driven by flywheel ring gear so fast that the armature windings would be thrown outward and the motor would be damaged. Therefore, provision must be made to engage the pinion gear into the flywheel ring gear when the engine is being cranked and to disengage the gears after the engine has started. This function is performed by the starter drive mechanism.

Generally speaking, there are two advantages from the starter drive mechanisms.

- When starting, it transmits motor torque to the engine flywheel and disconnects the starter motor from the flywheel after the engine has started.
- Provide gear reduction between the starter motor and the engine. Hence there will be sufficient torque to overcome the cranking resistance.

The common kinds of starter drives used on light vehicles are:

1. Bendix drive (Inertia drive)
2. Improved inertia type
3. Bendix Folo-thru type
4. Pre-engagement drive
5. Sliding armature type (Axial type)
6. Permanent magnet type

Most of these starter drives use a one-way clutch (overrunning clutch) to release the starter drive pinion from the flywheel after the engine has started. Some of the important *types* from the above mentioned will now be *discussed* in more detail.

## BENDIX DRIVE (INERTIA DRIVE)

This method was employed extensively in the past for small to medium size petrol engines. It is not generally used now as developments in technology have reduced the cost of pre-engaged starters.

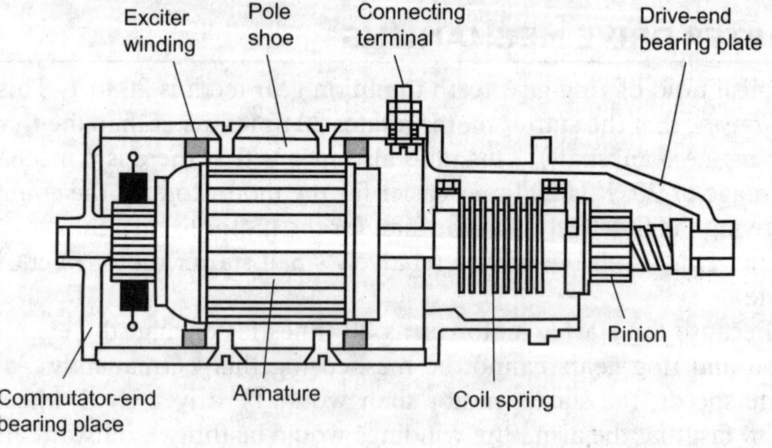


**Fig.5.20:** *Bendix inertia drive*

Inertia type drive works on the principle of inertia of unbalanced weight. In this type, pinion is mounted on threaded sleeve. When ignition switch is pressed, the starter motor starts rotating and this rotation wound the spring of the drive and also starts rotating the threaded sleeve. Now, the unbalance weight attached to the pinion, restricts the rotation of pinion and this action moves the pinion towards the ring gear. This action causes engagement of pinion with ring gear. The shock at the instant of engagement and disengagement is severe, so a large spring is fitted adjacent to the helix to act as a cushion. Pinion starts rotating ring gear and engine is cranked. The moment an engine starts running at the ideal speed, pinion is driven by the ring gear at high speed and it moves back on the sleeve and gets disengaged from ring gear by high speed inertia effect and starter motor is saved from the terrific centrifugal force.

## Disadvantage

The main disadvantage of this drive is lots of moving parts and engagement of pinion is done after starter motor starts. It causes rapid wearing of pinion teeth. To overcome this, pinion must be engaged before starter motor starts rotating. As a result, pre-engage drive is developed.

## IMPROVED INERTIA TYPE (BARREL TYPE)

Pinion is attached with a barrel and the weight of pinion and barrel assembly is much less than Bendix drive pinion. Due to less weight, inertia to move is lesser than Bendix drive, hence called as improved inertia type.

The control nut has internal helical splines matching with external helical splines of the screw in sleeve. The control nut has lugs matching with the slots of the barrel.

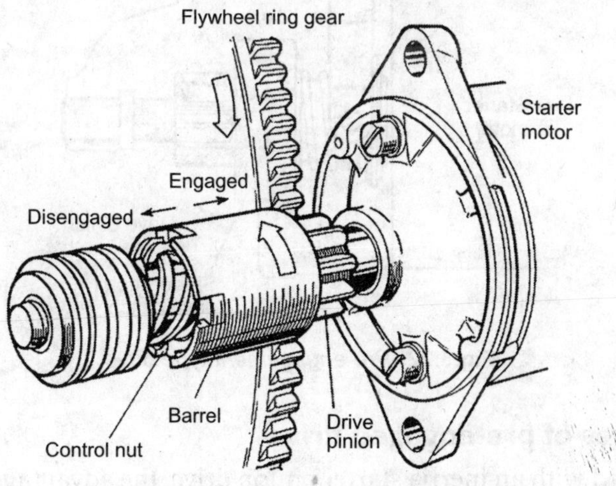


**Fig.5.21:** *Improved inertia type*

When the armature shaft rotates, the attached screw in sleeve rotates which makes the control nut to turn and axially move. Since the control nut is attached with the pinion and barrel assembly, this assembly rotates and move axially towards the flywheel ring gear to engage. Once the pinion got engaged, the starter motor delivers the torque to start the engine.

## PRE-ENGAGED DRIVE

The pre-engagement arrangement is used with both petrol and diesel engines and is now the preferred type.

The name itself suggests the working principle of pre-engage type drive. In this arrangement when the ignition key is turned to START position the drive pinion on the starter motor shaft is pushed axially into mesh with the flywheel ring gear before the starter motor is switched on. When the pinion meshes with flywheel ring gear, the starter motor is switched on by engagement relay. A solenoid is usually employed to move the lever and close the switch contacts. This method is quite simple and having the advantage not to damage gear teeth. As the engine is started, it is the time to disengage the pinion from the flywheel. This is done by the overrunning clutch which is attached to a pinion. Overrunning clutch make the pinion free from the starter motor shaft as the engine starts. And again motor is saved from the high speed damage.

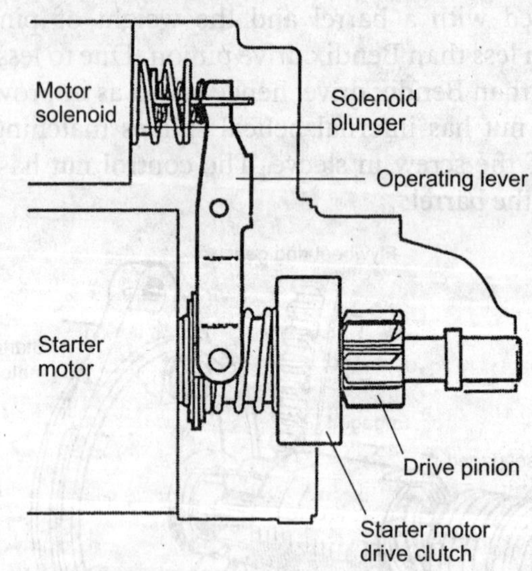


**Fig.5.22:** *Pre-engagement type drive*

### Advantages of pre-engaged drive

As compared with an inertia starter pinion drive, the advantages of the pre-engaged type of drive are as follows.

1. A smaller pinion can be utilised to provide a higher reduction ratio, so that cranking torque is increased to improve cold starting capability.
2. The pinion drive is not so highly stressed, because less energy has to be absorbed during pinion engagement, especially under difficult starting conditions.
3. Prolonged cranking of a diesel engine is possible without premature disengagement or the pinion repeatedly traveling in and out of mesh as resisting torque fluctuates.
4. Repeated travelling in and out of mesh of the pinion is also avoided when starting modern petrol engines with high compression ratios, and also those where flywheel inertia is reduced with automatic transmission installations.
5. A major source of excessive wearing and damage of the starter pinion and ring gear teeth is eliminated.

## SLIDING ARMATURE TYPE (AXIAL STARTER)

Starter motors using this system of pinion drive are also known as axial starter motors. They are now generally confined in application to the larger diesel engine used in heavy vehicles, where owing to the high inertia of the flywheel and crankshaft assembly, the starter pinion is required to engage the ring gear before the starter motor develops full cranking torque, thereby avoiding heavy engagement shock and excessive wear of the teeth.

In its rest position (Fig.5.23a), the armature is displaced axially out of the field windings. The pinion is driven from the armature shaft by way of a multi-plate clutch. The starter motor has three field windings: the auxiliary winding, the hold-in winding (shunt-wound) and the main winding (series-wound). To permit the armature to move axially, it has extra-long bearings and a wide commutator. There is also a return spring, a control relay with tumbler bridge and a locking pawl with release disc.

Engagement is in two stages (Fig.5.23b). In stage one, the armature is displaced axially by the magnetic field of the auxiliary and hold-in windings, so that the pinion engages in the flywheel ring gear, aided by the armature rotating slowly. The release disc on the commutator raises the release lever to switch on the second stage of the control relay. The main winding is then energised by way of the tumbler bridge in the control relay.

A multi-plate clutch is installed between the armature and the pinion. It ensures that the drive is taken up smoothly between armature and pinion after the pinion has engaged with the flywheel ring gear. If the flywheel ring gear begins to drive the pinion after the engine has started, the clutch is freed and acts like a freewheel. This prevents dangerous acceleration forces from reaching the armature of the starter motor as the engine speeds up. If a given torque is exceeded at the pinion, the multi-plate clutch interrupts the power flow between armature and flywheel ring gear (overload protection), this minimizes loads on the starter motor, pinion and ring gear.

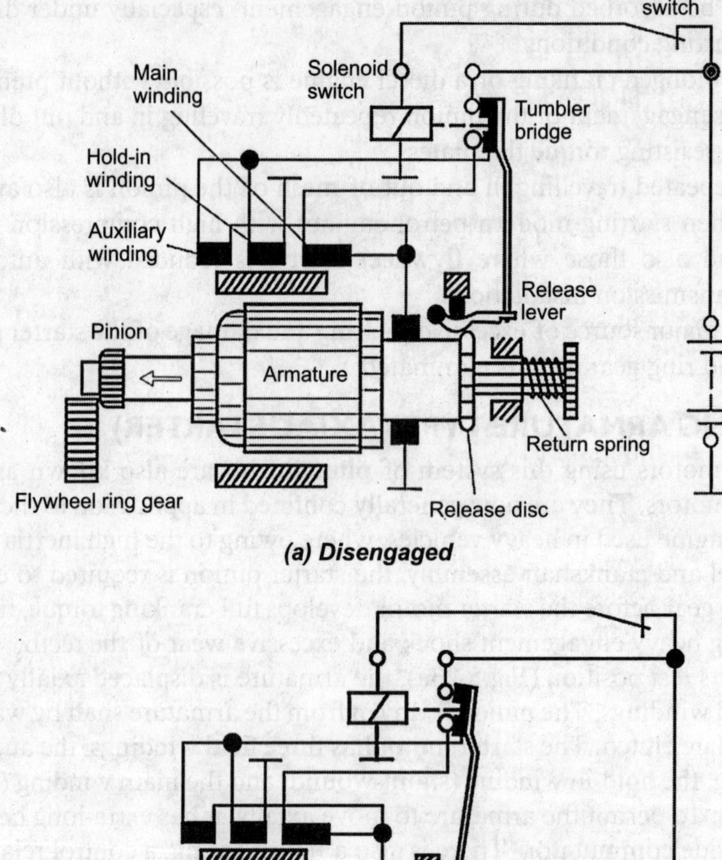


***(a) Disengaged***

***(b) Engaged***

**Fig.5.23:** *Sliding armature starter drive*

## OVERRUNNING CLUTCH

Both conventional and gear reduction starter motors are usually fitted with a one-way, over-running clutch.

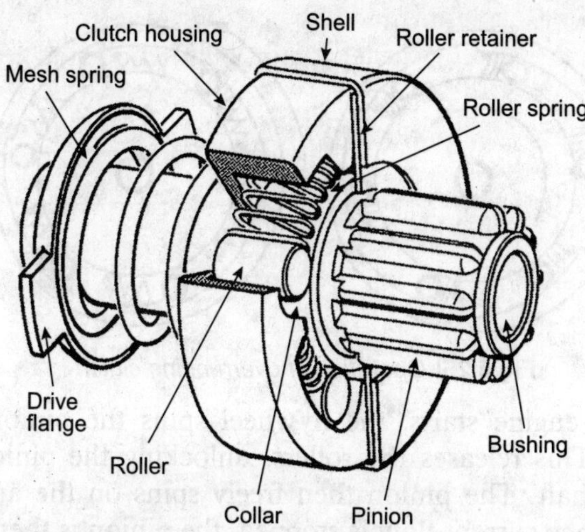


**Fig.5.24:** *Overrunning clutch*

To prevent damage to the starter motor, the starter drive must be disengaged as soon as the engine turns faster than the starter motor. This job is performed by the overrunning clutch.

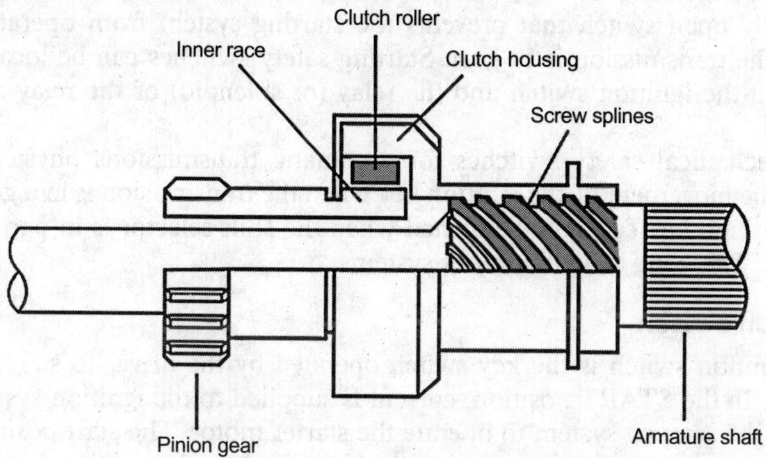


**Fig.5.25:** *Cut-view of overrunning clutch*

The overrunning clutch housing is internally splined to the armature shaft. The drive pinion turns freely on the armature shaft within the clutch housing. When the clutch housing is driven by the armature, the spring loaded rollers are forced into the small ends of their tapered slots and wedged tightly against the pinion barrel (Fig.5.26B). This locks the pinion and clutch housing solidly together, permitting the pinion to turn the flywheel and, thus, crank the engine.

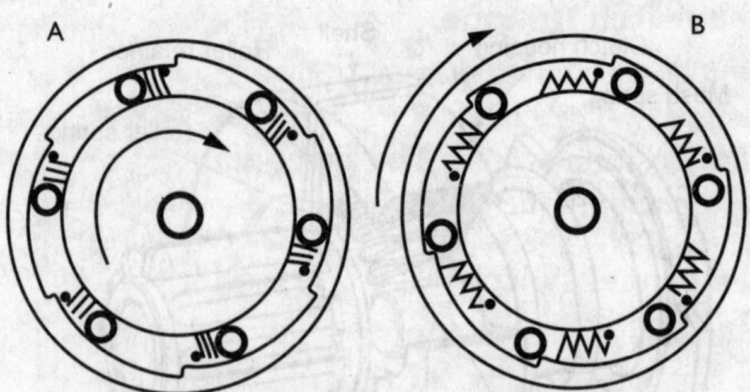


**Fig.5.26:** *Operation of overrunning clutch*

When the engine starts, the flywheel spins the pinion faster than the armature. This releases the rollers, unlocking the pinion gear from the armature shaft. The pinion then freely spins on the armature shaft (Fig.5.26A). Once current flow is stopped, the pinion is then pulled away from the flywheel. The overrunning clutch is moved in and out of mesh by the starter drive linkage.

## STARTER SWITCHES

The starting safety switch, often called the neutral safety switch, is a normally open switch that prevents the starting system from operating when the transmission is in gear. Starting safety switches can be located between the ignition switch and the relay (or solenoid) or the relay and ground.

Mechanical safety switches for automatic transmissions physically block the movement of the ignition key when the transmission is in a gear. The ignition key can only be turned when the shift selector is in park or neutral. These are called interlock systems.

### Ignition switch

The ignition switch is the key switch operated by the driver to start the engine. In the START position, current is supplied to the ignition system and to the starting system to operate the starter motor. The start position of the ignition switch is a momentary contact. In the start position, battery voltage is switched to the starter relay to operate the starter motor. When the engine starts, the driver releases the key and the switch automatically returns to the run (ON) position.

### Park/Neutral position switch (Automatic transmission)

The Park/Neutral position switch prevents operation of the starter motor unless the shift lever is in Park or Neutral. The switch contacts are in series with the starter control circuit. The switch is normally mounted on the transmission housing.

## Clutch start switch (Manual transmission)

The safety switches used with manual transmissions are usually controlled by the clutch pedal. The clutch start switch serves the same purpose as a Park/Neutral position switch. The clutch start switch keeps the starter control circuit open until the clutch pedal is depressed.

## Clutch start cancel switch

In some off-road situations it is advantageous to start a manual transmission vehicle while in gear with the clutch engaged. The driver-controlled safety cancel switch allows the driver to bypass the clutch start switch to make this possible. This feature is only available on some models.

# STARTER SOLENOID (MAGNETIC SWITCH)

Solenoids are electromagnetic switches with a movable core that converts current flow into mechanical movement.

In a 'pull-in' type solenoid, the magnetic field pulls a core into a coil. These solenoids are called magnetic switches on starters. A hold-in coil holds the core in place.

In a starting system, the solenoid uses the electromagnetic field generated by its coil to perform two distinct jobs. The first is to push the drive pinion of the starter motor into mesh with the engine's flywheel. This is the solenoid's mechanical function. The second job is to act as an electrical relay switch to energise the motor once the drive pinion is engaged. Once the contact points of the solenoid are closed, full battery current flows to the starter motor.

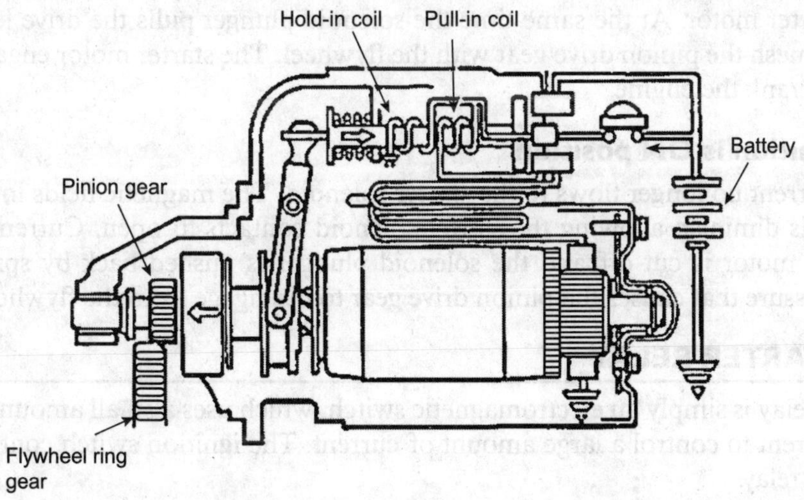


**Fig.5.27:** *Starter solenoid*

The solenoid assembly has two separate windings: a pull-in winding and a hold-in winding. The two windings have approximately the same number of turns but are wound from different size wires. Together these windings produce the electromagnetic force needed to pull the plunger into the solenoid coil. The heavier pull-in windings draw the plunger into the solenoid, while the lighter-gauge windings produce enough magnetic force to hold the plunger in this position.

Both windings are energized when the ignition switch is turned to the start position. When the plunger disc makes contact with this solenoid terminal, the pull-in winding is deactivated. At the same time, the plunger contact disc makes the motor feed connection between the battery and the starting motor, directing current to the field coils and starter motor armature for cranking power.

As the solenoid plunger moves, the shift fork also pivots on the pivot pin and pushes the starter drive pinion into mesh with the flywheel ring gear. When the starter motor receives current, its armature starts to turn. This motion is transferred through an overrunning clutch and pinion gear to the engine flywheel and the engine is cranked.

### Ignition is OFF position

Battery voltage is applied at all times from the positive battery terminal to the ignition switch and the normally open starter solenoid contacts.

### Ignition in START position

Current flows through the solenoid coil. The solenoid coil energizes, which close the starter solenoid contacts and the battery voltage is applied to the starter motor. At the same time the solenoid plunger pulls the drive lever to mesh the pinion drive gear with the flywheel. The starter motor engages to crank the engine.

### Ignition is ON position

Current no longer flows to the starter solenoid. The magnetic fields in the coils diminish allowing the starter solenoid contacts to open. Current to the motor is cut off and the solenoid plunger is pushed back by spring pressure that causes the pinion drive gear to disengage from the flywheel.

## STARTER RELAY

A relay is simply an electromagnetic switch, which uses a small amount of current to control a large amount of current. The ignition switch controls the relay.

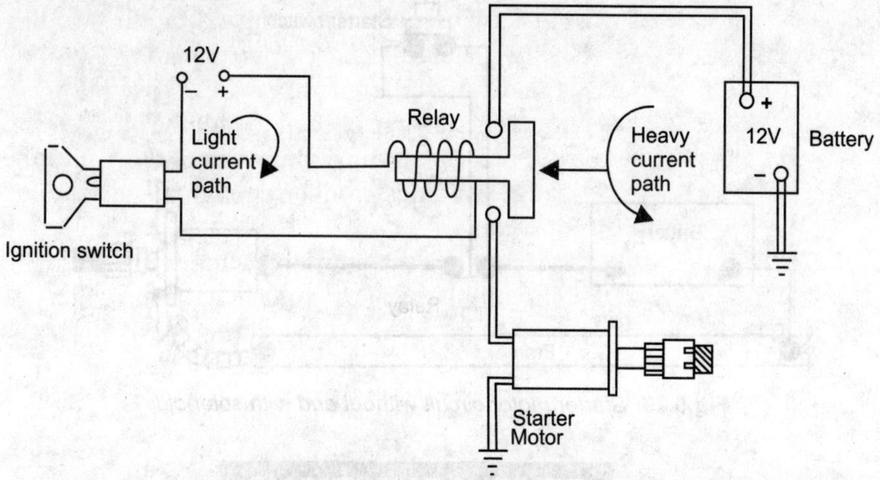


**Fig.5.28:** *Starting system with relay*

Relays are often used in starting circuits because the ignition switch contacts are not large enough to carry the starter motor current. They would overheat and burn. To design an ignition switch to carry this current would be impractical (the switch would be too large). Instead, the relay contacts are made large enough to carry the starter motor current. By using a relay, the small current to energize the relay coil can control the large starter motor current. Therefore, the ignition switch contacts that carry the current to control the relay can be small. Therefore, ignition switch is protected from being damaged. Moreover, the length of the starter cable can be reduced by using a solenoid as shown in Fig.5.29.

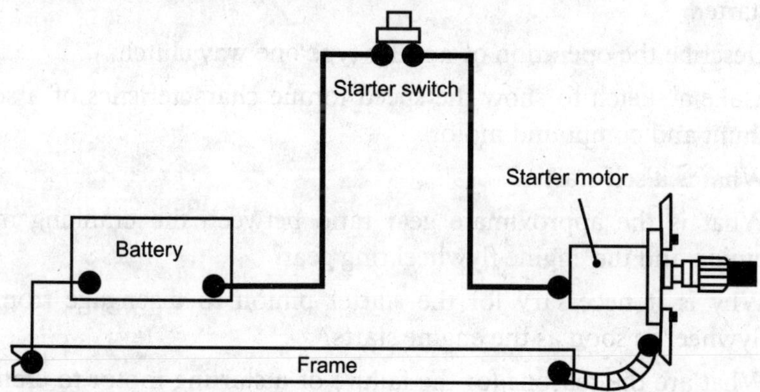

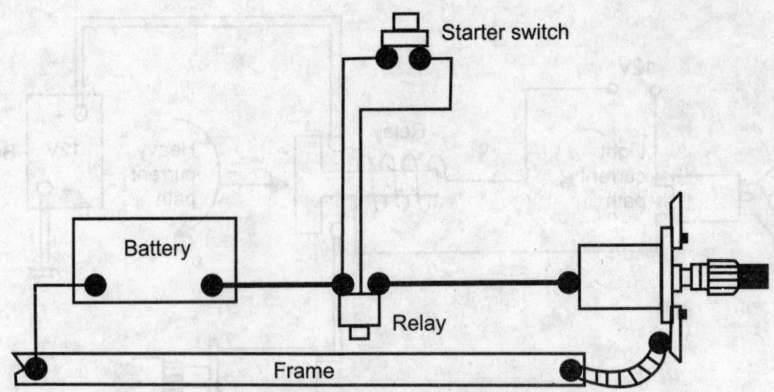


**Fig.5.29:** *Starter motor circuit without and with solenoid*

## REVIEW QUESTIONS

1. State four advantages of a pre-engaged starter when compared with an inertia type.
2. Describe the operation of the pull-in and hold-on windings in a pre-engaged starter solenoid.
3. Make a labeled sketch of the engagement mechanism of a pre-engaged starter.
4. Explain what is meant by 'voltage drop' in a starter circuit and why it should be kept to a minimum.
5. Describe the engagement and disengagement of a Bendix inertia starter.
6. Describe the operation of a roller-type one way clutch.
7. Make a sketch to show the speed torque characteristics of a series, shunt and compound motor.
8. What is a self-starter?
9. What is the approximate gear ratio between the cranking motor pinion and the engine flywheel ring gear?
10. Why is it necessary for the starter pinion to disengage from the flywheel as soon as the engine starts?
11. What are the reasons for the failure of a starting motor to crank an engine?
12. Explain the constructional details of starter motor.
13. Sometimes the starter motor spins but does not crank the engine, Why?

14. For what purpose is the solenoid used in a starting circuit?
15. Why is a relay used to operate the starter?
16. Why do we prefer series motor as automotive starter motor?
17. List the two starting system circuits.
18. List the components that make up the control circuit.
19. List the components that make up the motor circuit.
20. Explain in detail how a conventional starter differs from that of a gear reduction starter.
21. Explain why an overrunning clutch is needed and how it works.
22. Explain how the starter drive pinion engages (pushed out) with the ring gear when the ignition key is turned to the START position.
23. A self-starter and a dynamo look alike, then what is the difference between these two?
24. State the basic electrical principle governing the operation of a starting motor.
25. What is the purpose of the commutator in a starting motor?
26. Name two types of starter motor controls, state the difference between them, and name the type of starter drive each requires.
27. What is the function of a solenoid switch?

# UNIT 6 Ignition System

An internal combustion engine used in automobiles operates by the force of expansion obtained through combustion of the air-fuel mixture in the cylinders. However, the mixture will not ignite by itself. Therefore, in petrol engines, some means must be provided to ignite the mixture to start it burning. This function is performed by the ignition system. This system serves to distribute the ignition at the proper time to the proper cylinder in correct time relationship with the piston position and start of ignition.

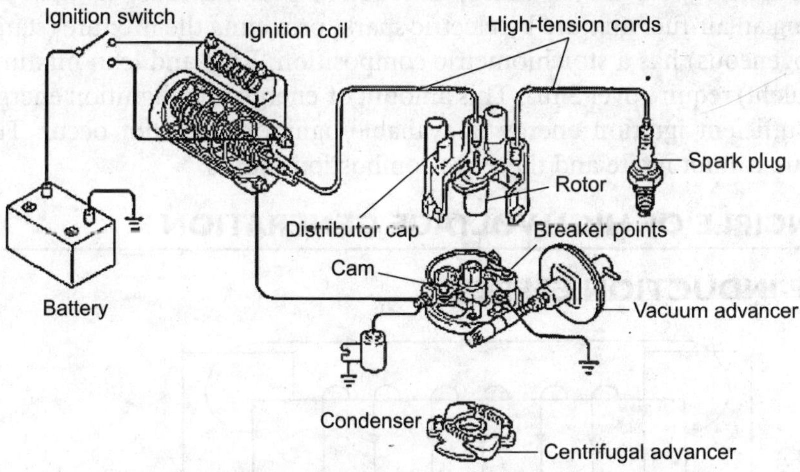


**Fig.6.1:** *Ignition system*

## FUNCTIONS AND REQUIREMENTS

The basic function of the ignition system is to generate sparks that can ignite the air-fuel mixture in the cylinders, so the following conditions must be satisfied:

### A STRONG SPARK

When the air-fuel mixture is compressed in the cylinders, it becomes difficult for the spark to pass through the air. This is because even air has electrical resistance and this resistance rises as the air is compressed. For this reason, the voltage that is supplied to the plugs must be high enough to ensure the generation of a powerful spark between the electrodes of the spark plug.

## PROPER IGNITION TIMING

In order to obtain the most effective combustion of the air-fuel mixture, there must be some means of varying the ignition timing in accordance with engine speed and load.

## SUFFICIENT DURABILITY

If the ignition system fails, the engine will stop running. The ignition system must therefore have sufficient reliability to endure the vibration and heat generated by the engine, as well as the high voltage in the ignition system itself.

## IGNITION ENERGY

Approximately 0.2 mJ of energy is required per individual ignition for igniting an air-fuel mixture by electric spark, providing the mixture (static, homogeneous) has a stoichiometric composition. Rich and lean mixtures (turbulent) require over 3 mJ. This amount of energy is the ignition energy. If insufficient ignition energy is available, ignition does not occur. The mixture cannot ignite and there are combustion misses.

## PRINCIPLE OF HIGH VOLTAGE GENERATION

## SELF-INDUCTION EFFECT

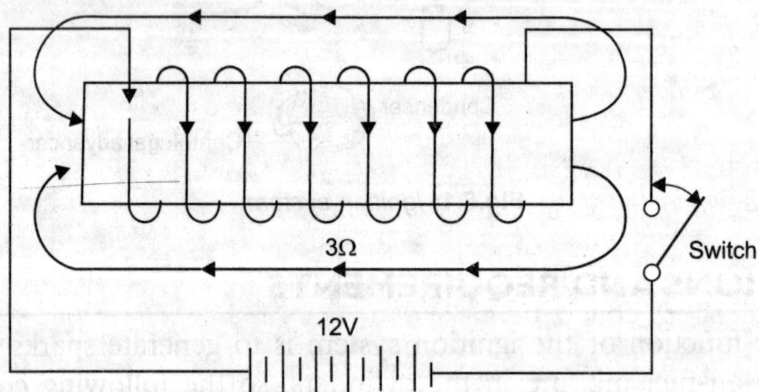


**Fig.6.2:** *Self-induction*

A magnetic field is generated when a current flows through a coil. As a result, an EMF (electromotive force) is generated which creates a magnetic field in a direction which impedes (obstruct) the generation of magnetic field in the coil. Therefore, current does not flow immediately when it is first introduced in the coil, but a certain period of time is required for the current to rise.

When current starts to flow in a coil or when current is cut off, the coil generates EMF which acts to impede changes in the coil's magnetic field. This is called the "self-induction effect".

## MUTUAL INDUCTION EFFECT

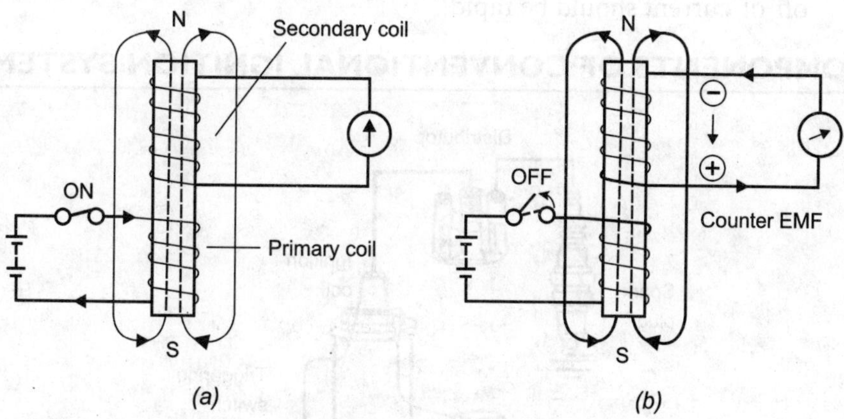


**Fig.6.3:** *Mutual induction*

When two coils are arranged in a line and the amount of current flowing in one of the coils (the primary coil) is changed, an EMF is generated in the other coil (the secondary coil) in a direction which impedes the change in the primary coil's magnetic field. This is called the "mutual induction effect".

In the Fig.6.3a, when a constant current is flowing in the primary coil, there is no change in the magnetic field, so no EMF is generated in the secondary coil. However, when the switch is turned off (Fig.6.3b), cutting off the flow of current in the primary coil, the magnetic field which has been generated up to this time suddenly disappears, so an EMF is generated in the secondary coil in a direction which impedes the decay of the magnetic flux.

The ignition coil generates a high-voltage current by means of the mutual induction that takes place between the primary coil and the secondary coil when the primary current is suddenly cut off by the opening of the breaker points.

The amount of EMF is determined by the following three factors:

1. **Amount of magnetic flux**

The more magnetic flux generated in a coil, the higher the amount of induced voltage.

2. **Number of coil windings**
The more turns in a coil, the higher the induced voltage.

3. **Rate at which magnetic flux changes**
The faster the change in the amount of magnetic flux generated in a coil, the higher the induced voltage. In order to obtain a large EMF due to mutual inductance (secondarily generated voltage), the current flowing in the primary coil should be as large as possible and the cut-off of current should be rapid.

## COMPONENTS OF CONVENTIONAL IGNITION SYSTEM

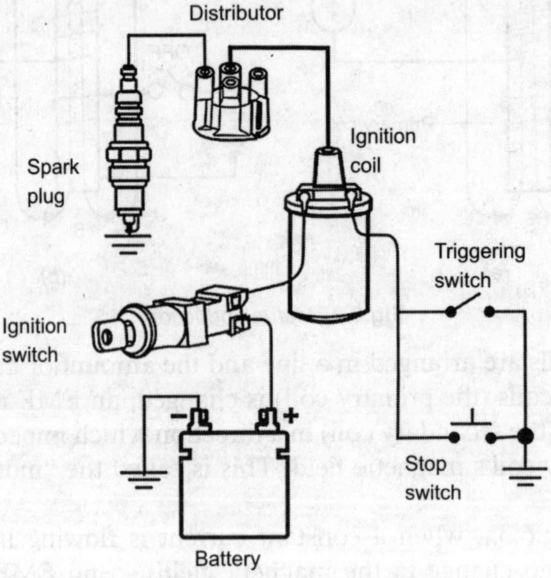


**Fig.6.4:** *Ignition system*

### Battery

The battery is the power source for current to primary circuit when starting the engine or when the engine is idling. When the engine speed rises, the generator becomes the source of power.

### Ignition coil

By passing the battery current through the primary winding of this coil, a high voltage is induced in the secondary winding that is large enough to discharge strong sparks across the spark plug electrode gaps.

### Distributor

The functions of the distributor are to switch off the primary circuit current so as to induce high voltage in the ignition coil and to distribute the induced voltage to the cylinders.

### Spark plugs

These are screwed into the combustion chambers, and form the gaps for the sparks. When the high voltage discharges across the gap, the resulting spark ignites the air-fuel mixture in the combustion chamber.

### Ignition switch

The switch is located either on the instrument panel in front of the driver's seat or on the steering column and serves to close the ignition circuit and to start and stop the engine.

### Ballast resistor

Ballast is a resistor that has the purpose of reducing the voltage to the coil. Ballast resistor in not used with all the ignition systems.

The starter motor draws a huge current from the battery leaving less energy to spark plugs. The result is a weaker than normal spark which is not ideal for starting an engine. To overcome such a problem, ignition systems were changed to run a lower voltage coil (usually 9V), and these coils could still give the same output as the original 12V coils. In order to run such a coil, the 12V ignition feed runs through a ballast, reducing it to 9V at the coil. To assist starting, a 12V feed (usually from the starter) bypasses the 9V ignition feed, giving the 9V coil a 12V feed. The result is a better than normal spark which is ideal for starting, particularly on cold damp mornings. As soon as the engine has started, the 12V feed is cut and the coil will run on the 9V ignition feed.

## IGNITION COIL

The ignition coil is the part where the high voltage necessary to make the spark jump across the spark plug gap is generated. As shown in the Fig.6.5, the ignition coil is made up of the core, primary coil, and secondary coil. The primary coil is made up of 300 to 400 turns of relatively heavy wire. The secondary coil is made up of 15,000 to 20,000 turns of fine wire wound in layers. Insulation paper is wound between the secondary coil layers to insulate the layers from each other. The secondary coil is wound over the core made of iron sheets having high permeability. The primary coil is wound over the secondary coil.

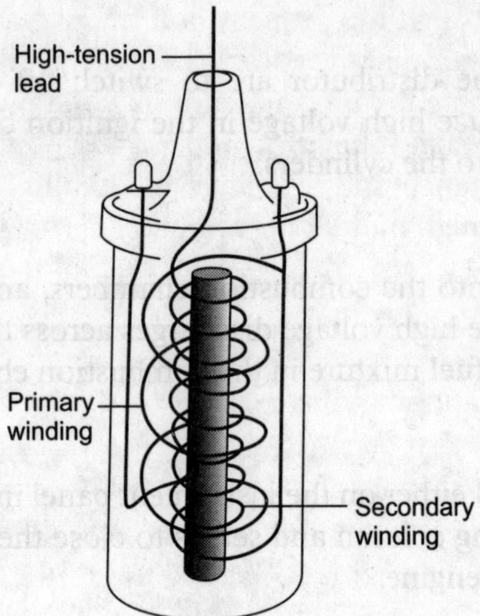


**Fig.6.5:** *Ignition coil*

The secondary coil has one end connected to the high voltage terminal and the other end to the primary coil. The primary coil ends are connected to the positive and negative terminals (battery) respectively. The coil assembly is contained in a case with the space between the coil and case filled with pitch for insulation. Oil may be added in the coil to dissipate heat rapidly, to provide greater insulation, and to reduce the possibility of insulation failure.

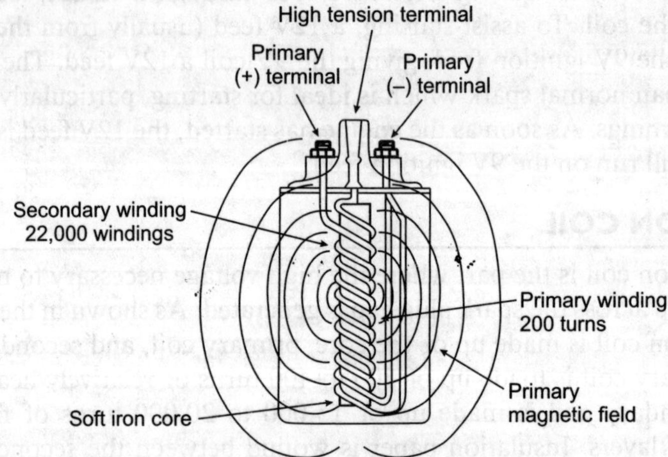


**Fig.6.6:** *Operation of ignition coil*

## Operation

When the contact points touch, the ignition circuit is complete. The current flowing through the primary coil magnetizes the core and builds up a magnetic field around the core. When the primary current is suddenly cut off by the opening of the contact points, the magnetic field collapses, resulting in producing voltage of 300 to 400 Volts in the primary coil through self-inductance and a voltage of 15,000 to 20,000 Volts in the secondary coil through mutual inductance.

## DISTRIBUTOR

The distributor can be divided into the following three parts, each performing important functions.

1. *Contact breaker* that opens and closes the ignition coil primary current.
2. *Distributor unit* that distributes the high voltage to the various cylinders.
3. *Ignition advancer* that automatically adjusts the ignition timing to conform with the engine speed and load.

## CONTACT BREAKER

The contact breaker, also called circuit breaker, is made up primarily of a cam, breaker arm, contact points, breaker plate, and condenser.

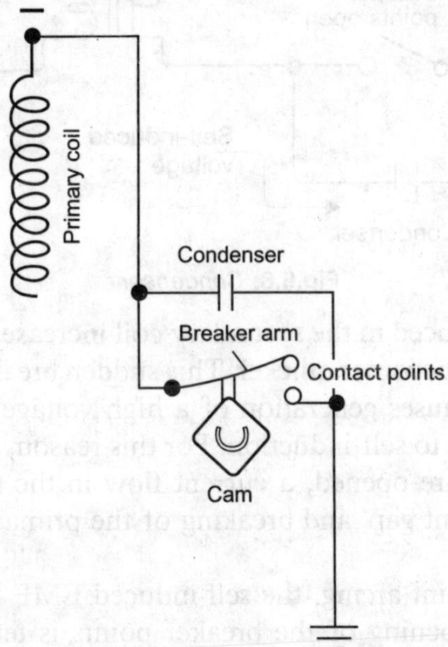


**Fig.6.7:** *Components of contact breaker*

The cam is installed to the distributor shaft driven by the camshaft gear, and rotates at one-half the engine speed. The cam has the same number of lobes as there are cylinders in the engines. When the cam turns, the breaker arm is raised and lowered to open and close the contact points, to make and break the primary current. For each cam rotation, high voltages are induced in the ignition coil, equal to the number of cylinders in the engine.

When the contact points are closed, the current flows through the ignition coil and builds up a magnetic field, and when the contact points are open, the primary current is cut out and high voltage is induced in the primary coil. The amount of contact points opening is made adjustable as it has a large effect on the induced high voltage.

## CONDENSER (CAPACITOR)

Generally, the condenser is installed on the outside of the distributor case and connected in parallel with the breaker points. It serves two purposes: It reduces arcing of the contacts.

It ensures a quick collapse of the magnetic field.

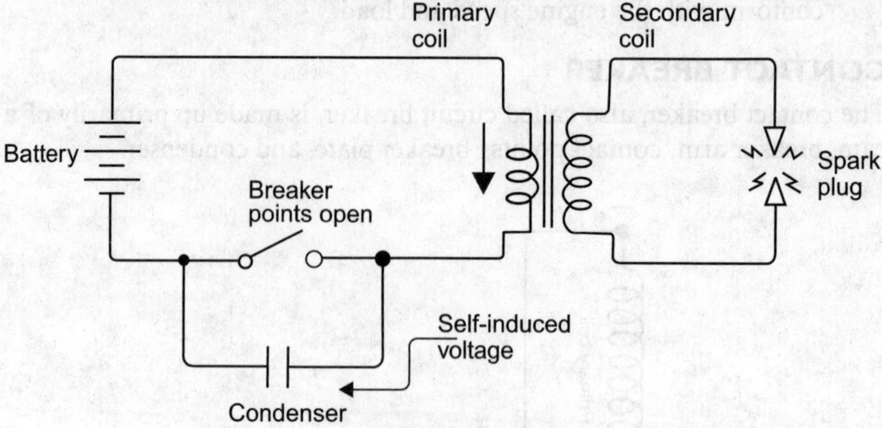


**Fig.6.8:** *Condenser*

The voltage induced in the secondary coil increases as the breaking of the primary current becomes quicker. This sudden breaking of the primary current, however, causes generation of a high voltage of about 500 V in the primary coil due to self induction. For this reason, at the moment that the breaker points are opened, a current flow in the form of an electric spark across the point gap, and breaking of the primary current does not immediately occur.

To minimise point arcing, the self-induced EMF in the primary coil that occurs upon opening of the breaker points is temporarily "stored" in the condenser to provide quick shut-off of the primary current and sends it back to the primary circuit when the points close. As a result, the current flow in the primary circuit is increased so that the ignition coil self induction becomes stronger to enable inducing higher voltage in the secondary coil.

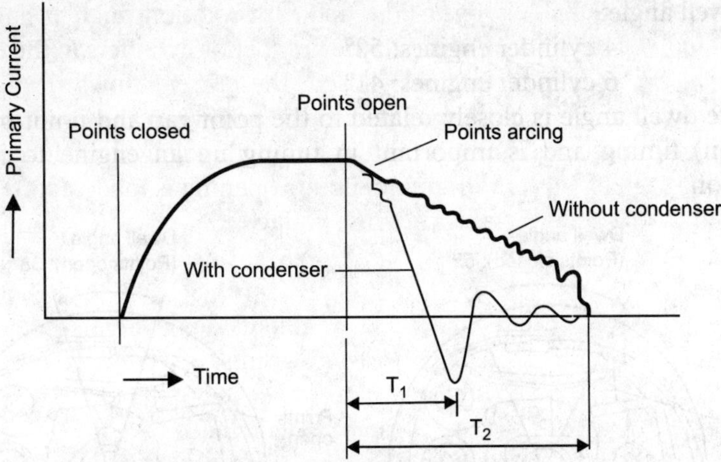


**Fig.6.9:** *Change in primary current due to condenser*

The graph 6.9 shows how the primary current behaves after the points open; when a condenser is provided, time $T_1$ is shorter than $T_2$, and point arcing is less likely to occur.

## Dwell angle

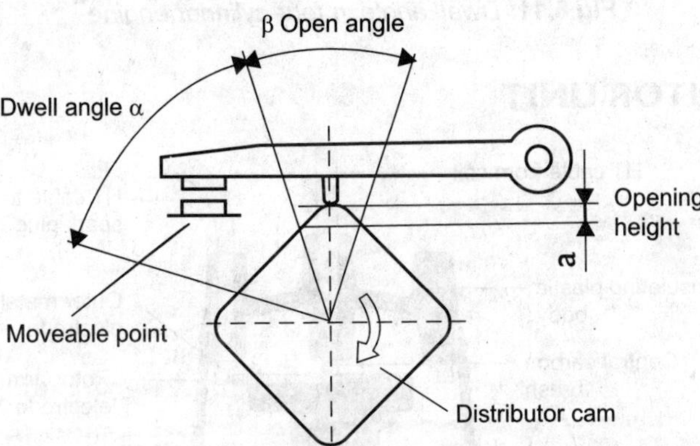


**Fig.6.10:** *Dwell angle*

The dwell angle refers to the angle of distributor shaft (cam) rotation between the time that the breaker points are closed by the breaker arm spring and the time that they are opened by the next cam lobe. If the gap of a 4-cylinder engine has been adjusted correctly to the standard value, the points should remain closed while the cam rotates 52° ± 6°.

Furthermore, the points should stay open until the cam rotates another 38° ± 6°. Since the combination of the point closing angle and opening angle is 90° (52° + 38°), the breaker points are opened and closed by every ¼ turn of the cam.

Dwell angle:

4-cylinder engines: 52°

6-cylinder engines: 41°

The dwell angle is closely related to the point gap and point opening (ignition) timing and is important in tuning up an engine to its best condition.

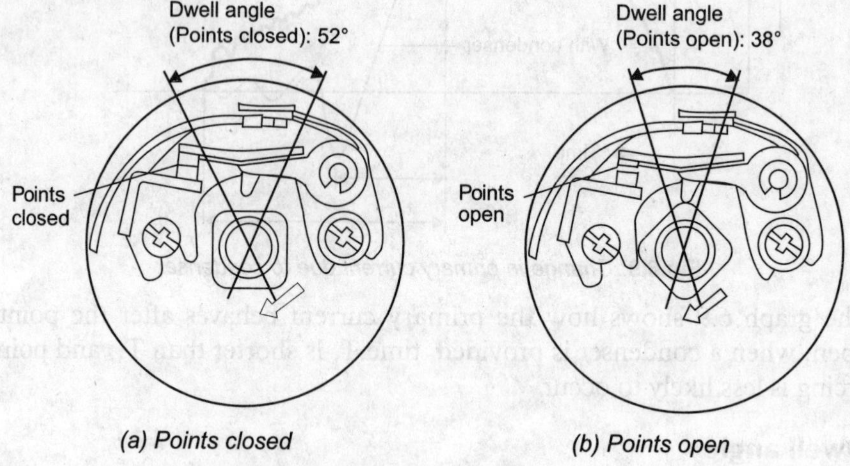


(a) Points closed (b) Points open

**Fig.6.11:** *Dwell angle in four-cylinder engine*

## DISTRIBUTOR UNIT

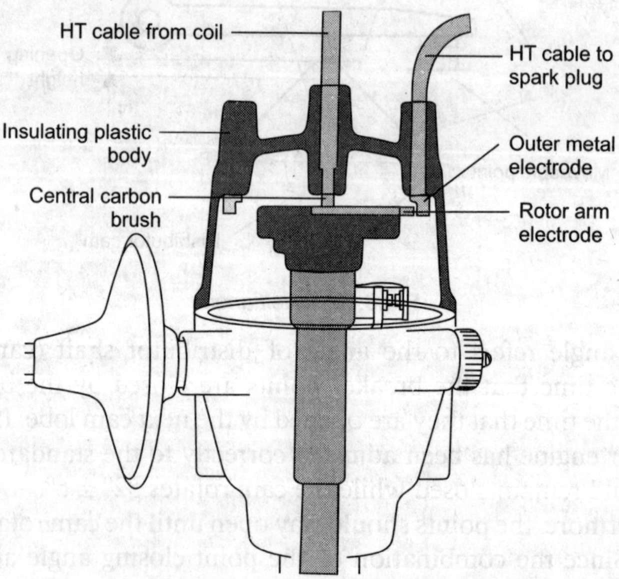


**Fig.6.12:** *Distributor cap and rotor*

This part distributes the high voltage correctly to the spark plugs in the various cylinders in accordance with the specified firing order. The distributor is comprised of a plastic body (cap) with terminals, and a rotor. Since ozone is generated due to ionization during high-tension discharge across these air gaps, small air vent holes are provided in the cap and housing for ventilation.

The high tension current generated in the secondary winding of the ignition coil passes from the secondary terminal of the ignition coil to the center electrode of the distributor cap through a high-tension card. The high tension current then passes from the center electrode through the center contact piece to the side electrodes in the form of an electric spark via the rotor, which rotates at half the crankshaft speed. The turning of the rotor distributes the high voltage to the spark plugs in firing order.

## IGNITION ADVANCE

The timing of the firing of the air-fuel mixture compressed in the cylinder with the spark plug is called ignition timing. It is necessary that the mixture be ignited at the correct time conforming with the engine load and speed in order to obtain large output from the engine with small fuel consumption.

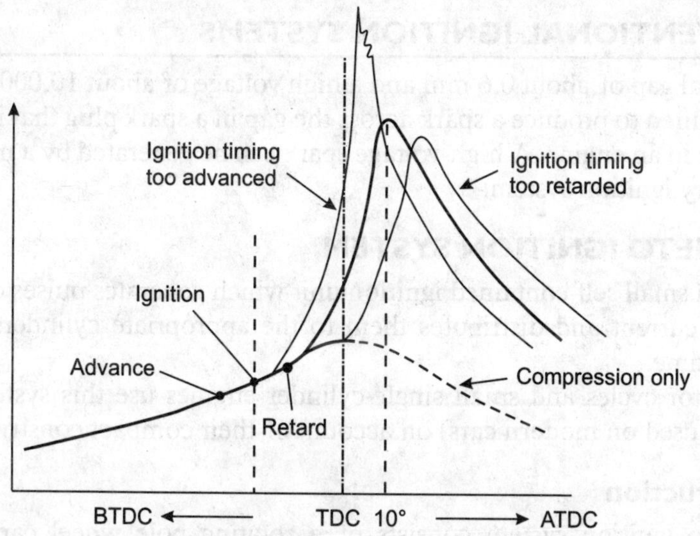


**Fig.6.13:** *Ignition timing advanced and retarded*

When the mixture is ignited, the flame must spread out successfully so that it takes time for the entire mixture to burn. The maximum gas force during the power stroke can be obtained when the piston has travelled slightly from the top dead center (about 10° crank rotation after top dead center). From these considerations, the mixture must be ignited slightly before the piston reaches its top dead center, that is, while still at the compression stroke. It is necessary to have some means to change (advance or retard) ignition timing so that it best matches engine speed, load, etc. The distributor is therefore provided with an ignition advancers, which consists of a governor advancer for controlling the ignition timing in accordance with engine speed and a vacuum advancer for controlling ignition timing according to the engine load.

### If the ignition timing is advanced too far

The maximum combustion pressure will occur before 10°ATDC. Since the pressure inside the cylinder will become higher than during ignition at the proper timing, spontaneous combustion of the air-fuel mixture will occur and knocking will also occur easily. Excessive knocking leads to burning of valves, spark plugs, pistons, etc.

### If the ignition timing is retarded too far

The maximum combustion pressure will occur after 10° ATDC (at a point when the piston has dropped considerably). Compared to ignition at the proper timing, the pressure inside the cylinder will be relatively low, so the engine output will drop and fuel economy and other problems will occur.

## CONVENTIONAL IGNITION SYSTEMS

A normal gap of about 0.6 mm and a high voltage of about 10,000-30,000 V is required to produce a spark across the gap in a spark plug that is under pressure in an engine. A high-voltage spark can be generated by a magneto or battery ignition system.

## MAGNETO IGNITION SYSTEM

This is a small self-contained ignition unit which generates pulses of high-tension current and distributes them to the appropriate cylinders at the correct time.

Motor cycles and small single-cylinder engines use this system (but it is not used on modern cars) on account of their compact construction.

### Construction

Magneto ignition system consists of a rotating pole wheel carrying a number of permanent magnets (Magneto rotor), the ignition armature with primary and secondary windings, the contact breaker points, the condenser and a cam rotating at crankshaft speed.

### Operation

The alternating north and south poles of the rotating permanent magnets induce an alternating current in the primary winding. When the contact breaker points are closed, a current flows in the primary winding of the ignition armature and generates a magnetic field. The points are opened at the moment when the current flow is at its greatest, that is to say when the magnetic field in the primary winding has reached a maximum. The change of pole of magneto rotor causes the direction of the magnetic field in the ignition armature to change suddenly. This sudden reversal of the magnetic field induces a high voltage in the secondary winding. The ignition condenser, which is wired in parallel with the contact breaker, suppresses spark formation and encourages the magnetic field to decay rapidly. Since the cam revolves at crankshaft speed, a spark is produced during each revolution of the engine. This is ideal for the requirements of a single-cylinder two-stroke engine. In a single-cylinder four-stroke engine, a 'false spark' occurs every second revolution of the crankshaft, but takes place during the exhaust stroke and therefore has no effect on the engine's actual performance.

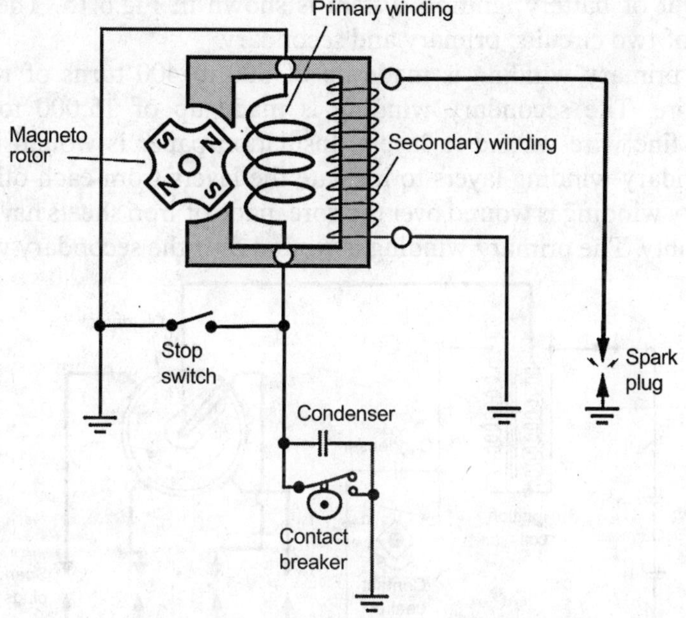


**Fig.6.14:** *Magneto ignition system*

## Advantages

1. More reliable
2. Economical
3. More compact
4. Good spark energies at low cranking speed
5. Reduced cable requirements due to the absence of low tension cables

### Disadvantages

1. Not suitable for multi-cylinder engines
2. Poor performance at starting

## BATTERY IGNITION SYSTEM

Battery coil ignition system is not self-contained, like a magneto ignition system. A separate source of supply (6, 12 or 24V battery) is required to feed the primary winding. Compared with a magneto system, the battery ignition system makes engine starting much easier. Also it is simpler to control the maximum voltage to suit the conditions.

This system uses electrical energy produced by a battery or alternator to supply the low-tension ignition primary current. The ignition coil transforms the voltage to that required to produce a high-voltage spark.

### Construction

The layout of battery ignition system is shown in Fig.6.15. The system consists of two circuits: primary and secondary.

The primary winding is made up of 300 to 400 turns of relatively heavy wire. The secondary winding is made up of 15,000 to 20,000 turns of fine wire wound in layers. Insulation paper is wound between the secondary winding layers to insulate the layers from each other. The secondary winding is wound over the core made of iron sheets having high permeability. The primary winding is wound over the secondary winding.

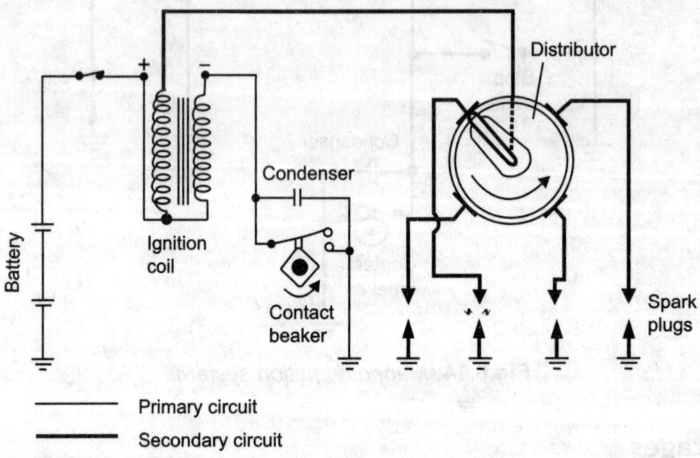


**Fig.6.15:** *Battery ignition system*

The primary winding is arranged in series with a battery, switch and contact breaker. A condenser is connected in parallel with the contact breaker. The secondary winding has one end connected to the high voltage terminal and the other end to the primary winding.

### Operation

The current flowing through the primary coil magnetises the core and builds up a magnetic field around the core. When the primary current is suddenly cut off by the opening of the distributor contact points, the magnetic field collapses, resulting in producing voltage of 300 to 400 Volts in the primary winding through self-inductance and a voltage of 15,000 to 20,000 Volts in the secondary winding through mutual inductance.

### Advantages

1. More reliable
2. Gives a much better spark for starting purposes
3. Easy maintenance
4. More suited for high speed engines

### Disadvantages

1. Rely upon the battery
2. Contact points wear, need adjustment, prone to failure, bounce at high RPMs, and become wet and dirty.

## COMPARISON OF BATTERY IGNITION AND MAGNETO IGNITION

| Sl.No | Battery ignition | Magneto ignition |
|---|---|---|
| 1. | Less moving parts, and requires less maintenance. | Most of the parts moving and requires more maintenance. |
| 2. | Less cost | More cost |
| 3. | Gives efficient spark at start | Gives weaker spark at start |
| 4. | Gives weaker spark at high speeds and very high speed is not desired. | Gives efficient spark at high speed |
| 5. | Efficient operation depends upon the condition of battery or generator. | Independent for its efficient operation |
| 6. | Battery maintenance is required. | No battery maintenance problems. |

## IGNITION ADVANCE MECHANISMS

Due to the time lag needed for flame propagation after ignition, the mixture must actually be ignited before TDC. It is necessary to have some means to change (advance or retard) ignition timing so that it best matches engine speed, load, etc. For this purpose, ignition advancers are provided.

Three types of spark advancing mechanisms are provided in general:

1. Governor advancer – advances in relation to engine speed.
2. Vacuum advancer – advances in relation to throttle valve opening.
3. Octane selector – advances in relation to kind of gasoline (petrol).

## CENTRIFUGAL (OR) GOVERNOR ADVANCER

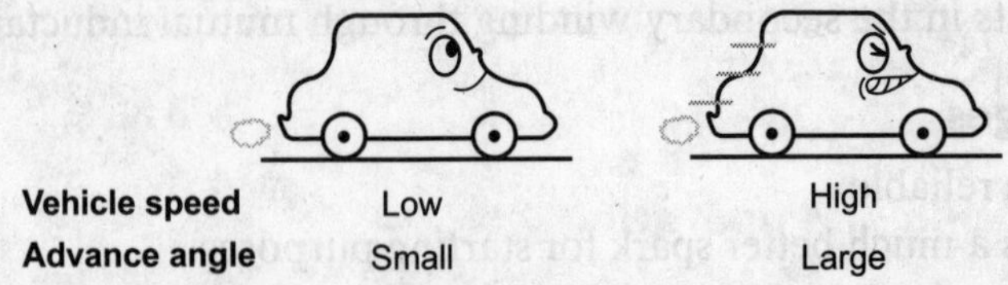


The spark advance during engine high speeds can be achieved by a centrifugal or governor advance mechanism. The governor advancer adjusts the ignition timing based on the engine speed. Since the flame propagation time is almost constant irrespective of the engine speed (as long as the air-fuel ratio is constant), the crankshaft angle during the flame propagation period increases as the engine speed rises.

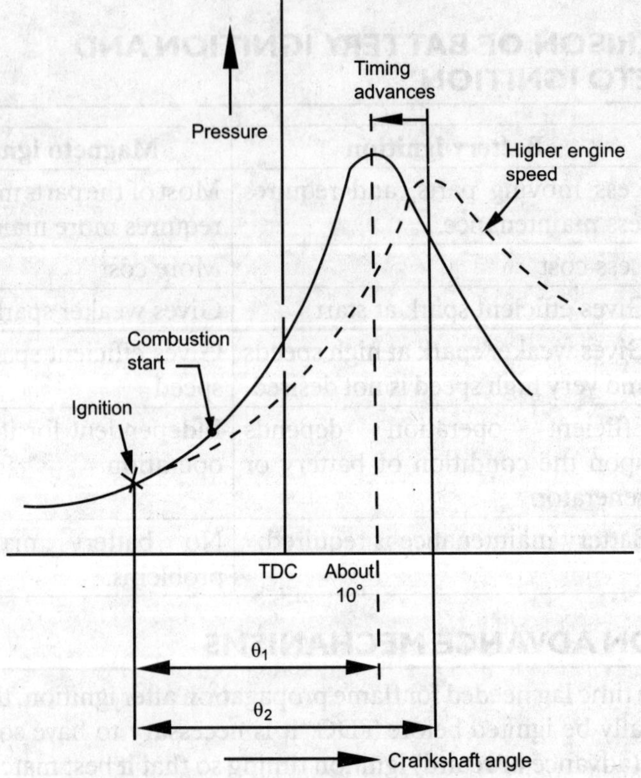


**Fig.6.16:** *Ignition advance at higher engine speed*

In other words, the flame propagation time gets relatively longer ($\theta_1 < \theta_2$) as the engine speed rises, so the curve indicating the crankshaft angle at maximum combustion pressure shifts to the right as shown in Fig.6.16. The governor therefore advances the ignition timing as the engine speed rises so that the maximum combustion pressure will always occur at a position 10° after TDC.

In other words, if the engine speed increase to higher rpm the time required for complete combustion of the air-fuel mixture is very less and hence it is necessary to increase the time to burn air-fuel mixture.

## Construction and operation

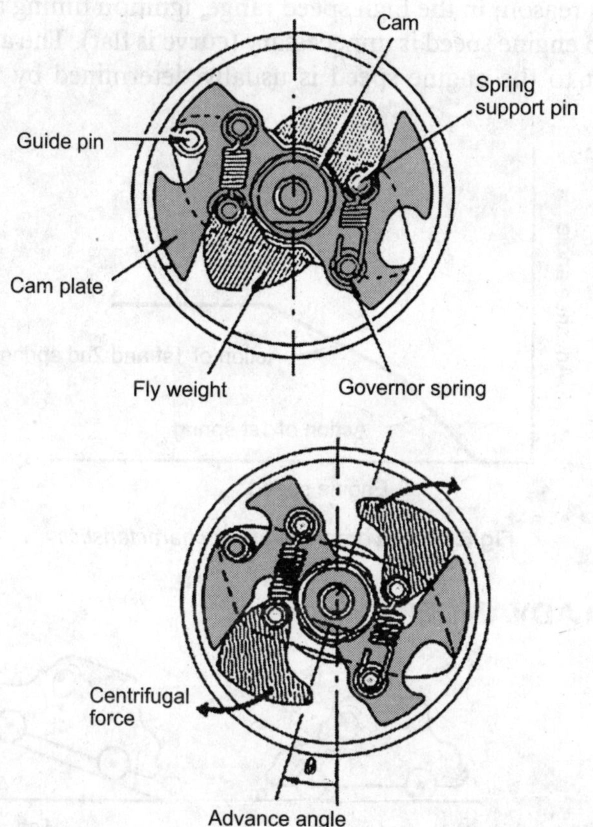


**Fig.6.17:** *Operation of governor advancer*

The Fig.6.17 shown above illustrates the construction of a centrifugal advance mechanism. It consists of two pivoted weights in the lower housing of the distributor assembly. These are held towards the center of the distributor shaft by two independent springs. As the rotation speed increases, the weights are thrown outwards, due to centrifugal force, against the spring tension. This will turn the center shaft cam lobe to open the contact points prior to the normal timing and cause the advance sparking.

As the speed reduces to idle running, the centrifugal forces also reduces, thus the weights are pulled back to their original position by the springs tension. Thereby the distributor shaft lobe turns back to the original position to provide the ignition at initial timing (normal timing, usually 10°BTDC).

## Governor advancer characteristics

When the engine speed rises considerably high, the turbulence of the mixture in the cylinder causes the flame propagation speed to accelerate.

For this reason, in the high speed range, ignition timing advancement according to engine speed is unnecessary (curve is flat). The advance angle with respect to the engine speed is usually determined by the governor springs.

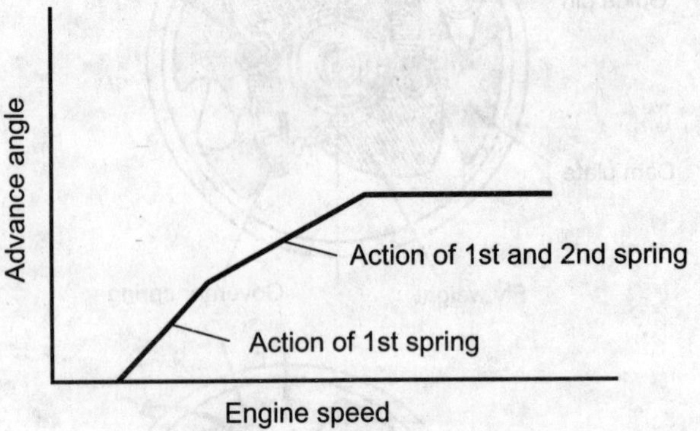


**Fig.6.18:** *Governor advancer characteristics*

## VACUUM ADVANCER

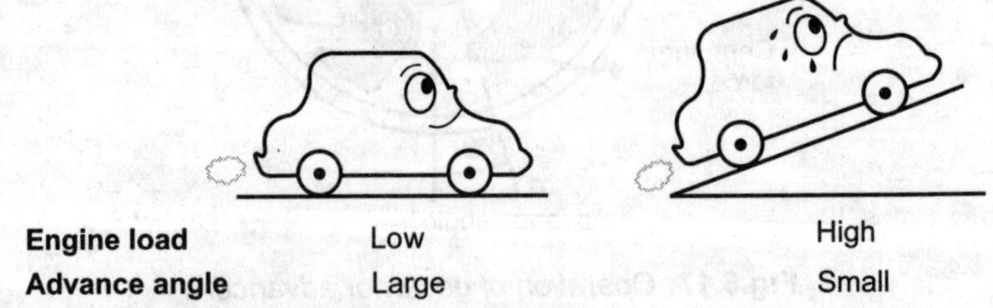


**Fig.6.19:** *Engine load Vs advance angle*

When the engine is running at partial throttle with average load, the air fuel ratio is very lean. As the mixture is lean which contains less fuel, it requires more time to complete the combustion process. In such case it is better to give ignition advance. The vacuum ignition advancer adjusts the ignition timing based on the varying vacuum in the intake manifold under varying engine loads, and advances the ignition timing accordingly.

When the load on the engine is light, the throttle valve opening is also small, so the vacuum in the intake manifold increases. A strong vacuum in the intake manifold lowers the air-fuel mixture intake efficiency, causing little air-fuel mixture to be drawn into the cylinders. This results in lowered flame propagation speed after ignition. However, when the load on the engine increases, the vacuum in the intake manifold decreases, improving the air-fuel mixture intake efficiency and resulting in an increase in the flame propagation speed after ignition.

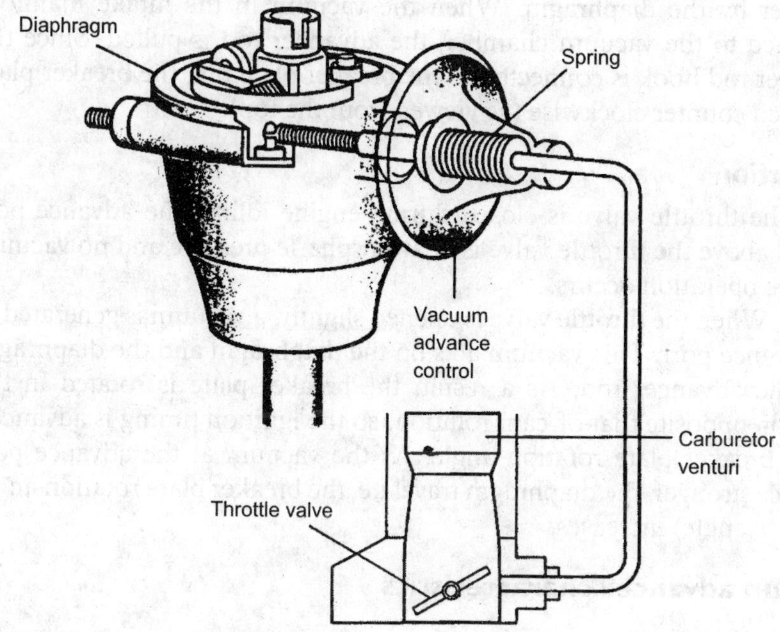


**Fig.6.20:** *Vacuum advance mechanism*

The vacuum ignition advancer therefore advances the ignition timing when the load is light to insure that the maximum combustion pressure will always occur at 10° after TDC.

**Note**: *Since the vacuum outlet port is above the throttle valve when the throttle is fully closed, the vacuum advancer does not advance during idling.*

## Construction

The vacuum ignition advance mechanism consists of the vacuum advancer (diaphragm, diaphragm spring, advancer rod, etc.) and the distributor breaker (breaker plates, breaker points, stationary plate, etc).

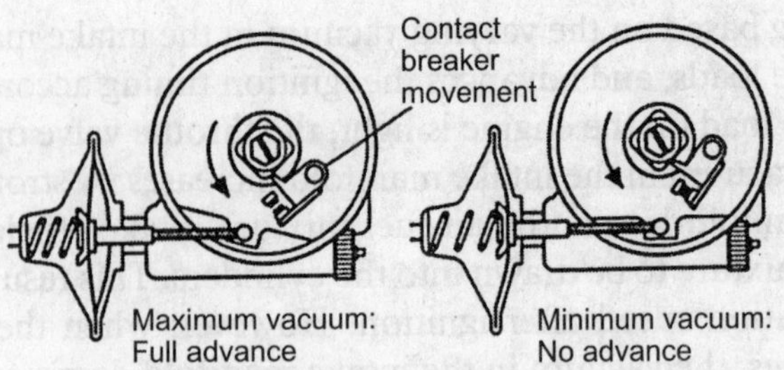


**Fig.6.21:** *Operation of vacuum advancer*

The vacuum advancer is divided into a vacuum chamber and an air chamber by the diaphragm. When the vacuum in the intake manifold is applied to the vacuum chamber, the advancer rod is pulled. Since the advancer rod hook is connected to the breaker plate pin, the breaker plate is rotated counter clockwise (as viewed from the top).

## Operation

Since the throttle valve is closed during engine idling, the advance port located above the throttle valve is at atmospheric pressure and no vacuum advance operation occurs.

When the throttle valve is opened slightly, a vacuum is generated at the advance port. This vacuum acts on the diaphragm and the diaphragm pulls the advancer rod. As a result, the breaker plate is rotated in the direction opposite that of cam rotation, so the ignition timing is advanced by the breaker plate rotation angle. As the vacuum at the advance port becomes stronger, the diaphragm travel i.e. the breaker plate rotation angle (advance angle) increases.

## Vacuum advancer characteristics

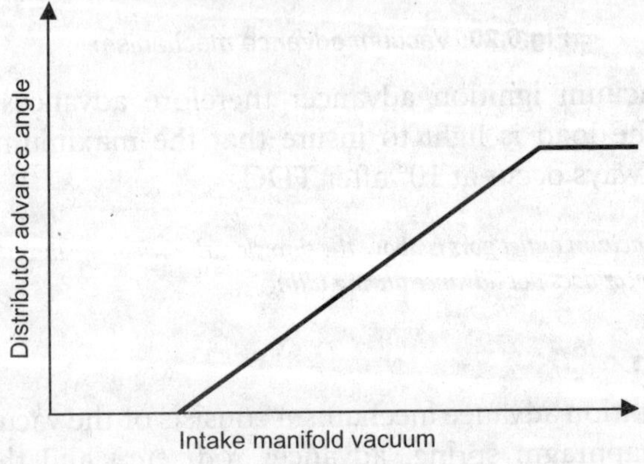


**Fig.6.22:** *Vacuum advancer characteristics*

The vacuum advancer characteristics, like the governor advancer characteristics, are determined on the basis of the required engine performance. Advancement is controlled by the strength of the vacuum advancer diaphragm spring as shown in the graph below.

## OCTANE SELECTOR

The octane selector makes fine adjustments of the ignition timing possible by changing the vacuum advancer characteristics in accordance with the octane value of the petrol.

The combustion rate of the air-fuel mixture (flame propagation rate) differs depending on the octane rating of petrol. Therefore, to utilize the explosive force in the cylinder to the maximum advantage which results in larger output and greater fuel economy, the ignition timing must be adjusted in accordance with the octane rating of petrol.

If petrol with a low octane rating is used, the ignition point (temperature) of the petrol is lower than for normal petrol, so the time from spark to ignition and combustion is short and the combustion rate (flame propagation rate) is high. Therefore, the maximum combustion pressure occurs before 10° ATDC. For that reason, not only is the engine prevented from displaying its full potential performance, but the pressure inside the cylinder becomes too high, making it easy for knocking due to spontaneous combustion to occur.

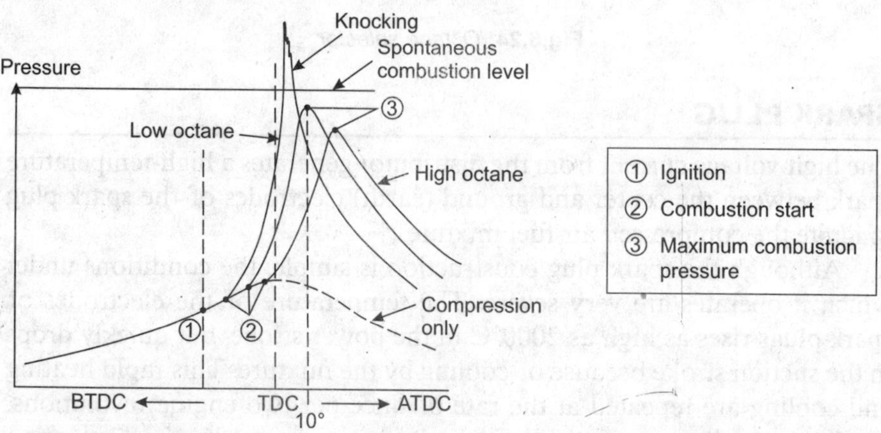


**Fig.6.23:** *Flame propagation time differs depending on differences in the octane rating*

When high octane petrol is used, the ignition point is higher than with ordinary petrol, so more time is needed from spark to ignition and combustion, and the combustion rate is slow. Therefore, the time until the maximum combustion pressure is achieved may be too long, causing the combustion rate to be retarded beyond 10° ATDC. Since the piston will be quite far down in the cylinder in such a case, the combustion pressure will be too low and the engine will not be able to display its full potential performance.

As can be seen from the Fig.6.23, the ignition timing is retarded when low octane petrol is used and advanced when high octane petrol is used.

In areas where the petrol sold has higher octane values, ignition angle adjustment using the octane selector is necessary.

## Operation

Turning the octane selector knob changes the hook position ($\ell$) relative to the diaphragm rod. As a result, the breaker plate setting position relative to the distributor cam also changes, altering the vacuum advancer characteristics as shown below. Normally one turn of the octane selector knob advances or retards the ignition timing by about 4°.

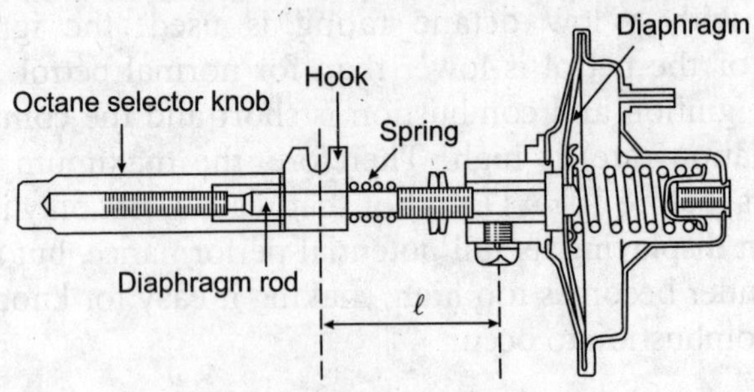


**Fig.6.24:** *Octane selector*

## SPARK PLUG

The high-voltage current from the distributor generates a high-temperature spark between the center and ground (earth) electrodes of the spark plug to ignite the compressed air-fuel mixture.

Although the spark plug construction is simple, the conditions under which it operates are very severe. The temperature of the electrodes of spark plugs rises as high as 2000°C in the power stroke, but quickly drops in the suction stroke because of cooling by the mixture. This rapid heating and cooling are repeated at the rate of once per two engine revolutions. Furthermore the pressure in the cylinder is less than 1 bar in the suction stroke, but rises to 45 bar in the power stroke. The spark plugs must be able to maintain their ignition performance for long periods of time while being exposed to such high temperatures and pressure changes, and while maintaining insulation resistance against voltages of as high as 10,000 to 30,000 Volts.

## SPARK PLUG LOCATION

There are three locations for the spark plugs which are to be considered while designing an engine.

Spark plug is to be located at the center of the cylinder head in order to have shorter flame travel which ultimately reduces detonation even at high compression ratios.

It is to be located near to the exhaust valve region so that the flame front travels from the hottest to the coolest part which ensure lower end charge temperature and hence no self ignition of end charge.

And finally it should be near inlet valve to combust more intake.

## PARTS OF A SPARK PLUG

The spark plug consists mainly of the insulator, casing, center electrode, and ground electrode.

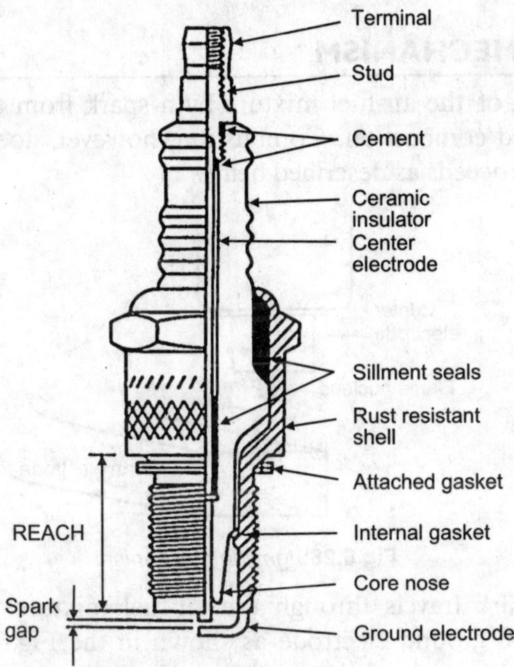


**Fig.6.25:** *Parts of a typical spark plug*

### Insulator

The insulator holds the center electrode and serves as the insulation between the center electrode and casing. The insulator is made of high-purity alumina porcelain having excellent heat resistance, mechanical strength, dielectric strength at high temperatures, and thermal conductivity.

### Casing

The casing supports the insulator while at the same time serving as the means for mounting the spark plug to the engine.

### Center electrode

The center portion of the spark plug consists of the terminal, upper stem, glass seal, resistor, copper core and center electrode. The center electrode generates the spark with the ground electrode.

### Ground electrode

The ground electrode is made of the same material as the center electrode.

### Resistor

The resistor filters out excessive electrical noise generated by the ignition system.

## IGNITION MECHANISM

The explosion of the air-fuel mixture by a spark from the spark plug is generally called combustion. Combustion, however, does not occur in a moment but proceeds as described below.

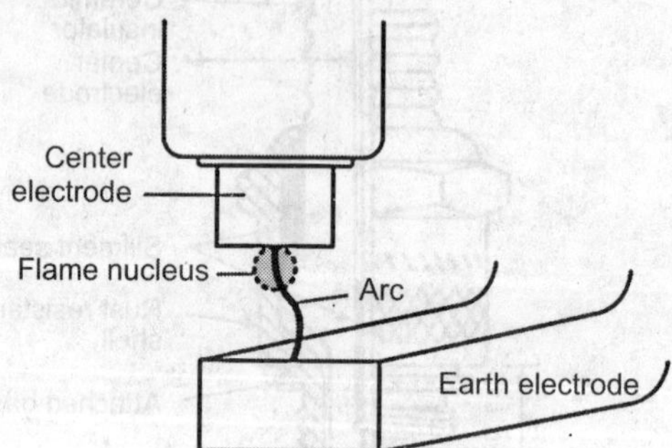


**Fig.6.26:** *Ignition mechanism*

As the spark travels through the air-fuel mixture from the center electrode to the ground electrode as shown in the Fig.6.26, the air fuel mixture along the path of the spark is activated and a so-called "flame nucleus" is formed. The molecules of the air-fuel mixture in the immediate vicinity of this flame nucleus are activated by the propagation of the heat of the spark and are pushed outward away from the nucleus by the shock wave created by the spark, activating the molecules further out from the center of combustion. These molecules then become part of the flame nucleus, which now has enough energy to continue to propagate outwards in this way by itself.

## ELECTRODE QUENCHING

The outward propagation of the flame in the ignition process is resisted by the relatively low temperature of the electrodes, which absorb the heat produced by the spark, thus tending to extinguish the flame nucleus before it can get started. This is called "electrode quenching" and must be overcome by each new spark as it attempts to ignite the air-fuel mixture and create a steadily propagating flame in the combustion chamber.

If the flame nucleus is too small, it will be extinguished by electrode quenching and the air-fuel mixture will not be activated. The flame will therefore go out and misfiring will occur. Suppressing electrode quenching and facilitating flame nucleus expansion and growth will enhance flame nucleus heat generation and flame propagation, thus improving combustion of the mixture.

## SPARK PLUG RESISTORS

When a spark is formed, it acts as a miniature radio transmitter and causes interference to nearby radio, television and mobile telephones etc. All cars are required by law to be fitted with resistors (suppressors), which limit interference. Many modern ignition systems require resistor plugs to stop electrical noise from interfering with the vehicle's on-board electronic control units (ECUs). Resistor spark plugs should always be fitted where specified.

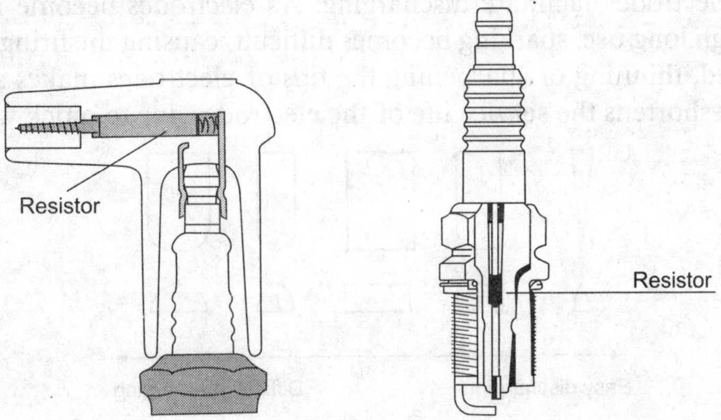


**Fig.6.27:** *Spark plug resistor*

Automotive resistor spark plugs contain a ceramic monolithic resistor. The most effective place to situate a resistor in the high tension circuit is as close to the spark plug as possible making the spark plug an ideal place to house the resistor.

## ELECTRODE MATERIALS

The electrodes in a spark plug typically consist of high-nickel alloys, while the insulator is generally made of aluminum oxide ceramic and the shell is made of steel wire.

Diverse materials are used for the center electrodes. Special nickel-based alloys and copper-core electrodes offer good thermal conductivity and high corrosion resistance. Silver has even higher thermal conductivity. Platinum and Iridium offer excellent resistance to erosion, so it extends the length of time between replacements. The design of the earthing electrode is just as important. Its geometry influences mixture accessibility, wear, heat dissipation and ignition voltage requirement, among other things. Titanium, platinum and iridium offer particularly long operating lives, for the same electrode gap.

## IGNITION PERFORMANCE

The high voltage generated in the secondary winding of the ignition coil is discharged between the center and ground electrodes of the spark plug. The sparking performance depends on many factors. The main factors affecting sparking performance are as explained below.

### Electrode shape and discharge performance

Rounded electrodes make discharging difficult, while squared-off or pointed electrodes facilitate discharging. As electrodes become rounded off through long use, sparking becomes difficult, causing misfiring. On the other hand, thinning or sharpening the tips of electrodes makes sparking easier but shortens the service life of the electrodes due to quick wearing.

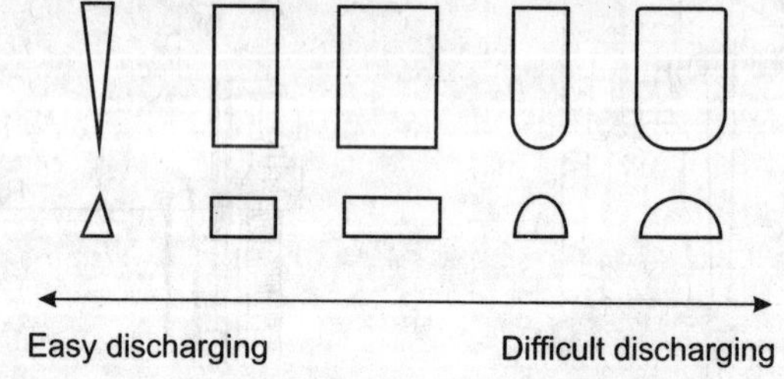


**Fig.6.28:** *Electrode shape and discharge performance*

### Spark gap and required voltage

Discharging becomes harder and the required voltage increases as the air gap becomes wider. When electrodes are worn, thus increasing the gap, sparking becomes difficult, causing misfiring.

If the electrode gap is too small this may cause misfiring, noisy idling and poor exhaust gas quality levels.

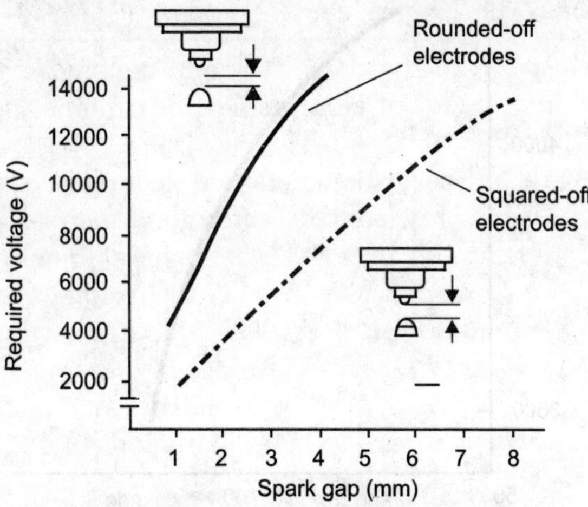


**Fig.6.29:** *Spark gap Vs Required voltage*

## Compression pressure and required voltage

Discharging becomes difficult and the required voltage increases as the compression pressure increases. This occurs mainly under a high engine load when the vehicle is running at a low speed with the throttle valve fully opened. The required voltage also increases as the temperature of the air-fuel mixture decreases.

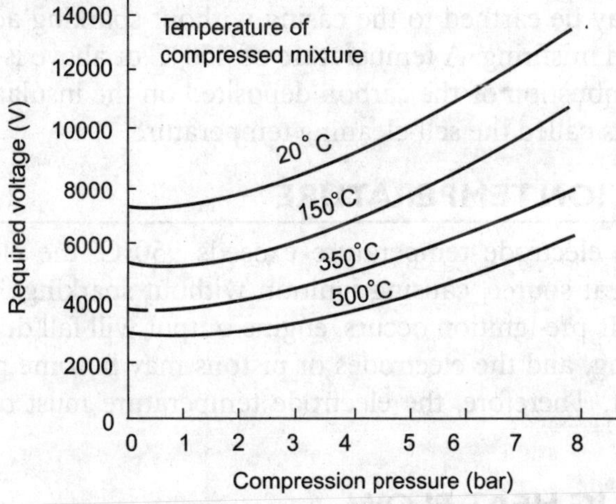


**Fig.6.30:** *Compression pressure Vs Required voltage*

## Electrode temperature and required voltage

The electrode temperature generally rises as the engine speed increases. However, the required voltage for discharge decreases as the electrode temperature rises.

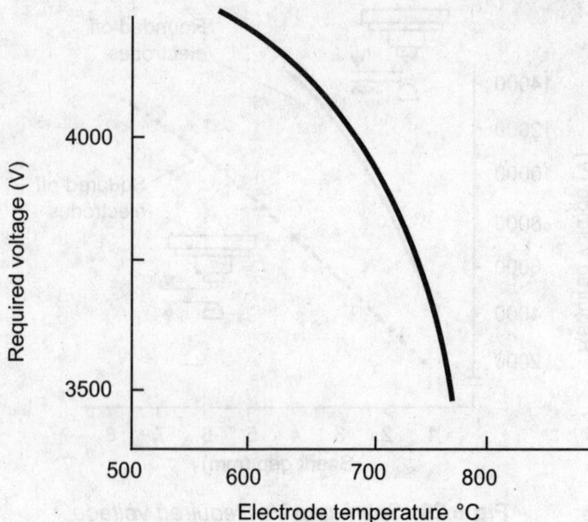


**Fig.6.31:** *Electrode temperature and required voltage*

## SELF-CLEANING TEMPERATURE

If the center electrode temperature is less than 450°C, the carbon generated by incomplete combustion of the fuel adheres to the surface of the porcelain insulation, lowering the insulation resistance between the center electrode and casing. As a result, the high voltage applied to the electrodes may be earthed to the casing without sparking across the gap. This results in misfiring. A temperature of 450°C or above is necessary for complete combustion of the carbon deposited on the insulator nose. This temperature is called the self-cleaning temperature.

## PRE-IGNITION TEMPERATURE

If the center electrode temperature exceeds 950°C, the electrode itself becomes a heat source, causing ignition without sparking. This is called pre-ignition. If pre-ignition occurs, engine output will fall due to incorrect ignition timing, and the electrodes or pistons may become pitted or even partially melt. Therefore, the electrode temperature must be kept below 950°C.

## SPARK PLUG HEAT FLOW

The routes of heat radiation and the percentages of heat radiated by the spark plug are shown in Fig.6.32.

The self-cleaning temperature of 450°C and the pre-ignition temperature of 950°C occur near the tip of the spark plug. The temperature at 'A' is important to the performance of the spark plug and this depends on the temperature of the gas in the combustion chamber and on the spark plug design.

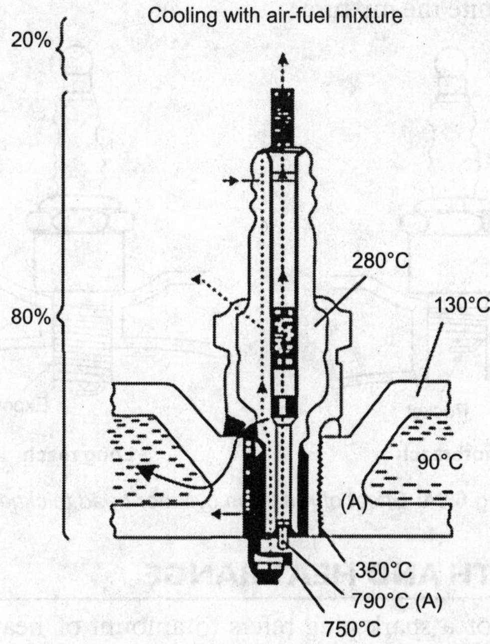


**Fig.6.32:** *Heat flow in spark plug*

The temperature of the spark plug center electrode depends on the absolute amount of heat radiated to the outside through the routes.

## SPARK PLUG REACH

Spark plug reach refers to the length of the shell from the contact surface at the seat to the bottom of the shell, including both threaded and non-threaded sections.

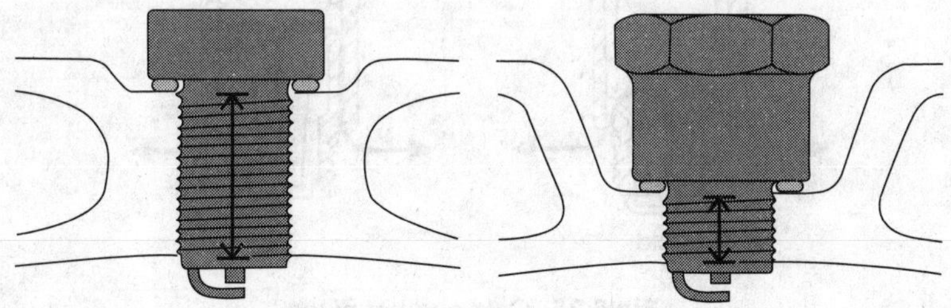

**Fig.6.33:** *Spark plug reach*

Long-reach plug is used where the cylinder head section is very deep. In a thinner head such a plug would project too far into the combustion chamber. The exposed plug threads act as hot spot and cause preignition.

Short-reach plug is used where the cylinder head section is thin. In a deeper cylinder head, its electrodes are in a pocket and the arc is not able to adequately ignite the mixture.

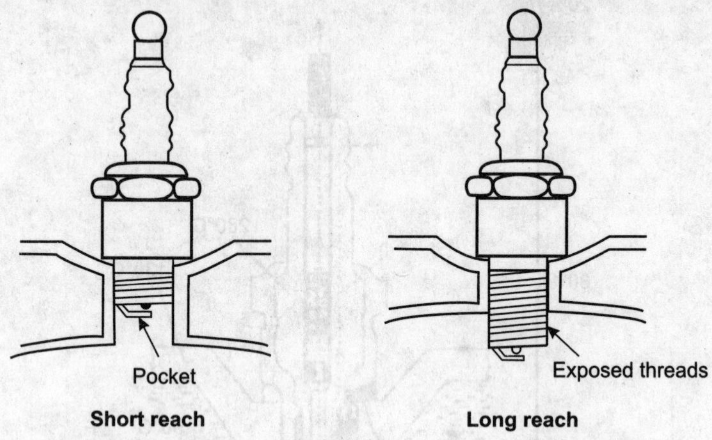


**Fig.6.34:** *Effect of reach on cylinder head thickness*

## NOSE LENGTH AND HEAT RANGE

The heat range of a spark plug refers to amount of heat radiated by the spark plug. It is determined by the heat radiated distance before escaping to the spark plug shell. Spark plug that radiates more heat is called "cold plug", because the plug itself stays cooler while one that radiates less heat is called a "hot plug", because it keeps its heat in.

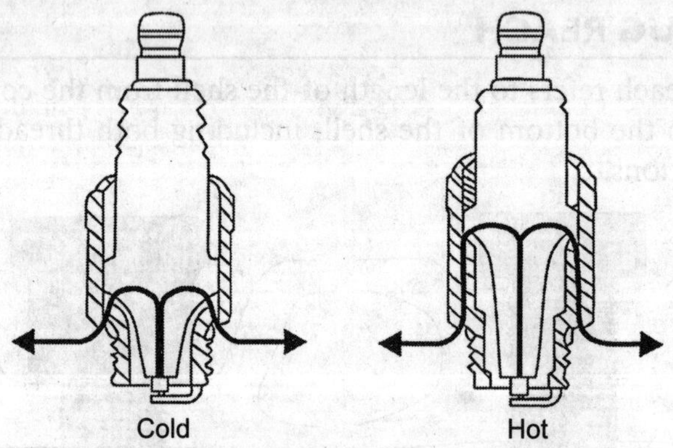


**Fig.6.35:** *Cold and hot plugs*

The lower operational limit of a spark plug is the self-cleaning temperature (450°C), while the pre-ignition temperature (950°C) is the upper limit. Spark plugs perform best when the center electrode temperature is between about 450 and 950°C.

The insulator nose length (T) of cold and hot type plugs differs as shown in Fig.6.35. The cold plug has a shorter insulation nose. Since the surface area exposed to the flame is small and the heat radiation route is short, heat radiation is excellent and the center electrode temperature does not rise very high. For this reason, if the cold spark plug is used, it is more difficult for pre-ignition to occur.

On the other hand, since the hot type plug has a longer insulator nose, the surface area exposed to the flame is large, the radiation route is long and heat radiation is small. As a result, the center electrode temperature rises fairly high and the self-cleaning temperature can be attained more quickly in the low-speed range than in the case of the cold type.

Use of excessively cold plugs promotes sluggish combustion and spark plug fouling. Use of excessively hot plugs leads to pre-ignition and rapid erosion of the electrodes.

The heat range is affected by the reach of a spark plug. Long-reach spark plugs run hotter and short-reach plugs run colder.

**Note:**

a. The 'Hot' Plug is used in cold running engines.
b. The 'Cold' Plug is used in hot running engines.

## PROJECTING-ELECTRODE SPARK PLUG

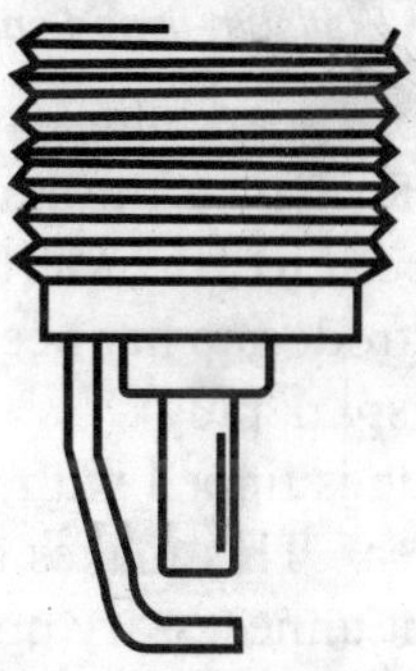

**Fig.6.36:** *Projecting electrode spark plug*

A spark plug which has longer electrodes shown here is called a projecting-electrode spark plug. Since this type of spark plug projects into the combustion chamber, the probability of exposure to gasoline molecules in the air-fuel mixture increases, thus improving ignition performance.

## PLATINUM-TIPPED SPARK PLUG

The ignition performance items mentioned previously can also be used for platinum-tipped spark plugs. The tip of the center electrode and the opposing earth electrode are covered with thin platinum films to extend the service life of the plug. This type of spark plug is adopted for some engines equipped with emission control devices.

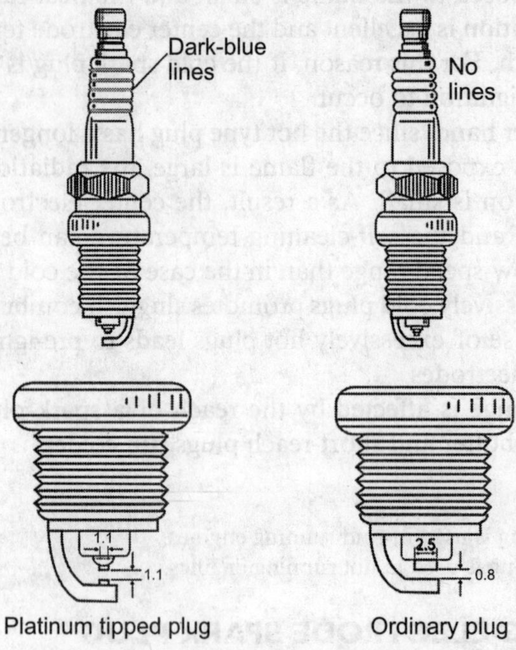


**Fig.6.37:** *Platinum-tipped spark plug*

The features of platinum-tipped spark plugs are as follows:

1. To improve ignition performance, the diameter of the tip of the center electrode has been reduced to 1.1 mm (from 2.5 mm of an ordinary spark plug) and the electrode gap has been increased to 1.1 mm (from 0.8 mm of an ordinary spark plug).
2. The end of the electrode is tipped with platinum, thus reducing the amount of electrode wear. This makes checking and adjustment of the gap and replacement unnecessary up to the 10,000 km mark.
3. The width across flats of the hexagonal part has been reduced from the 20.6 mm of ordinary spark plugs to 16 mm, both to reduce size and weight and to improve cooling of the spark plug.
4. To make it easier to distinguish platinum tipped spark plugs from ordinary plugs without removing them from the engine, there are dark-blue lines around the insulator.

## TWIN TIP SPARK PLUG

The large center and ground electrodes seen on a standard plug increases "quenching" of the explosion, meaning the spark explosion is restricted by the electrodes above and below the gap, allowing it to spread to either side, but not grow with full range.

The Twin-Tip (TT) technology delivers far less obstruction to the spark allowing an Omni-directional explosion. This minimises the quenching effect and results in a more effective, more complete combustion. Better combustion means more power and better fuel economy, as fuel is burned more efficiently with each piston firing.

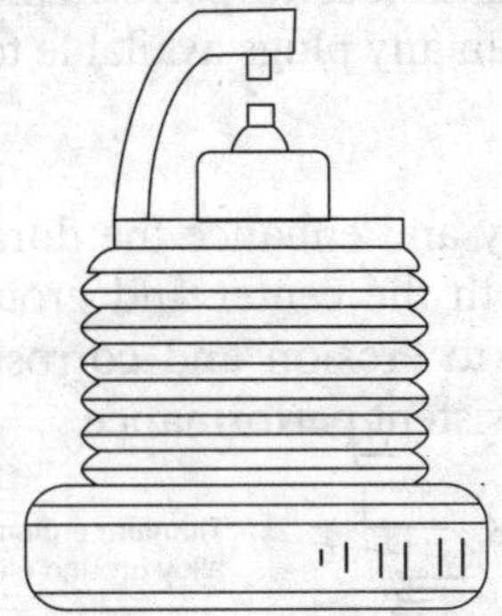

**Fig.6.38:** *Twin-tip spark plug*

## Advantages

1. Lower emissions
   Significantly reduces $CO_2$ and HC emissions.
2. Superb fuel economy
   Excellent fuel efficiency due to enhanced ignitability.

## IRIDIUM TT

For complete combustion, the Iridium TT's two tips allow the spark plug to create the largest spark concentration and explosion, which translates into more power and superior fuel economy than any other plug.

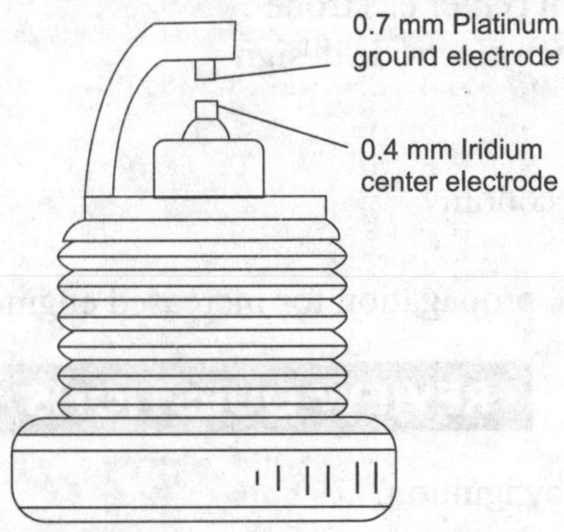


**Fig.6.39:** *Iridium TT spark plug*

With this unique design, there are literally two tips: a 0.4-millimeter Iridium center electrode, and a 0.7-millimeter Platinum ground electrode. The 0.4-millimeter center electrode has a low required voltage and high ignitability, that means vehicle experiences less interference to spark growth (quenching), fewer misfires and significantly improved combustion, optimising power output and fuel economy.

The unparalleled Iridium TT spark plug is tested to last well beyond 100,000 miles, with minimal loss to performance, contributing to more overall lifetime savings than any plugs available today.

## PLATINUM TT

To maximise the longevity and enhance the durability of the spark plug, platinum is applied to both the center and ground electrodes. Platinum offers superior resistance to erosion and corrosion enabling to create a spark plug capable of consistent performance.

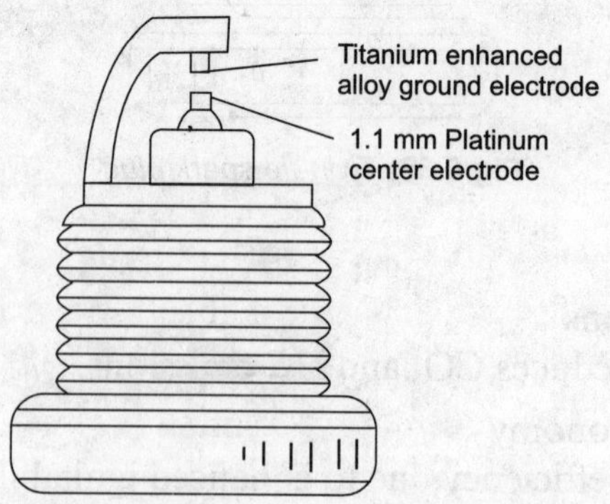


**Fig.6.40:** *Platinum TT spark plug*

### Features

1. Titanium-enhanced alloy ground electrode
2. 1.1mm platinum center electrode
3. Minimum-quench electrode design

### Advantages

1. Improved fuel economy
2. Faster starts
3. Enhanced flame propagation for increased engine efficiency

## REVIEW QUESTIONS

1. What is meant by ignition?
2. What is pre-ignition?

3. What are the classifications of an ignition system?
4. Why is a resistor used in the primary ignition circuit?
5. How does ignition take place in a petrol engine?
6. How many volts of current reaches at the spark plug?
7. What is magneto ignition system and how it works?
8. What are the special features of the magneto ignition system?
9. Where magneto ignition system is being used?
10. How many kinds of magneto ignition systems are there?
11. What is the principle of current generation?
12. What other functions are performed by magneto except ignition?
13. What is battery ignition system?
14. What is the function of a distributor?
15. Which are the main parts of a distributor?
16. What is the function of contact breaker point?
17. How much gap is maintained between contact breaker points?
18. What will happen if the contact breaker points gap is less or more?
19. What is ignition timing and how is it set? What is the effect of too early ignition in the working of an engine?
20. Describe the factors which affect ignition timing in an engine.
21. What do you understand by advance or retard of the ignition system?
22. What is the main benefit of advance ignition?
23. What are the different types of advance or retard mechanisms?
24. What is the function of spark plug?
25. Which are the main parts of a spark plug?
26. What should be the characteristics of a good spark plug?
27. What is the importance of spark plug gap? How much gap should be maintained?
28. What will be the effects if the spark plug gap is less or excess?
29. How many types of spark plugs are there?
30. What could be the possible faults in a spark plug?
31. What should be done for the maintenance of the spark plug?
32. What is the correct gap between the cut-out points?
33. What are the causes of the engine backfire?
34. What is meant by reach of the spark plug?

35. What is a hot plug?
36. What is a cold plug?
37. What is the effect of too hot and too cold plugs in the working of an engine?
38. What will happen if a long reach plug is replaced with a short reach plug by mistake?
39. What will happen if a short reach plug is replaced with a long reach plug by mistake?
40. What are the reasons for no spark?
41. What is an engine misfire?
42. What are the causes for misfires at slow speed?
43. What are the causes for misfires at high speed?
44. What are the reasons for misfires in a scooter?
45. When magneto does not generate current, what are the possible faults in it?
46. What could be the possible faults when magneto misfires at slow speed?
47. When should we do spark maintenance?
48. What happens when the condenser gets discharged?
49. What is the essential difference between the magneto and the coil ignition system?
50. Why is it necessary to retard the spark for full throttle operation under load?
51. Why is it desirable to advance the ignition timing when the engine speed increases?
52. Why are advance mechanisms necessary in the ignition system?
53. What is referred to as vacuum advance? How does this mechanism works?
54. What is referred to as centrifugal advance? How does this mechanism works?
55. When does a condenser get charged?
56. Trace the electrical path through (a) the primary circuit, (b) the secondary circuit.
57. How is the distributor shaft driven?
58. Why is a condenser used?

59. Why are the distributor cap and rotor necessary?
60. What type of high-tension wire is most commonly used today? Why?
61. Explain why the different heat ranges of spark plugs are necessary?
62. Why are extended nose spark plugs used?
63. How are spark plug sizes determined?
64. Explain what is meant by fouling of a spark plug.
65. Explain the effects of (a) over-tightening, (b) under-tightening a spark plug.
66. Describe the working of battery coil ignition system in detail. What is the function of CB point?

# Electronic Ignition System

In case of conventional contact breaker point ignition system, large amount of current is allowed to pass through the contacts. Thereby there is a chance of spark (arcing) at contacts causing damage to the contacts and also loss of high amount of current through the contacts. The spark produced also not powerful to ignite the lean mixture. The coil used in such system has to carry high amount of current to produce a powerful spark. This makes the size of the coil bulk and heavy. Later the electronic ignition has been introduced to avoid the problems in the conventional type of ignition.

Electronic ignition system uses electronic switching devices instead of mechanical contact breaker points to send a high voltage to the spark plug.

## ADVANTAGES

The advantages of electronic ignition are:

1. It is small and light weight
2. It gives higher ignition voltage
3. It gives longer duration of spark
4. It has very accurate control of timing
5. It improves cold weather starting
6. Less maintenance
7. Longer service life
8. Less operating cost
9. It helps vehicles meet exhaust emission regulations and reduces fuel consumption.

## DISADVANTAGES

The common problems with electronic ignition systems are:

1. Hard to diagnose
2. Adjustment is not as accessible
3. More specific expertise is necessary
4. More susceptible to heat damage

## TYPES OF ELECTRONIC IGNITION SYSTEM

Electronic ignition systems may be categorised as capacitor discharge or transistorised or distributor less.

## CAPACITIVE DISCHARGE IGNITION SYSTEM

This system is not widely used and differs considerably from the standard and inductive systems. One of the problems with using a coil to provide the high voltage is the time required to build up and collapse the magnetic field before a spark can be produced. On eight or twelve cylinder engines this can be a major problem.

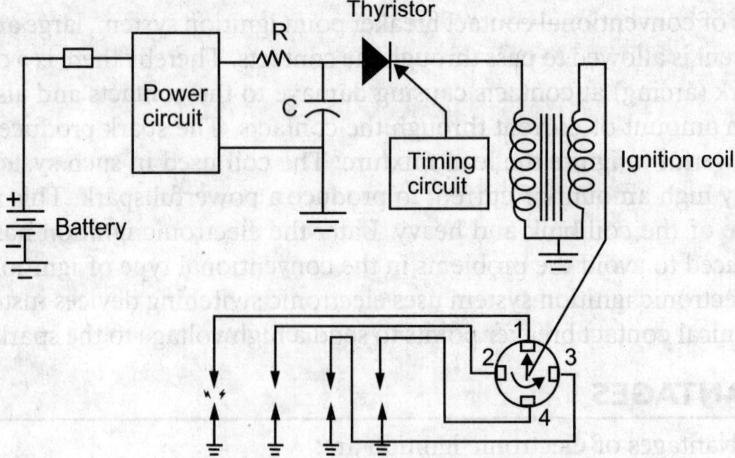


**Fig.7.1:** *Capacitive discharge ignition system*

In the capacitor discharge system the supply to the coil is stored in a capacitor (C) inside an electronic control unit. The timing circuit (trigger) in Fig.7.1 could be either mechanical breaker points or a pick-up coil. A trigger suddenly releases the charge stored in the capacitor to the primary winding of the coil.

### Operation

In operation, an inverter converts the AC (Alternating Current) supply to DC (Direct Current) and feeds it to a step-up transformer to provide the 400 V needed by the capacitor. When a spark is required, the trigger releases the energy from the capacitor to the coil by activating a thyristor (an electronic switch). The sudden discharge of high voltage electrical energy causes a rapid build-up of the magnetic field around the coil and induces very high voltage in the secondary windings. This results in a spark of high intensity but short duration. Voltage transformation time is about one hundredth of the time taken by the standard system. This allows a much higher rate of sparks per second. However, the short duration of each individual spark renders the system unsuitable for most engines.

### Advantages

1. Capacitor can be fully charged in a very short time (typically 1ms), only 20 microseconds (μs) are required to fully charge a capacitor.
2. Fast voltage rise (between 3 ~ 10 kV/μs) compared to typical inductive systems (300 ~ 500 V/μs).
3. Decreases spark plug fouling.
4. Increases sparkplug life.
5. Uses less current. This means less demand on the battery during starts and a potentially longer battery life.
6. Provides improved cold starting ability.

## SOLID STATE TRANSISTORISED IGNITION SYSTEM

This is widely used type of electronic ignition system. It is similar to the conventional system but the big difference is that the primary current is switched on and off by a transistor (an electronic switch). It therefore eliminates the problems associated with contact breaker points. The system does, however, still use the mechanical centrifugal advance mechanism to sense engine speed and a vacuum advance to sense engine load.

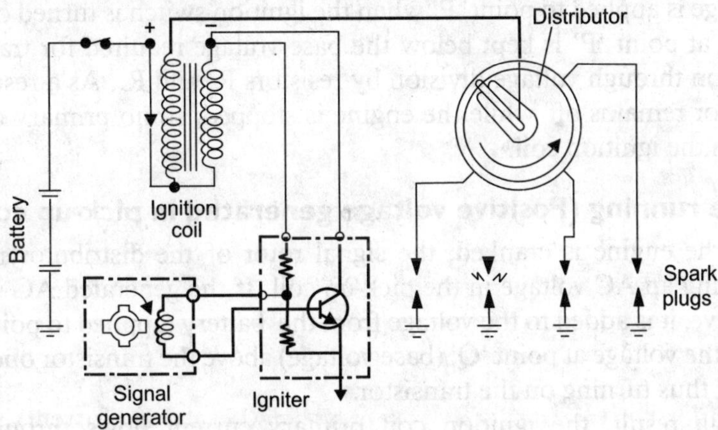


**Fig.7.2:** *Transistorised ignition system*

## Operation

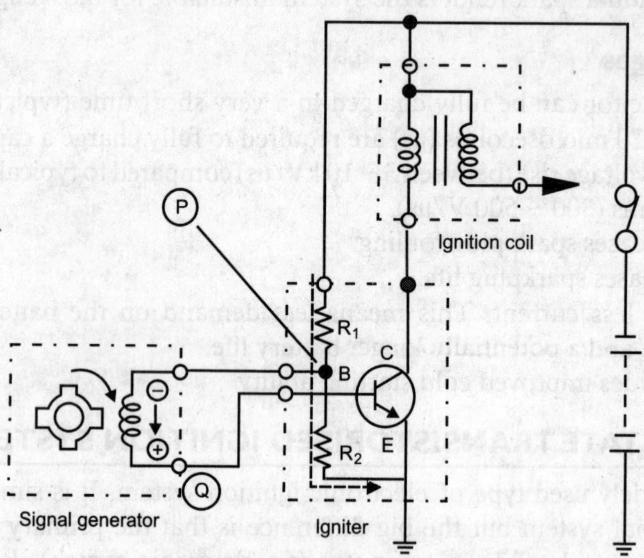


**Fig.7.3:** *Circuit of transistorised ignition system*

### Engine stopped

A voltage is applied to point 'P' when the ignition switch is turned on. The voltage at point 'P' is kept below the base voltage required for transistor operation through voltage division by resistors $R_1$ and $R_2$. As a result, the transistor remains off while the engine is stopped, so no primary current flows in the ignition coil.

### Engine running (Positive voltage generated in pick-up coil)

When the engine is cranked, the signal rotor of the distributor rotates, generating an AC voltage in the pick-up coil. If the generated AC voltage is positive, it is added to the voltage from the battery (applied to point 'P'), raising the voltage at point 'Q' (base voltage) above the transistor operating voltage, thus turning on the transistor.

As a result, the ignition coil primary current flows through the transistor from the collector (C) to the emitter (E).

### Engine running (Negative voltage generated in pick-up coil)

When the AC voltage in the pick-up coil is negative, this voltage is added to the voltage at point 'P', so the voltage at point 'Q' decreases below the transistor operating voltage, turning off the transistor. As a result, the ignition primary current is turned off and a high voltage is induced in the secondary coil.

### Advantages

1. Delivers much higher voltage
2. Plug life at least doubles
3. Because the points are scrapped, the ignition timing is stable over the full operating range.
4. No moving parts to fail or bounce.
5. Longer service between tune-ups
6. Better fuel economy
7. Safer and more reliable.

## DISTRIBUTOR LESS IGNITION (DLI) SYSTEM

In past transistorised ignition systems, the high voltages generated by a single ignition coil were distributed to each spark plug through the distributor. DLI is a spark-ignition engine ignition system which uses no distributor. This system uses one ignition coil for every two spark plugs. The electronic control unit (ECU) distributes the primary current to each ignition coil directly, causing each spark plug to spark.

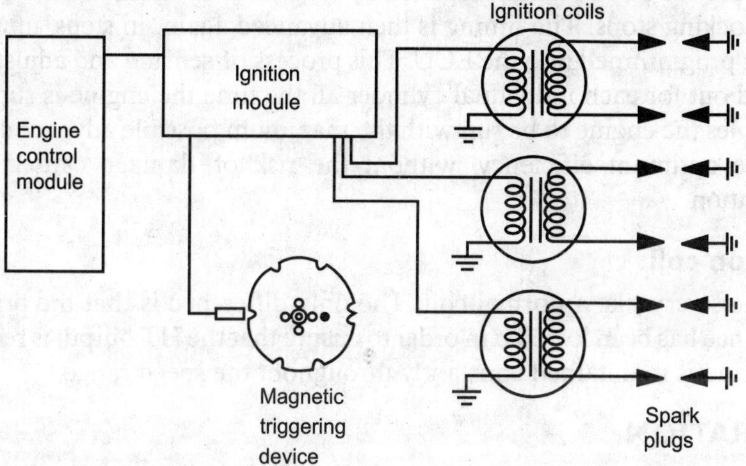


**Fig.7.4:** *Distributor less ignition system*

## COMPONENTS OF DLI

This system consists of the following major components.

### Electronic Control Unit (ECU)

The ECU operates as a pre-programmed computer. It gathers information from a variety of sensors interprets that information and adjusts the ignition timing accordingly. Its two main functions are to control the ignition timing electronically using information from sensors and to control the ignition coil output. The main advantage is that the ignition timing is controlled by the ECU instead of the centrifugal and vacuum advance mechanisms.

### Crankshaft sensor

This determines the crankshaft speed and position. As the crankshaft rotates, the sensor produces pulses, which are fed to the ECU. From this the ECU can determine the TDC position and the engine speed. These inputs are used by the ECU to calculate basic ignition timing.

### Coolant temperature sensor

The ECU uses the information from coolant temperature sensor to adjust ignition timing according to engine operating temperature.

### Knock sensor

This provides a small input voltage to the ECU, which is proportional to the vibration sensed in the engine block. The vibrations set up by combustion detonation are detected by the sensor and the information is fed to the ECU. The ECU responds by retarding the ignition timing until the knocking stops. The timing is then advanced again, in steps, up to the setting programmed into the ECU. This process of sensing and adjusting is carried out for each individual cylinder all the time the engine is running. It enables the engine to be run with the maximum possible advance setting to give optimum efficiency, without the risk of damage caused from detonation.

### Ignition coil

This acts very like a normal coil. The only difference is that the primary resistance has been lowered in order to ensure that the HT output is reached faster and is maintained constantly throughout the speed range.

## OPERATION

The distribution of the spark is obtained by using double-ended coils which are fired alternately by the ECU. The timing is determined by signals from sensors. When one of the coils is fixed a spark is delivered to two engine cylinders. For the majority of engine layouts where an even number of cylinders is used, two of the pistons will rise and fall within the cylinders at the same time. The spark delivered to the cylinders on the compression stroke will ignite the mixture as normal. The spark produced in the other cylinder will have no effect, as this cylinder will be just completing its exhaust stroke.

Because of the low compression and the exhaust gas in the waste spark cylinder, the voltage used for the spark to jump the gap is only about

3000 Volts. The spark produced in the compression cylinder is therefore not affected.

## ADVANTAGES

This system has the following advantages:

1. High tension cords can be shortened, thus reducing noise.
2. Reduction in the number of mechanical components improves reliability.
3. Ignition timing can be controlled over a wider range.
4. More precise timing
5. Timing curve is more adjustable
6. More powerful spark
7. Fewer moving parts
8. No distributor to accumulate moisture and cause starting problems
9. No distributor to drive thus providing less engine drag

## DIRECT IGNITION SYSTEM

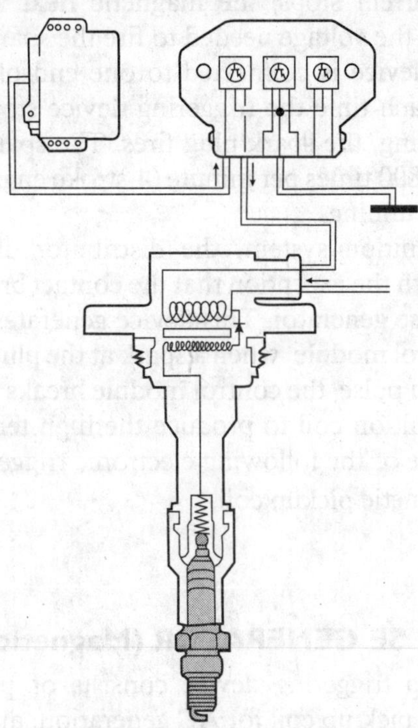

**Fig.7.5:** *Direct ignition system*

Direct ignition is in a way the follow on from distributor less ignition. This system utilises an inductive coil for each cylinder. These coils are mounted directly on the spark plugs. The use of an individual coil for each plug ensures that the rise time for the low inductance primary winding is very fast. This ensures that a very high voltage, high energy spark is produced. This voltage, which can be in excess of 400kV, provides efficient initiation of the combustion process under cold starting conditions and with weak mixtures. Some direct ignition systems use capacitor discharge ignition.

### Features

1. Mounted directly on the plugs
2. No Sparkplug wires

## TRIGGERING DEVICES (SIGNAL GENERATORS)

All ignition systems use some type of triggering devices to turn the primary coil winding on and off. A triggering device works much like a switch. The ignition system's circuit is closed when the switch closes. When the switch closes, current flows from the power source to the ignition coil. When the switch opens, the circuit is opened and the flow of current immediately stops. When the current stops, the magnetic field in the ignition coil collapses, producing the voltage needed to fire the spark plug.

The triggering device is connected to one end of the ignition coil's primary winding. Each time the triggering device stops the current flow in the primary winding, the spark plug fires. The spark plug keeps firing continually, about 1,800 times per minute (4-stroke engine with 3600 rpm), so the engine keeps running.

In electronic ignition system, the distributor unit is similar to a conventional unit with the exception that the contact breaker is replaced by a sensor called a pulse generator. This device generates an electrical pulse to signal to the 'control module' when a spark at the plug is required. When the trigger produces a pulse, the control module breaks the primary circuit, which causes the ignition coil to produce the high tension (HT) voltage. Most engines use one of the following electronic triggering devices:

- Inductive (magnetic pickup coil)
- Hall
- Optical

## INDUCTIVE PULSE GENERATOR (Magnetic pickup coil)

The magnetic-pickup triggering device consists of permanent magnets which magnetise the pick-up coil for AC generation, and the trigger wheel (signal rotor) which induces an AC voltage in the pick-up coil in accordance with the ignition timing. The number of teeth on the trigger wheel is equal to the number of engine cylinders (i.e., 4 teeth for a 4-cylinder engine and 6 teeth for a 6-cylinder engine).

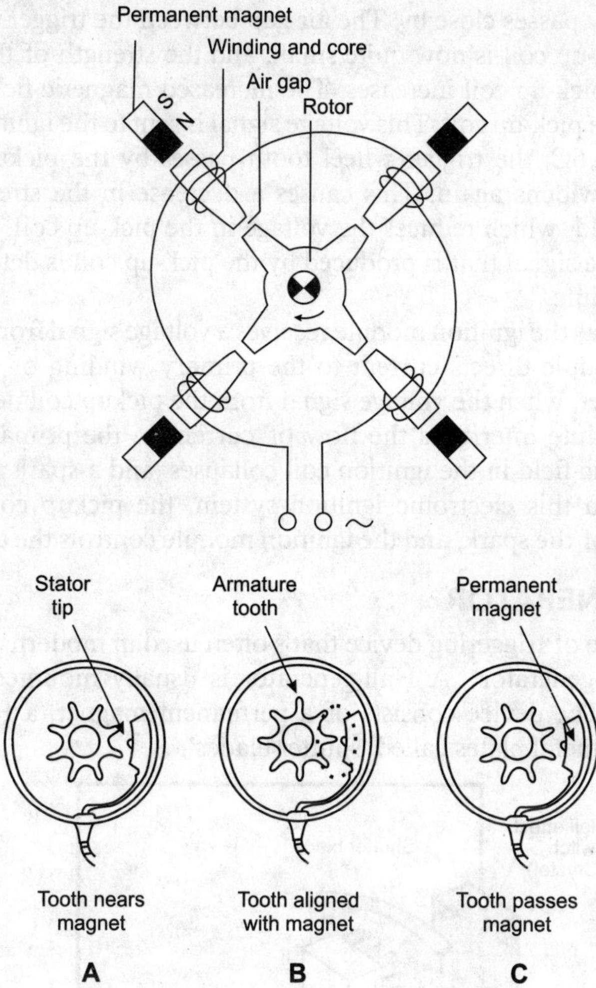


**Fig.7.6:** *Trigger wheel position relative to pickup coil*

## Operation

When the ignition switch is on, current from the battery reaches the magnetic pick-up coil, and as the distributor shaft turns the trigger wheel, a magnetic field is induced at the pickup coil. In addition, a magnetic field develops between the pickup coil and the trigger wheel. The strength of the magnetic field is influenced by the trigger wheel as it revolves.

In Fig.7.6A, note that none of the teeth on the trigger wheel are aligned with the pick-up coil. Because of the wide air gap between the pick-up coil and the trigger wheel at this moment, the magnetic field is quite weak.

In Fig.7.6B, a trigger wheel tooth approaches the pick-up coil. Note that the trigger wheel tooth does not come into contact with the pickup coil-it merely passes close by. The air gap between the trigger wheel tooth and the pick-up coil is now quite small, and the strength of the magnetic field in the pick-up coil increases. The increased magnetic field induces a voltage in the pick-up coil. This voltage signal is sent to the ignition module.

In Fig.7.6C, the trigger wheel tooth passes by the pickup coil, and the air gap widens again. This causes a decrease in the strength of the magnetic field, which reduces the voltage in the pick-up coil. The change in the voltage signal that is produced by the pick-up coil is detected by the ignition module.

As long as the ignition module receives a voltage signal from the pickup coil, the module directs current to the primary winding of the ignition coil. However, when the voltage signal from the pickup coil decreases, the ignition module interrupts the flow of current in the primary winding. The magnetic field in the ignition coil collapses, and a spark is produced. Therefore, in this electronic ignition system, the pickup coil times the occurrence of the spark, and the ignition module controls the actual spark.

## HALL GENERATOR

Another type of triggering device that's often used in modern automobiles is the 'Hall-generator'. A Hall-generator is usually mounted inside the distributor. The device consists of a permanent magnet, a Hall IC, and several thin metal plates called 'shutter blades'.

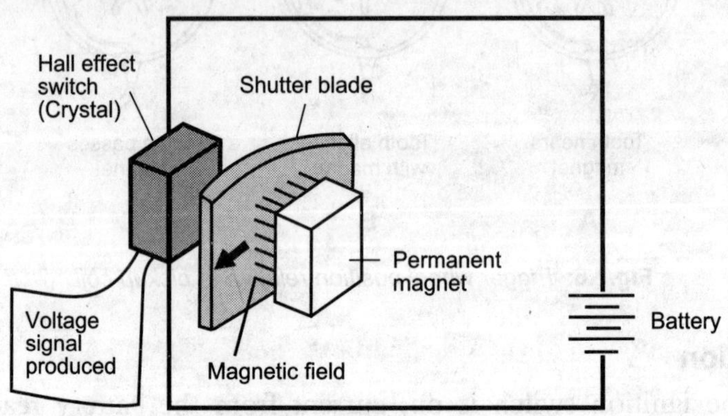


**Fig.7.7:** *Hall effect*

Since the Hall IC is placed next to a permanent magnet, the Hall IC has a magnetic field around it. As long as a magnetic field is present around the device, no electrical signal is produced. However, when there is no magnetic field around the switch, the Hall IC sends a signal to the ignition module.

In a Hall generator, shutter blades are attached to a round disk that rotates with the distributor shaft. This disk may be a separate item or the blades may be attached to the distributor rotor. In either case, the operation of the device is the same. The shutter blades rotate with the distributor shaft as the engine operates. Each time a shutter blade passes between the permanent magnet and the Hall IC, the shutter blade prevents the magnetic field from reaching the Hall IC (Fig.7.7). This causes the Hall IC to produce a voltage signal that is sent to the ignition module. The signal tells the ignition module that a cylinder is ready to receive a spark. A typical Hall-generator has one shutter blade for each cylinder.

## OPTICAL PULSE GENERATOR

The Inductive pulse generator and Hall generator use the forces of electromagnetism to signal the ignition module as to the position of the cylinders. Each time a cylinder is ready to fire, the triggering device signals the ignition module to turn off the voltage in the primary winding of the ignition coil. The distributor rotor and cap then direct the high voltage to the proper spark plug.

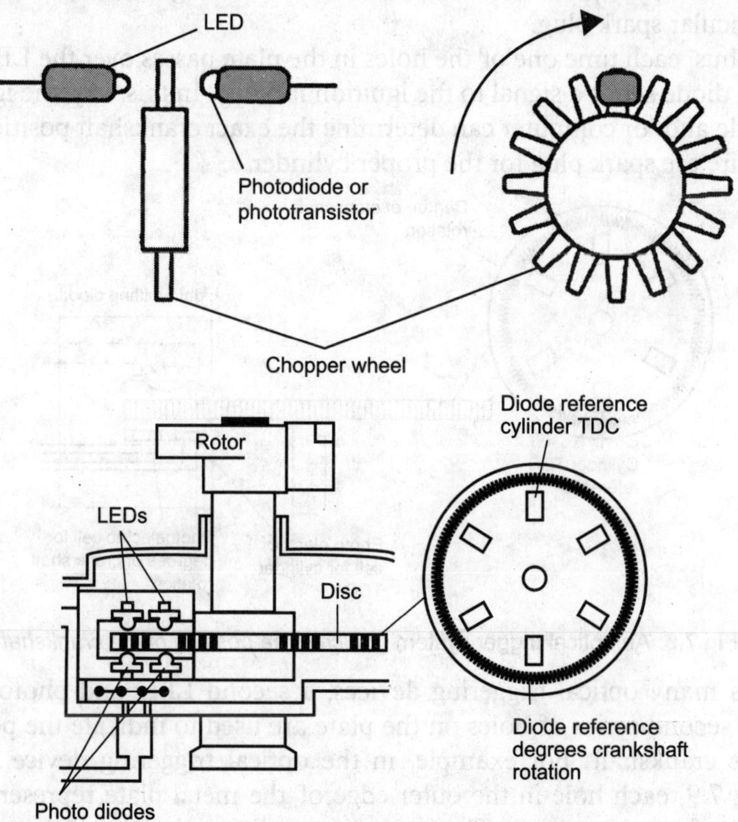


**Fig.7.8:** *An optical trigger system*

Fig.7.8 shows an optical trigger. An optical or LED triggering device performs the same function as an inductive pulse generator or a Hall generator. However, an optical triggering device uses a light source and a thin, flat metal plate for triggering. Holes in the metal plate represent the TDC point of each cylinder.

The optical system consists of a disc (chopper), which is connected to the distributor shaft cam. Either side of the disc is a light-emitting diode (LED) and a phototransistor (a semiconductor switch that is sensitive to light intensity).

## Operation

In operation, the metal plate is attached to the distributor shaft, so the plate rotates when the engine is running. When the light from the LED shines against the surface of the plate, no light reaches the photo diode on the other side of the plate. However, whenever one of the holes in the plate passes over the LED, the light passes through the hole and reaches the photo diode on the other side of the plate. When the light reaches the photo diode, the photo diode sends a signal to the ignition module to fire a particular spark plug.

Thus, each time one of the holes in the plate passes over the LED, the photo diode sends a signal to the ignition module. In this way, the ignition module and/or computer can determine the exact crankshaft position and then fire the spark plug for the proper cylinder.

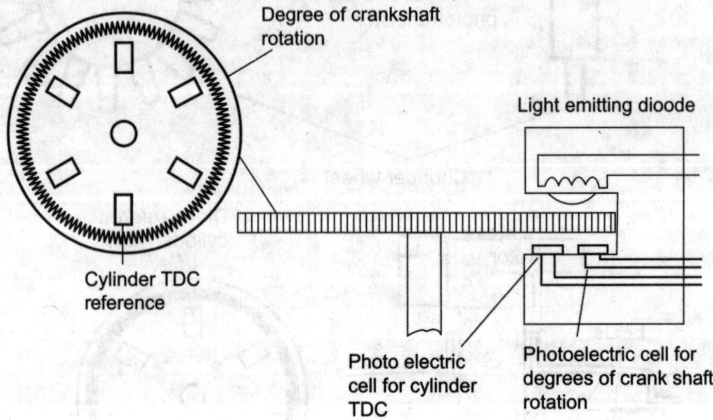


**Fig.7.9:** An optical trigger system *(indicate the position of the crankshaft)*

In many optical triggering devices, a second LED and photo diode and a second series of holes on the plate are used to indicate the position of the crankshaft. For example, in the optical triggering device shown in Fig 7.9, each hole in the outer edge of the metal plate represents one degree of crank rotation. The larger, inner slits in the plate represent the

TDC points of each cylinder. The second LED and photoelectric cell monitor the positions of the holes in the outer edge of the plate. The additional information provided by the second LED and photo-electric cell helps the computer system determine the ignition timing. Therefore, if a computer determines that a cylinder should be fired a few degrees of crankshaft rotation before it reaches TDC, the computer can simply monitor the information from the second LED and photoelectric cell so it knows exactly when to fire the spark plug.

## IGNITER

The igniter consists of a detector, which detects the EMF generated by the signal generator, a signal amplifier, and a power transistor, which carries out accurate interruption of the ignition coil primary current according to the amplified signal. Dwell angle control for correction of the primary signal according to increases in engine speed is also incorporated into the igniter.

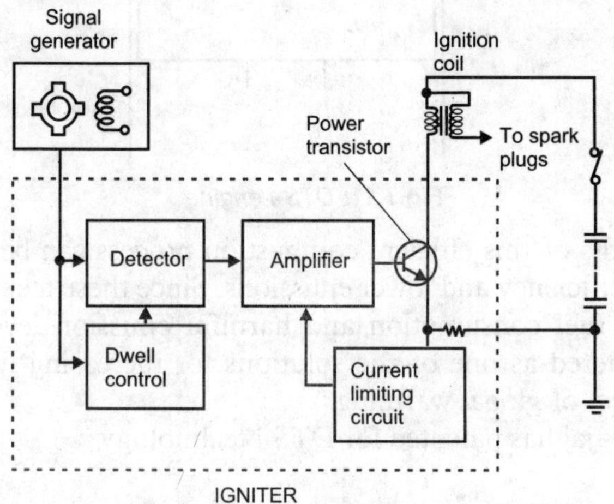


**Fig.7.10:** *Igniter circuitry*

## DIGITAL IGNITION SYSTEM

The intelligent electronic ignition system is programmed for precise ignition timing at all operating speeds, from idle to full power. It ensures easy starting every time, delivers optimum fuel efficiency and automatically limits the maximum engine revolutions.

## DIGITAL TWIN SPARK IGNITION SYSTEM (DTSI)

In actual practice, perfect combustion is not at all possible due to various losses in the combustion chamber as well as design of the internal combustion engine. Moreover the process of burning of the fuel is also not instantaneous. However an alternate solution to it is by making the combustion of fuel as fast as possible. This can be done by using two spark plugs which spark so as to increase the diameter of the flame and burn the fuel instantaneously. This system is called Digital Twin Spark Ignition (DTSi). In this system, due to twin sparks, combustion will be complete. The engine combust fuel at double rate than normal.

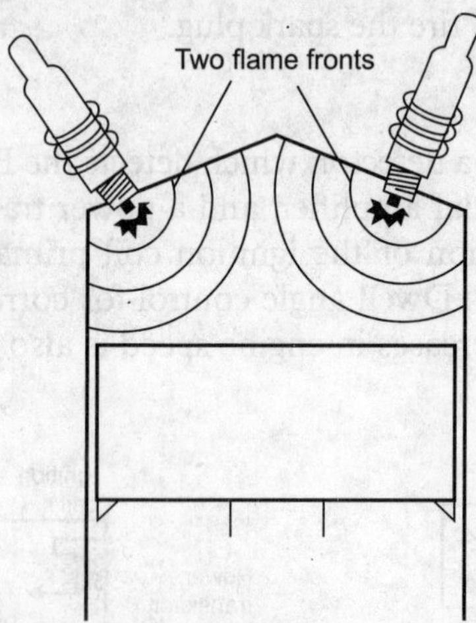


**Fig.7.11:** *DTS-i engine*

The benefits of this efficient combustion process can be felt in terms of better fuel efficiency and lower emissions. Since these technologies also minimize the fuel consumption and harmful emission levels, they can also be considered as one of the solutions for increasing fuel costs and increasing effect of global warming.

In India Bajaj has patented for DTS-i technology.

## Circuit

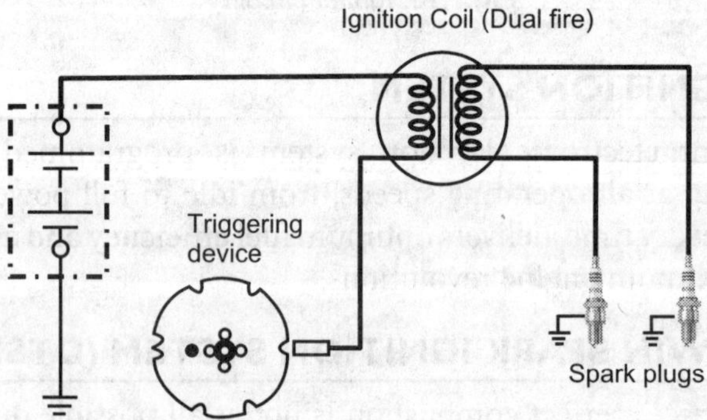


**Fig.7.12:** *Circuit of twin spark ignition*

Digital Twin Spark ignition engine has two spark plugs located at opposite ends of the combustion chamber and hence fast and efficient combustion is obtained. It is mapped by the digital electronic control box which also handles fuel ignition and valve timing.

This system uses a special configuration of the hemispherical combustion chambers and piston heads, ensures a fast, wide flame front when the air-fuel mixture is ignited, and therefore less ignition advance, enabling relatively lean mixtures to be used. A microprocessor continuously senses speed and loads of the engine and respond by altering the ignition timing there by optimizing power and fuel economy.

## Working

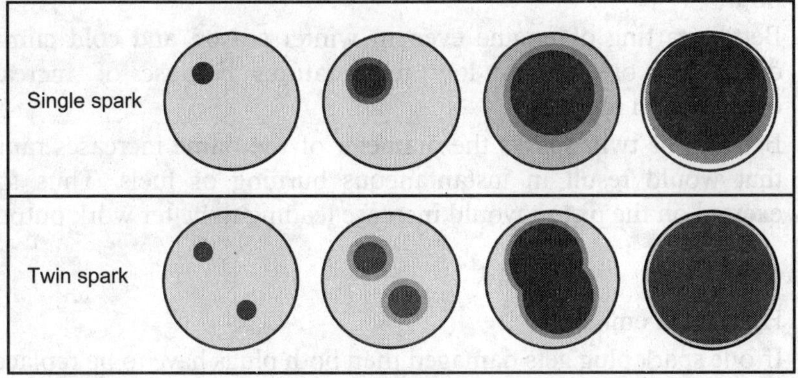


**Fig.7.13:** *Flame radius starting from the initiation of spark*

In a traditional single spark plug cylinder the combustion process is very slow. This is because the spark plug is located at one end of the combustion chamber, which sparks the air-fuel mixture and the flame inflates like a balloon that takes some time to reach the remote corner of the combustion chamber that delays the combustion process. Whereas, in the DTS-i technology equipped engine the cylinder head is fitted with two spark plugs at two of corners of the combustion chamber, which ensures that the sparks emit from two ends of the chamber and the air fuel mixture ignites to produce two flame fronts thus, the process of combustion is faster (reduction in flame travel of the order of 40% is achieved) and also the pressure rises at a good pace. This technology in the engine promotes greater torque, fuel-efficiency and low emission. The ignition system on the twin spark is a digital system with static spark advance and no moving parts subject to wear.

## Main characteristics

- Digital electronic ignition with two spark plugs per cylinder

- Twin overhead cams with camshaft timing variation
- Injection fuel feed with integrated electronic twin spark ignition
- A high specific power
- Compact design and superior balance

## Advantages

- Less vibrations and noise
- Long life of the engine parts such as piston rings and valve stem
- Decrease in the specific fuel consumption
- No over heating
- Increase thermal efficiency of the engine and even bear high loads on it.
- Better starting of engine even in winter season and cold climatic conditions or at very low temperatures because of increased compression ratio.
- Because of twin sparks the diameter of the flame increases rapidly that would result in instantaneous burning of fuels. Thus force exerted on the piston would increase leading to better work output.

## Disadvantages

- High NOx emission
- If one spark plug gets damaged then both plugs have to be replaced
- Cost is relatively more

## Applications

The use of twin spark plugs has been around for decades mostly in the engine of Alfa Romeo, Honda and on large capacity Ducati engine etc. Bajaj however have patented the technology for small capacity engines. At present Platina, XCD125, 135, Discover150, Pulsar135, 150, 180, 200, 220 etc. are using this technology. Hence the application of this technology will give the present generation what they want i.e. power bikes with fuel efficiency.

# DIGITAL TWIN SPARK-SWIRL INDUCTION (DTS-SI)

The DTS-Si technology gives the highest possible fuel efficiency by introducing 'Swirl induction' to the DTS-i engine to create turbulence in order to achieve extremely efficient combustion.

Like DTS-i the engine has 2 spark plugs, but, instead of conventionally positioned straight ports, the offset positioning of the ports generate high swirl and turbulence of the air fuel mixture in the combustion chamber.

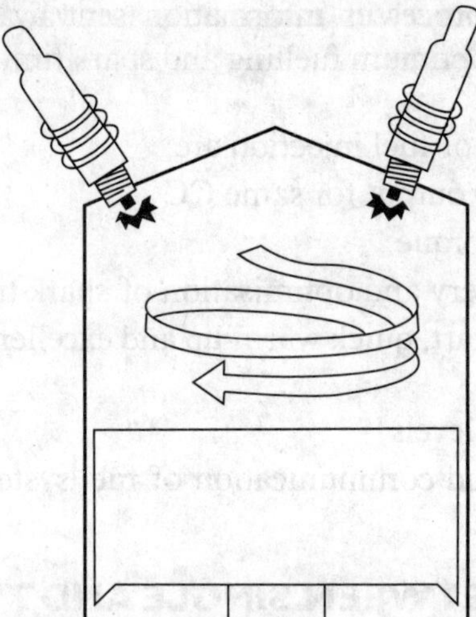

**Fig.7.14:** *DTS-Si engine*

Incorporated in the new engine, this technology results in a swirling motion of the incoming charge, which decays itself into turbulence as the piston moves in the induction and compression strokes. This results in the air-fuel mixture being more thoroughly mixed and spread around the combustion chamber. Sparks provided by the twin spark plugs ignite this highly turbulent and compressed air-fuel mixture, leading to a flame front with high surface area, resulting in a rapid rise of pressure due to rapid combustion. The values of turbulence achieved now, are substantially higher than that of a straight port cylinder head, such as in Pulsar. A combination of DTS-i and Swirl induction thus provides extremely rapid combustion, resulting in high efficiency.

### Application

Bajaj Auto new 125cc engine with DTS-Si technology will give an amazing mileage of 109 kilometers per liter under ideal test conditions surpassing the mileage of all current 100cc motorcycles. With this breakthrough there is a huge potential and opportunity to upgrade the 100cc customer with a engine which offers 100cc mileage and 125cc performance.

## DIGITAL TWIN SPARK-FUEL INJECTION (DTS-FI)

DTS-Fi stands for "Digital Twin Spark Fuel Injection", a 'Bajaj Patented Technology'. In fuel injection the conventional carburetor has been replaced by injector which injects fuel in to the engine in a spray form based on the instructions of the engine Electronic Control Unit (ECU) which is a part of the Engine Management System (EMS). The Electronic Control Unit (ECU) is microprocessor based and is the brain of the fuel injection system. It processes information sent by various sensors and instantly determines optimum fuelling and spark timing for various engine operating conditions.

The main advantages of fuel injection are:

- Increased power output for same CC
- Better low end torque
- Lower fuel delivery and optimisation of spark timing
- Improved cold start, quick warm-up and excellent response to sudden acceleration
- Lower emission levels
- Self detection and communication of fuel system malfunctioning if any

## COMPARISON BETWEEN SINGLE AND TWIN SPARK TECHNOLOGY

| Sl.No | Single spark technology | Twin spark technology |
|---|---|---|
| 1 | One spark plug used | Two spark plugs used |
| 2 | Slower and incomplete combustion of fuel | Faster and complete combustion of fuel |
| 3 | Increased rate of emission | Emission reduced considerably |
| 4 | Engine design is simple | Engine design is complex |
| 5 | Less rise in pressure and hence less torque generated | More rise in pressure and hence more torque generated |
| 6 | Less efficient | More efficient |

## DTS-I TRIPLE SPARK ENGINE

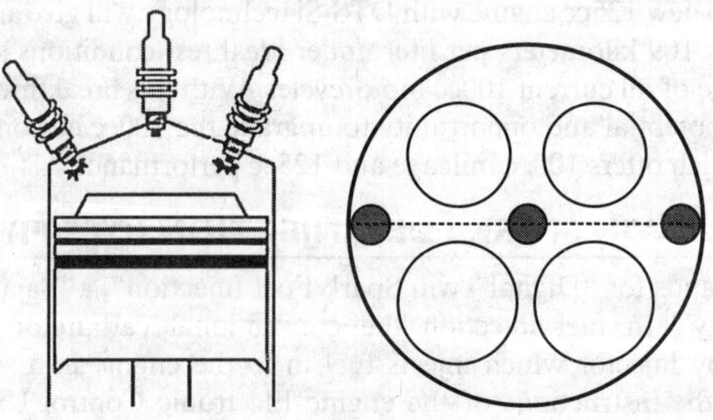

**Fig.7.15:** *Triple spark engine*

Here DTS-i Stands for "Digital Triple Spark Ignition". Three spark plugs provide unprecedented performance and efficiency. Fast and optimal combustion at part load conditions results in better fuel efficiency and ensures lower emissions. Combustion in a triple spark engine is 27% faster than a twin spark and 50% faster than a single spark engine.

To make use of 3 spark plugs, the pulsar engine houses a pent roof combustion chamber which in turn allows to house 3 spark plugs. Out of the three plugs, the primary plug is the center one and the other two secondary plugs are mounted below, each opposite each other.

The secondary plugs fires a bit after the primary one has fired and the timings are controlled by the ECU depending on various parameters like throttle position, engine speed, load on engine, etc. According to Bajaj, these plugs gain an advantage in low speed riding condition where it extracts the best economy. Compared to KTM Duke 200 in similar conditions it gives as much as 10-13 kmpl more, however the difference vanishes at higher speed.

## INTELLIGENT-DUAL SEQUENTIAL IGNITION (I-DSI)

The dual sequential ignition system optimizes the timing of each spark plug based on engine speed and engine load. The intensive combustion at all engine speeds not only controls knocking, but also permits a much higher compression ratio (10.8:1) to achieve a higher output with less fuel consumed compared to a conventional design.

The i-DSI engine provides rapid, intensive combustion, using two spark plugs in each cylinder mounted diagonally opposite one another. A very compact, high-swirl combustion chamber is made possible by the narrow (30°) valve angle and SOHC single pivot cylinder head.

Each pair of spark plugs is fired sequentially with the interval between the two depending on engine rpm and load. The intake side plug ignites first, then as the flame propagates, the exhaust side plug is fired (before top-dead-centre), and the flame expands rapidly into the whole area to achieve complete combustion. This arrangement provides much faster combustion and higher cylinder pressures to provide high engine output.

The programming of the ignition timing map achieves a careful balance between economy and power output. At large throttle openings, up to engine speeds of around 2600 rpm, intake side ignition is advanced and the exhaust side slightly retarded. In the mid-range the exhaust side is further retarded, optimising output. At high engine speeds ignition is almost simultaneous for optimum power and torque. At part-throttle, the simultaneous ignition phase is more extensive, commencing as low as 3500 rpm.

**Features of i-DSI**

- Twin Spark plug layout
- 2 valves per cylinder
- One coil per plug
- Optimum swirl formation
- Compact combustion chamber
- Compact SOHC cylinder head
- Chain driven cam
- Low friction design

## ELECTRONIC SPARK ADVANCE (ESA)

The electronic spark advance (ESA) system is a system in which an ECU (rather than a mechanical advancer) controls the ignition timing of the ignition system.

In early systems, the timing is advanced and retarded by a governor advancer in the distributor. Furthermore, ignition must also be advanced when the manifold pressure is low (i.e. when there is a strong vacuum). However, optimal ignition timing is also affected by a number of other factors besides engine speed and intake air volume, such as the shape of the combustion chamber, the temperature inside the combustion chamber, etc. For these reasons, electronic control provides the ideal ignition timing for the engine.

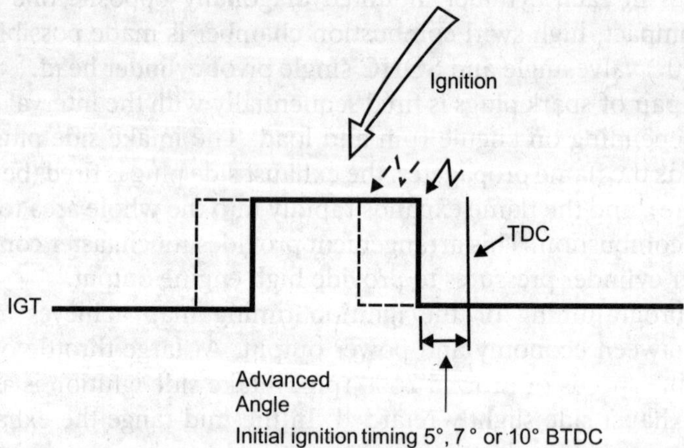


**Fig.7.16:** *Ignition advance angle*

As engine rpm increases, there is less time for the mixture to complete its combustion at the proper time because the piston is travelling faster, ignition must begin earlier (Fig.7.16). The ECM controls when the spark occurs through the IGT signal. By varying the time the IGT signal is turned off, the ECM changes ignition spark timing.

## Operation

In the electronic spark advance (ESA) system, the engine is provided with nearly ideal ignition timing characteristics. The Electronic Control Module (ECM) determines ignition timing based sensor inputs and on its internal memory, which contains the optimal ignition timing data for each engine running condition. After determining the ignition timing, the ECM sends the ignition timing signal (IGT) to the igniter. When the IGT signal goes off, the igniters will turn on, shut off primary current flow in the ignition coil producing a high voltage spark in the cylinder.

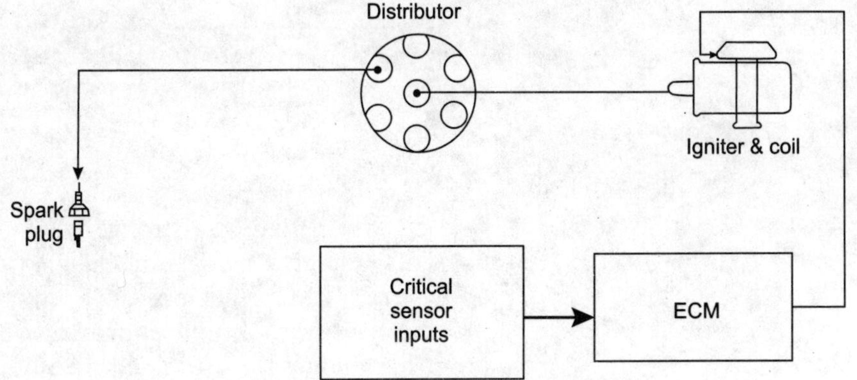


**Fig.7.17:** *Electronic spark advance*

Since the ESA always ensures optimal ignition timing, emissions are lowered and both fuel efficiency and engine power output are maintained at optimal levels.

## REVIEW QUESTIONS

1. Describe the transistor assisted ignition system.
2. What is CDI system? Explain.
3. List the disadvantages of contact-breaker point ignition system.
4. Compare breaker point and breakerless ignition system.
5. Point out the advantages of electronic ignition systems.
6. What replaces the ignition points in a breakerless ignition system?
7. Explain the transistorised ignition system.
8. Explain the electronic ignition system with a sketch.
9. What is DLI? How it works?

10. List out the merits of twin-spark technology.
11. Give the importance of triple spark ignition.
12. Compare single spark ignition with that of two-spark ignition.
13. Mention the advantages of twin-spark fuel injection.
14. What is i-DSI? What are its features?
15. Write a note on electronic spark advance.

# Automotive Wiring

The required electric current in the automobile flows from the battery to the various electric parts and then returns to the battery. The paths through which the electric current flows are called electric circuits.

The wiring that serves these paths may be cords bundled together, called wiring harness, or may be individual cords, called single wire.

## INSULATED EARTH RETURN

Generally, insulated earth return system is very useful nowadays for the purpose of safety in the vehicle. Some special vehicles such as petrol and chemical tankers have what is known as an insulated earth return. This system use insulated cables to return current to the battery instead of using the chassis or body.

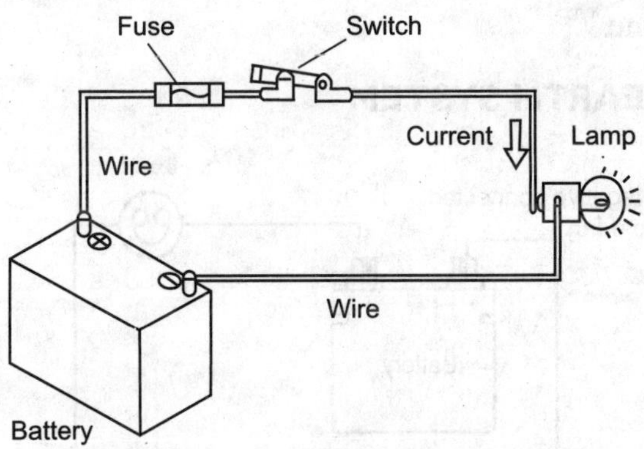


**Fig.8.1:** *Insulated earth return*

This type of return is used to minimise the build-up of static electricity on the vehicle body; this might otherwise cause a spark, which could ignite hazardous substances being carried by the vehicle. Also, some vehicles' bodies are made out of fibreglass; this is a very poor electrical conductor, and so an insulated earth cable is provided, which runs along the entire length of the vehicle's body to complete the electrical circuit.

## EARTH RETURN SYSTEM

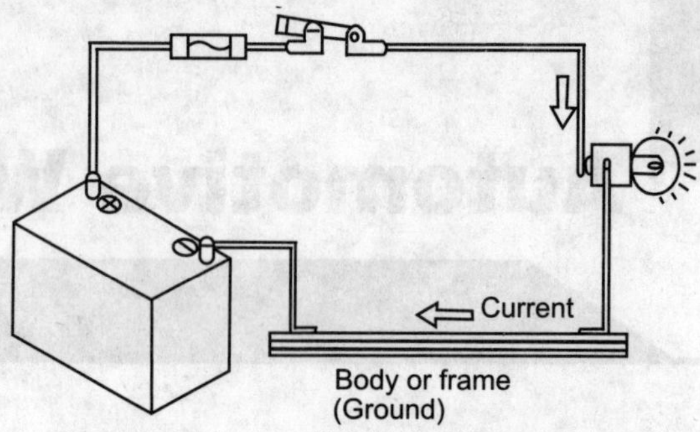


**Fig.8.2:** *Earth return system*

The automotive wiring system in automobiles is a complex system. Numerous electrical components are interconnected by a multitude of wires making up a system of their own. On modern vehicles the body and chassis forms one route for the electricity. Therefore no separate wires for the return of electrical current to the battery or alternator are used. Hence the type of electric circuit used in automobiles is earth return system (or) one wire system. This reduces the amount of wiring in the vehicle and makes circuit simple. This also saves considerably on manufacturing costs and weight. Fault finding also easier with this system.

An important point with respect to earth return connections is the polarity. That is, which of the two battery terminals will be connected to the vehicle frame.

## NEGATIVE EARTH SYSTEM

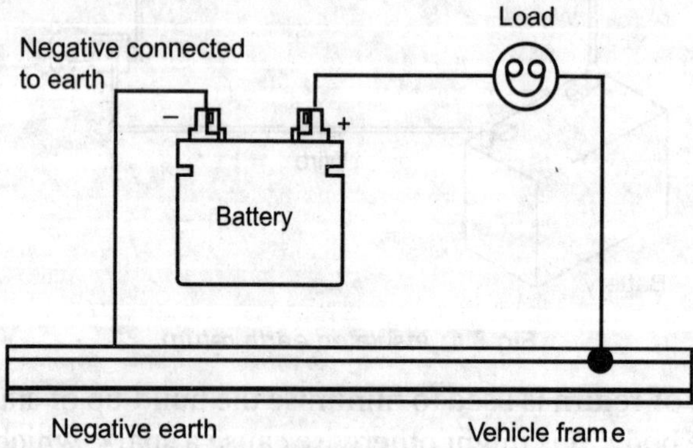


**Fig.8.3:** *Negative earth systems*

The common practice is to connect the negative terminal of the battery to earth, so that positive current is supplied to the electrical components.

## POSITIVE EARTH SYSTEM

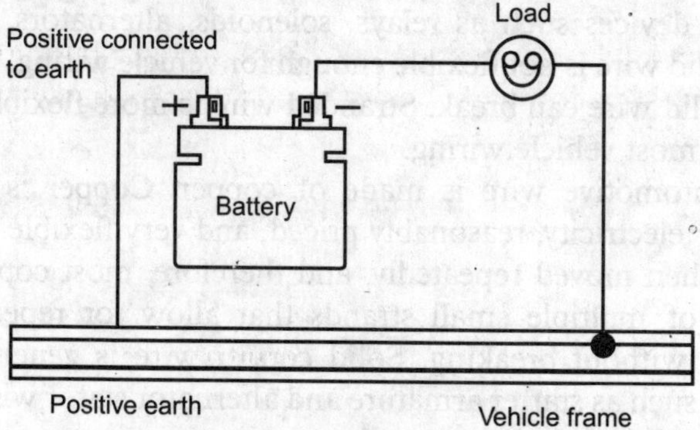


**Fig.8.4:** *Positive earth system*

In positive earth system, the positive terminal is connected to the chassis and giving the following advantages:

1. Lower spark plug voltages required.
2. Wear of plug point reduced.
3. Less wear on the electrodes in the distributor.
4. By having the electrodes in the distributor cap positive, burning is spread between them rather than being concentrated on the single rotor arm.
5. Reduced battery terminal (positive) corrosion.

Since the introduction of alternators and transistorised electronic ignition circuits it has become necessary to revert to the negative earth return system.

## WIRE

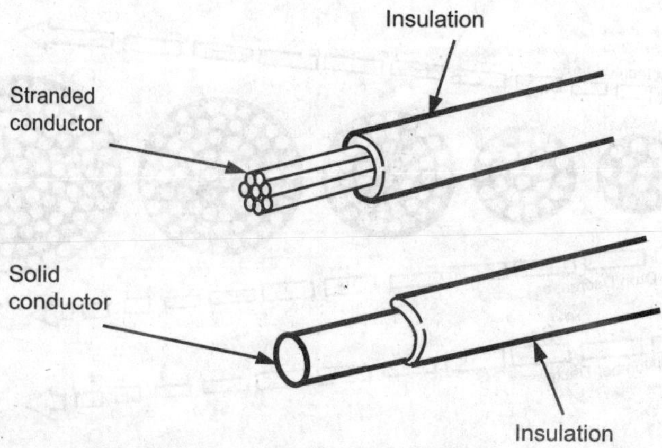


**Fig.8.5:** *Automotive wires*

Automotive wiring is either solid or stranded. Stranded wire is made of a number of small solid wires (strands). Solid wire is used in automobile in magnetic devices such as relays, solenoids, alternators, and motors. However, solid wire is not flexible enough for vehicle wiring. Under severe vibration, solid wire can break. Stranded wire is more flexible. Therefore, it is used for most vehicle wiring.

Most automotive wire is made of copper. Copper is an excellent conductor of electricity, reasonably priced, and very flexible. Even copper can break when moved repeatedly, and therefore, most copper wiring is constructed of multiple small strands that allow for repeated bending and moving without breaking. Solid copper wire is generally used for components such as starter armature and alternator stator windings which do not bend or move during normal operation.

## WIRE GAUGE (WIRE SIZE)

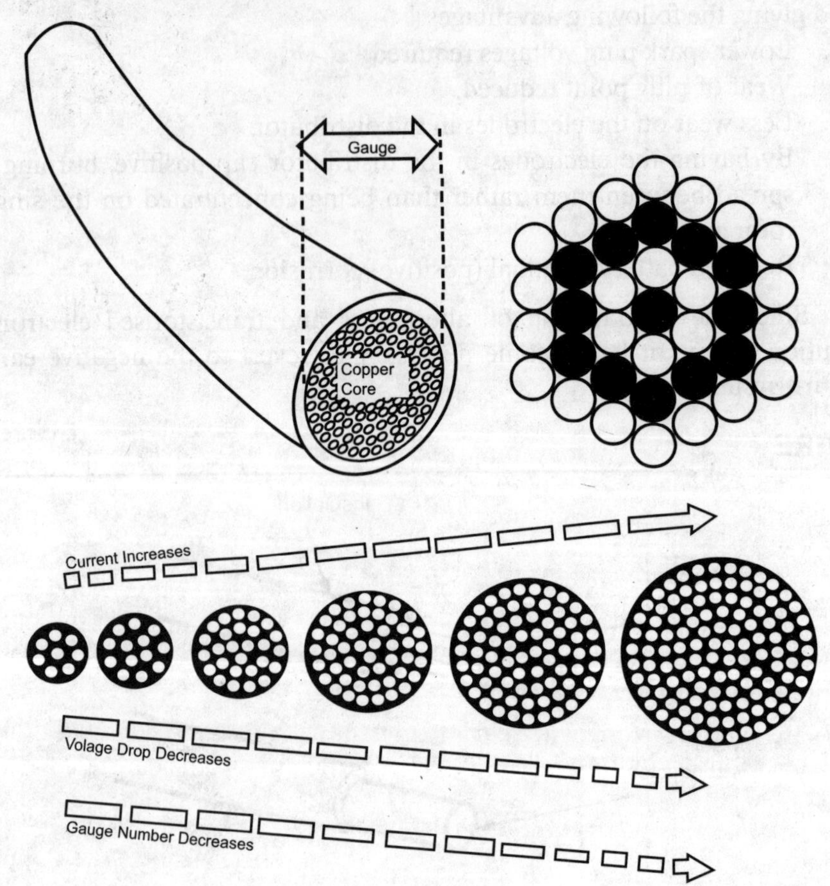


**Fig.8.6:** *Wire gauge*

The size of a wire determines how much current it can safely carry without overheating. The larger the wire diameter, the less resistance the wire offers to the flow of current and therefore, the greater the current carrying capacity. The wire or cable size is expressed in terms of standard gauge numbers. Size refers to the measure of the cross-sectional diameter of all the separate strands.

Wire size or gauge can be expressed either by metric system or by American wire gauge (AWG) number system.

## AMERICAN WIRE GAUGE SYSTEM

Wiring is sized and purchased according to gauge size as assigned by the American wire gauge (AWG) system. AWG numbers can be confusing because as the gauge number increases, the size of the conductor wire decreases. Therefore, a 14-gauge wire is smaller than a 10-gauge wire. The greater the amount of current (in amperes), the larger the conducting wire (the smaller the gauge number) required. The wire used in automotive wiring is usually between 10 and 18 gauge. Battery cables are between 2 and 6 gauge.

| Metric ($mm^2$) | AWG Number (gauge) |
|---|---|
| 0.22 | 24 |
| 0.5 | 20 |
| 0.8 | 18 |
| 1.0 | 16 |
| 2.0 | 14 |
| 3.0 | 12 |
| 5.0 | 10 |
| 8.0 | 8 |
| 13.0 | 6 |
| 19.0 | 4 |
| 32.0 | 2 |

## METRIC WIRE SYSTEM

In the metric system, wire size is expressed as the cross-sectional area of the wire in square millimetre ($mm^2$). This designation is the true measurement of area.

## NECESSITIES OF SELECTING WIRE GAUGES

For any wire system it is necessary to select the correct wire size or gauge for the following reasons.

- To keep the voltage drop in the wire as low as possible
- To make the voltage used by the consumer as high as possible

## CRITERIA FOR SELECTING WIRES

The criteria to select wires are as follows:

- The maximum permissible voltage drop
- The permissible heat of the conductor during operation
- The cost of the material

## CURRENT DENSITY

When a new cable or wire is installed, one has to take into consideration of the components, which get heated up by current flowing through them when in operation. If the current density is too high, this might cause damage to the insulation of wires or cables. This leads to burning of the insulation of wires or cables. This may cause burning of the whole vehicle.

### Approximate values of current densities

1. Wire size up to 6 $mm^2$ can be loaded 10 amperes/$mm^2$
2. Wire size up to 35 $mm^2$ can be loaded 6 amperes/$mm^2$
3. Wire sizes up to 120 $mm^2$ can be loaded 4 amperes/$mm^2$

But for short period of time all electrical lines can be loaded with current density of 20 amperes/$mm^2$

$$\text{Current density} = \frac{\text{Current}}{\text{Cross sectional area}}$$

$$I_{cd} = \frac{I}{A}$$

The standard wire sizes manufactured for automotive service use are available in 1, 1.5, 2.5, 4, 6, 10, 16, 25, 35, 50, 90, 120 $mm^2$.

## MEASUREMENT OF THE GAUGE OF A CABLE

To determine the gauge of a cable use the table as follows. This table is for copper wire and lists wire gauge, wire diameter, cross-sectional area in circular mills. A mill is 0.001 inch (one thousandth of an inch). A circular mill is the square of the diameter of the wire.

(a) Count the number of strands of wire.

**Size and Area of wires**

| (A) Wire diameter in inches | (B) American wire gauge | (C) Cross-sectional area in circular mill |
|---|---|---|
| 0.4600 | 0000 | 211600 |
| 0.4096 | 000 | 167800 |
| 0.3648 | 00 | 133100 |
| 0.3249 | 0 | 105500 |
| 0.2893 | 1 | 83690 |
| 0.2576 | 2 | 66370 |
| 0.2294 | 3 | 52640 |
| 0.2043 | 4 | 41740 |
| 0.1620 | 6 | 26250 |
| 0.1285 | 8 | 16510 |
| 0.1019 | 10 | 10380 |
| 0.0808 | 12 | 6530 |
| 0.0640 | 14 | 4107 |
| 0.0508 | 16 | 2583 |
| 0.0403 | 18 | 1624 |
| 0.0319 | 20 | 1022 |
| 0.0284 | 21 | 810.1 |
| 0.0253 | 22 | 642.4 |
| 0.0225 | 23 | 509.5 |
| 0.0201 | 24 | 404.0 |
| 0.0179 | 25 | 320.4 |
| 0.0159 | 26 | 254.1 |
| 0.0142 | 27 | 201.5 |
| 0.0126 | 28 | 159.8 |
| 0.0112 | 29 | 126.7 |
| 0.0100 | 30 | 100.5 |
| 0.0089 | 31 | 79.70 |
| 0.0079 | 32 | 63.21 |
| 0.0070 | 33 | 50.13 |
| 0.0063 | 34 | 39.75 |
| 0.0056 | 35 | 31.52 |
| 0.0050 | 36 | 25.00 |

(b) Measure the diameter of a single strand in mill.
(c) In column A of the table, find the diameter of the wire, which you have measured and on the same line in column C, find area.
(d) Multiply the area of a single wire by the number of strands to get the total area.
(e) In column C, find the figure closest to the area obtained in step (d) and on the same line in column B note the gauge.

### Example

A cable is found to have 19 strands of wire, the individual strands being 0.0112 inch in diameter. The table (column C) shows the circular mill area of each strand to be 126.7. Multiplying this by the number of strands 19, results in 2407.3 total circular miles and the closest figure in column C is 2583. On the same line in column B it is found 16 is the nearest cable gauge number.

## COLOUR CODING OF ELECTRICAL WIRES

| Code | Colour |
|---|---|
| BK | Black |
| BR | Brown |
| GY | Grey |
| O | Orange |
| P | Purple |
| PK | Pink |
| R | Red |
| T | Tan |
| W | White |
| Y | Yellow |
| DB | Dark blue |
| LB | Light blue |
| DG | Dark green |
| LG | Light green |

Automotive electric wires are colour coded to make fault finding in the electrical circuit and for further distinction according to its purpose. Electric wires in automotive vehicles are additionally provided with colour trailers or strips. The first colour mentioned is always the base colour.

For example, a Red wire with a white strip may be designated as RED/WHT, RO/WH, **R/W**, RED/W or RD/W.

## WIRING HARNESS

A wire harness is a group of individually insulated wires and cables, connecting parts, circuit protecting parts, etc., all grouped together for easy connection between electrical components of a vehicle. The wires are colour coded for tracing ease. Wire ends are equipped with lugs or connectors, which makes it easier to connect the wires to the various electrical system components. Each wire harness consists of the following items:

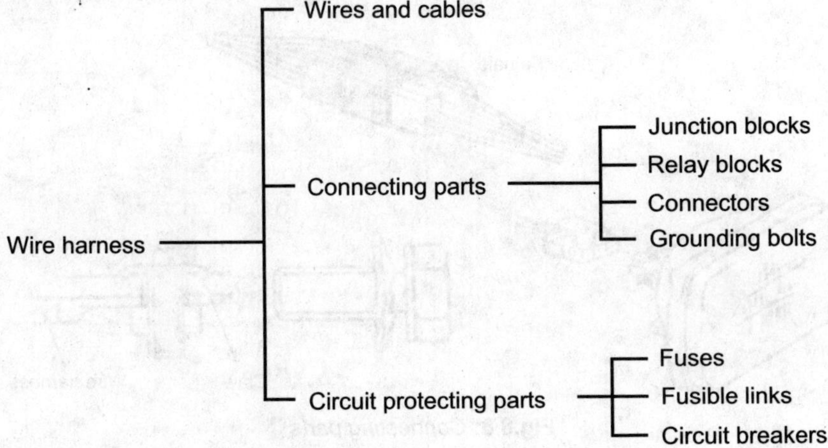


## WIRES AND CABLES

Various types of wires and cables are made for selective use in varying conditions (electric current, temperature, application, etc). Mainly three kinds of wires and cables are used in a vehicle.

1. Low-voltage wires
2. High-tension cords (High-tension wires)
3. Shielded cables

Most of the wires and cables found in a vehicle are low-voltage wires. Each low-voltage wire consists of the wire elements and the insulation. High-tension cords carry the high-tension current from the ignition coil to the spark plugs, via the coil cord and spark plug cord.

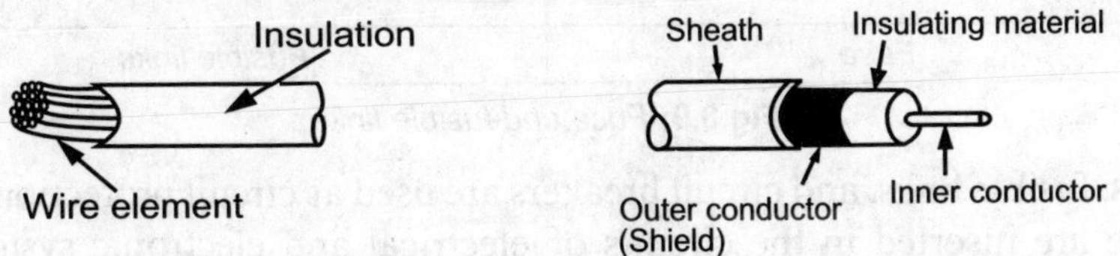


**Fig.8.7:** *Wires and cables*

Shielded cables are used as radio antenna feeder cables, ignition signal lines, oxygen sensor signal lines, etc. Since electricity of only very low voltage and current is allowed to flow through these signal lines, signals can be affected easily by inductive interference (switch on/off noises, ignition noises, etc.). For this reason, shielded cables were designed to prevent inductive interference from external sources and are used for signal lines.

## CONNECTING PARTS

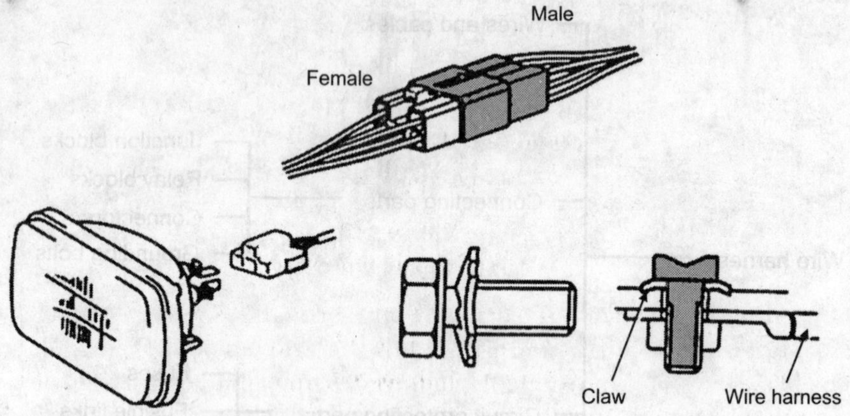


**Fig.8.8:** *Connecting parts*

A wiring harness is divided into several sections to fit more easily into the vehicle. The wiring harness sections are connected to one another by connecting parts, such as junction blocks, relay blocks, connectors and grounding bolts, so that electronic and electrical components can function as desired.

## CIRCUIT PROTECTING PARTS

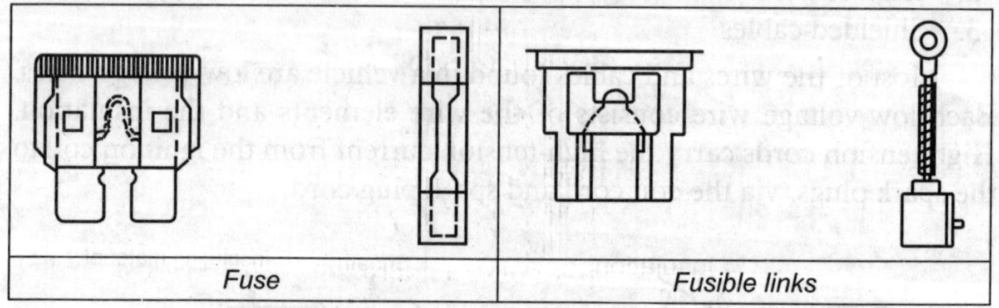


**Fig.8.9:** *Fuse and fusible links*

Fuses, fusible links, and circuit breakers are used as circuit protection parts. These are inserted in the circuits of electrical and electronic systems to protect the cables and connectors used in the circuit from being burnt out by an over-current or short-circuit.

The function of a fusible link is similar to that of a fuse. The main difference between the two is that the fusible link can be used for larger currents as it is larger in size and has a thicker element.

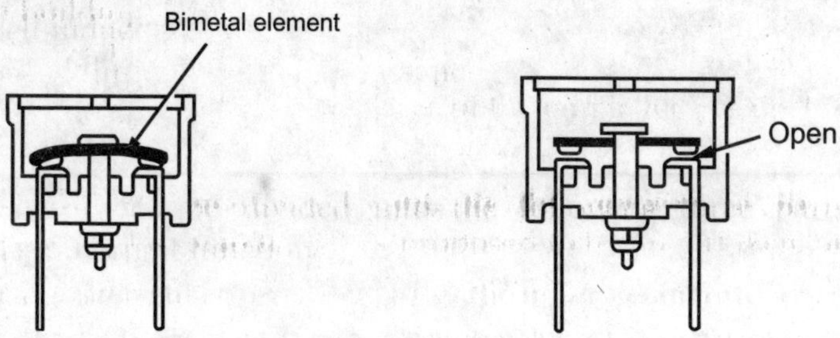


**Fig.8.10:** *Circuit breaker*

Circuit breakers are used in place of fuses for the protection of complicated power consuming circuits such as the power window, sun roof and heater circuits. When an over-current flows through the circuit breaker, the bi-metal strip heats up. This causes the strip to bend or warp outwards, tripping the circuit breaker (i.e., opening the contact and cutting off the flow of current). The circuit breaker can be reset after it is tripped.

## REVIEW QUESTIONS

1. What are various sizes of wire used in automotive wiring circuits?
2. Why are wires of different colours used in an automobile?
3. What does the gauge number of a wire refer to?
4. What are the gauges fitted on the instrument panel?
5. Why earthing is important?
6. What is positive earthing? What are its advantages?
7. What is insulated and earth return system?
8. Why are nearly all automobile electrical components wired on the earth return system?
9. Name and describe the operation of two types of circuit protectors used in automotive electric circuits.
10. Explain why different sizes of cable are used in vehicle wiring systems.
11. State the purpose of a fuse, and explain how it works.
12. Differentiate fusible link from fuse.

10. [illegible] in many cases, lengths of cable are used in vehicle wiring systems.


UNIT

9

# Lighting System

The lighting system of a vehicle has the following functions:

- Illumination of the road
- Defining the vehicle outline
- Indicating intended movements
- Warning other road users
- Indicating operating states.

The lighting system in automobiles includes exterior lights such as head lights, parking lights, turn signal lights, side marker light, back up lights, tail lights, number plate lights and interior lights such as instrumental-panel lights, map lights, various warning and indicator lights and courtesy lights which turn on when the car door is opened.

## HEADLIGHT

The headlight system is a lighting system to illuminate the road in front of the vehicle. Generally, high-beam and low-beam lights are provided and can be switched from one to the other by the dimmer switch. Two types of headlights are used in vehicles: sealed-beam type and semi-sealed beam type.

## SEALED BEAM HEADLIGHT

In a sealed beam headlight, no separate light bulb is used. Instead, the whole assembly itself is a light bulb. The filament is placed in front of a reflector, to which a glass lens is welded. The reflector directs the lights from the filament through the lens. The lens focuses the light from the reflector into the desired pattern.

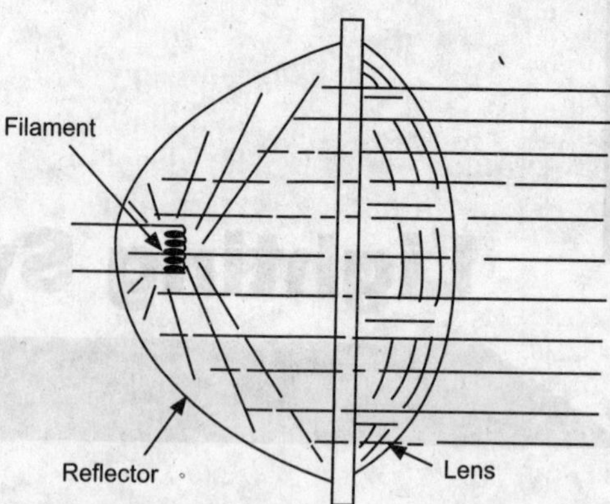


**Fig.9.1:** *Sealed beam headlight*

## SEMI-SEALED BEAM HEADLIGHT

The difference between the sealed beam and semi-sealed beam headlight is its design, which allows bulb replacement. Since the bulb can easily be replaced, it is not necessary to replace the whole headlight assembly if a filament burns out. Furthermore, when a bulb is replaced, the orientation (direction and angle) is not likely to change.

## HALOGEN HEADLIGHTS

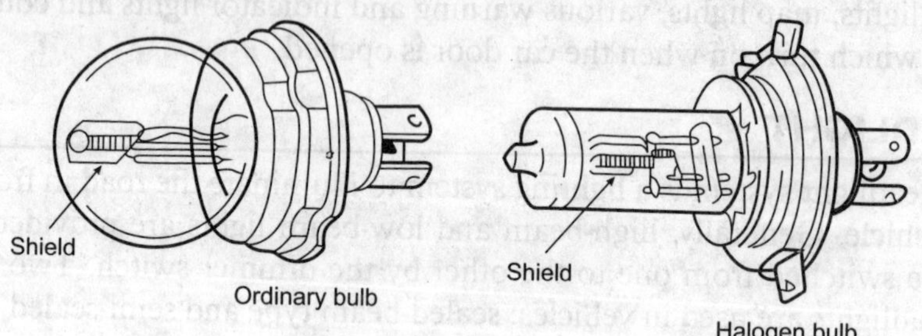


**Fig.9.2:** *Halogen headlight*

Halogen headlights are used on some vehicles. Halogen is the name given to a certain group of chemically related non-metallic elements, which include chlorine, fluorine and iodine. Iodine is the halogen most commonly used in automotive headlights. A halogen headlamp consists of a small bulb filled with iodine vapour located inside the light housing. The filament of the bulb is made of tungsten and is designed to operate at a higher temperature than the filaments of conventional headlights.

In any lamp, the heating of the filament causes atoms of tungsten to be released (3) from the surface of the filament. These atoms are deposited on the inside of the glass envelope. This produces a blackening effect, which reduces the light output of the lamp.

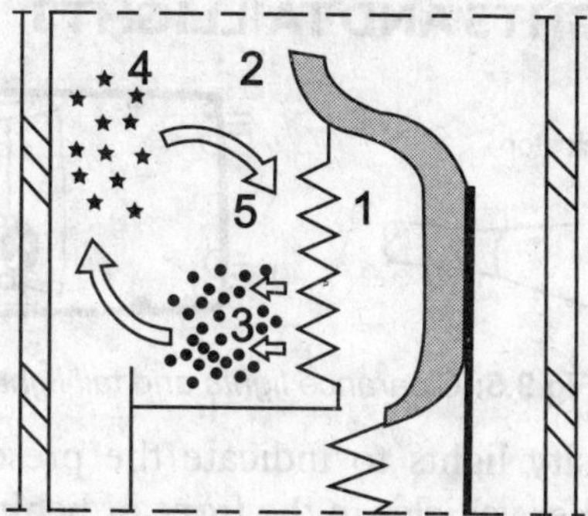

**Fig.9.3:** *Principle of halogen bulb*

With a halogen lamp, the iodine vapour causes the tungsten atoms to be redeposited (5) back on the filament. This almost eliminated blackening of the lamp. Therefore, the lamp produces a greater light output. They emit much whiter and brighter light than other sealed beam headlights.

## SIGNAL LIGHTS

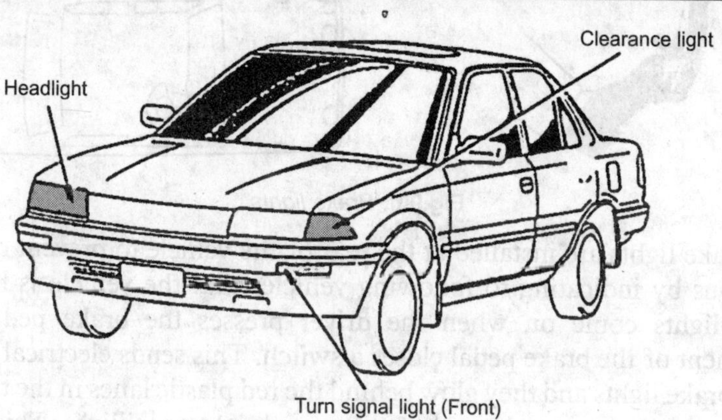


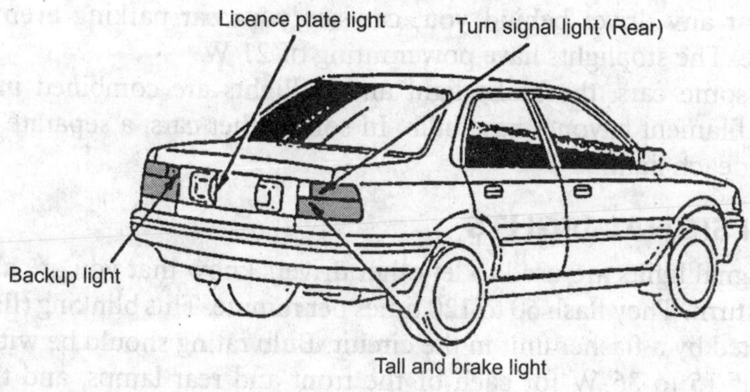

**Fig.9.4:** *Signal lights*

## CLEARANCE LIGHTS AND TAILLIGHTS

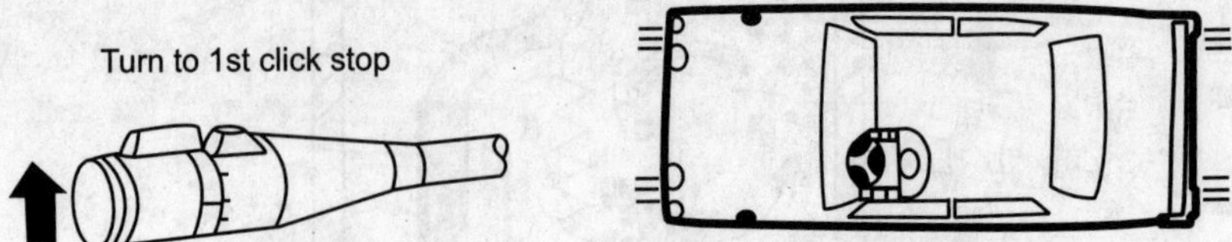


**Fig.9.5:** *Clearance lights and tail lights*

These are low-intensity lights to indicate the presence and width of a vehicle at night to other vehicles in the front or behind. Those in the front are called clearance-lights and those in the back are called taillights. The power rating of these lights is 5 W.

## BRAKE LIGHTS (STOP LIGHTS)

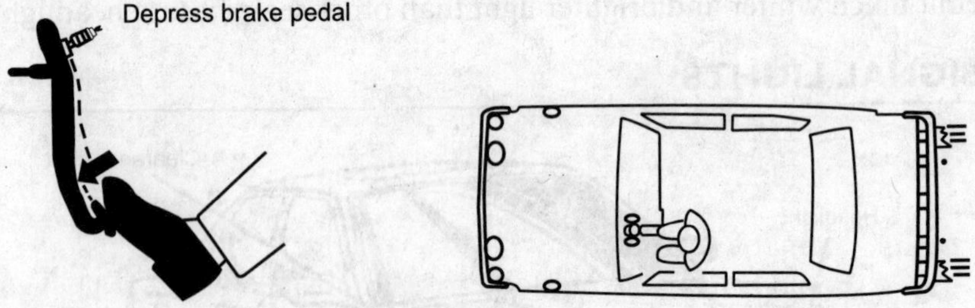


**Fig.9.6:** *Brake lights*

Brake lights are installed at the rear of the vehicle to prevent rear-end collisions by indicating to following vehicles that the vehicle is braking. Brake lights come on when the driver presses the brake pedal. The movement of the brake pedal closes a switch. This sends electrical current to the brake lights and they glow behind the red plastic lanes in the taillight.

The stoplights are about three times as bright as taillight. This makes sure that any driver behind you, can see your car parking even in the daytime. The stoplights have power rating of 21 W.

In some cars, the brake light and taillights are combined into one double filament bayonet base bulb. In some other cars, a separate bulb is used for each light.

## TURN SIGNAL LIGHTS

Turn signal lights are used to let other drivers know that you are about to make a turn. They flash 60 to 120 times per minute. This blinking (flashing) controlled by a flasher unit in the circuit. Bulb rating should be within the range of 15 to 36 W for each of the front and rear lamps, and the side mounted repeater lamp is about 6 W.

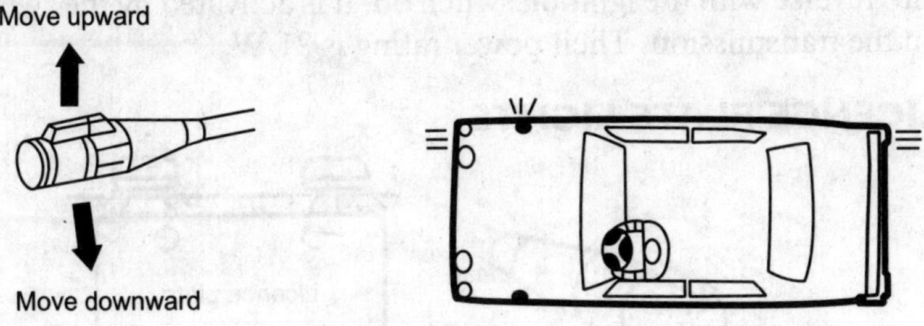


Fig.9.7: *Turn signal lights*

## HAZARD WARNING LIGHTS

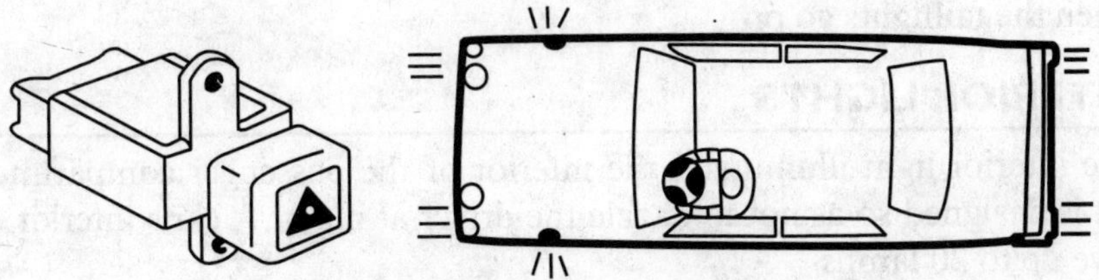

Fig.9.8: *Hazard warning lights*

In the event of an emergency, such as a roadside breakdown, all the lamps should be capable of working simultaneously, when they are called "hazard warning lights". The hazard warning lights indicate the existence of the vehicle to the front, rear and both sides during emergency stopping or parking. The emergency warning lights are turned on with a separate switch that causes all four turn signals to flash at once. The power rating of turn signal lights is 21 W.

## BACK-UP LIGHTS

Fig.9.9: *Back-up lights*

Back-up lights are installed at the rear of the vehicle to provide extra illumination to allow the driver to see behind the vehicle when backing up at night, and to warn following vehicles that the driver intends to back up or is backing up. The back-up lights go on when the transmission is shifted into reverse with the ignition switch on. It is activated by the switch located on the transmission. Their power rating is 21 W.

## LICENCE PLATE LIGHTS

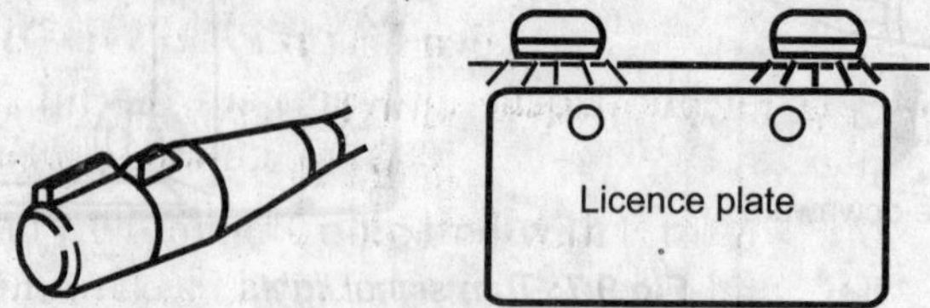


**Fig.9.10:** *Licence lights*

These lights illuminate the rear licence plate. The licence plate lights go on when the taillights go on.

## INTERIOR LIGHTS

The interior light illuminates the interior of the passenger compartment, but is designed so as not to dazzle the driver at night. A car's interior can have up to 20 lamps.

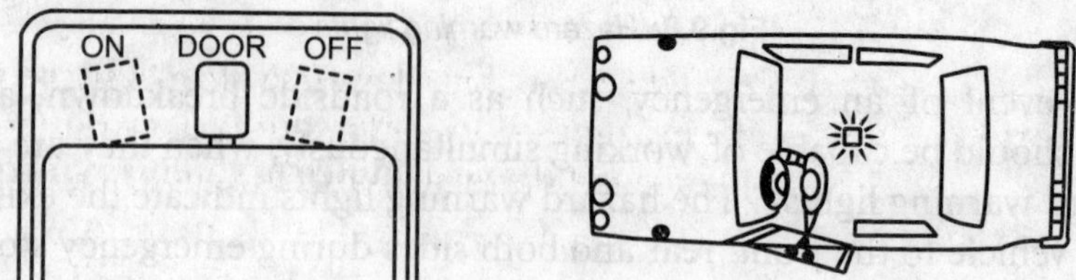


**Fig.9.11:** *Interior lights*

The bulbs that light when a door is opened are called courtesy lights. Instrument panel lights are used to illuminate the meters on the instrument panel at night, to enable the driver to read the meters and gauges quickly and easily while driving. The instrument panel lights go on when the taillights go on.

The Instrumental panel warning lights on the dashboard are used to warn the driver of a car with problem. These indicates unclosed door, low engine oil pressure, an under charging alternator, high coolant temperature, problems in the brake system and other potentially harmful situations.

## LOW BEAM AND HIGH BEAM OPERATION

### Low beam operation (Dipper)

The dipped or low beam produces a light pattern that illuminates the road a certain distance (about 40 m) from the vehicle. The low beam is low enough to prevent glare to oncoming vehicles. The low beam filament is located in front of the focal point of the parabolic reflector, so that all the light rays it emits are inclined to the mirror axis. A screen plate or cap or shield is located under the dipped beam filament. It prevents light rays from reaching the lower half of the reflector and being deflected upwards through the glass. It is the screen or cap, which provides the sharp boundary between the light and dark area visible.

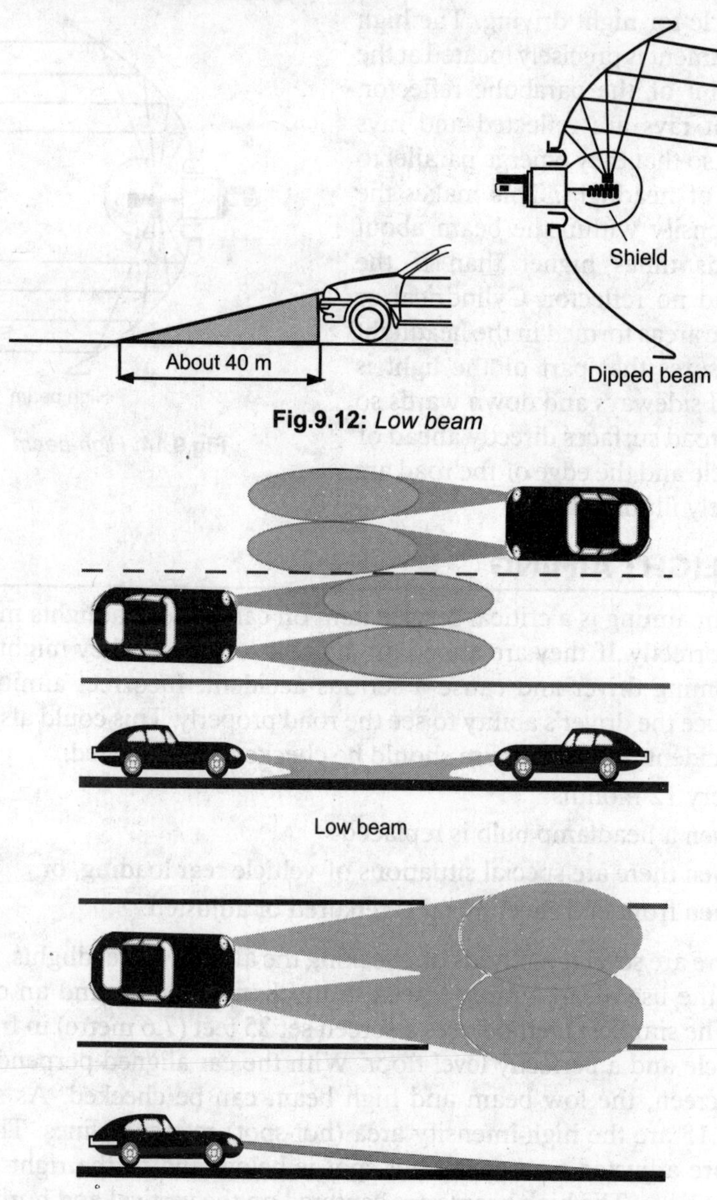


**Fig.9.12:** *Low beam*



**Fig.9.13:** *Low beam and high beam operation*

### High beam operation

The upper or far or high beam filament is focused to produce a high intensity light pattern to illuminate the road a greater distance (about 100 m) from the vehicle for night driving. The high beam filament is precisely located at the focal point of the parabolic reflector. The light rays are reflected and rays grouped so that they emerge parallel to the axis of headlight. This makes the light intensity within the beam about thousands times higher than if the bulbs had no reflector. Cylindrical or prismatic areas formed in the headlight glass ensures that part of the light is deflected sideways and down wards so that the road surfaces directly ahead of the vehicle and the edge of the road are adequately illuminated.

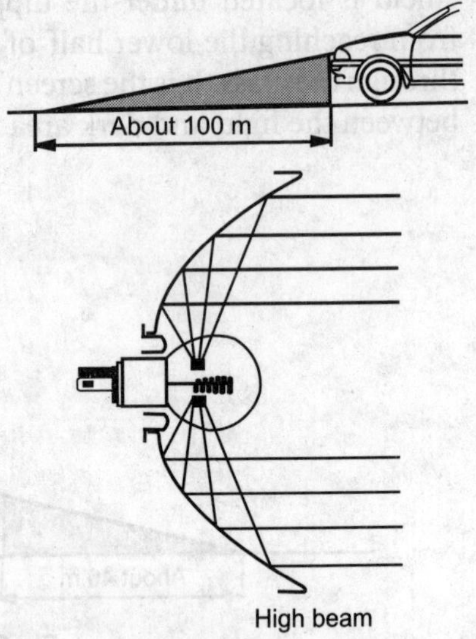


**Fig.9.14:** *High-beam*

## HEADLIGHT AIMING

Headlight aiming is a critical service item on cars. The headlights must be aimed correctly. If they are aimed too high or to the left, they might blind an oncoming driver and cause a serious accident. Incorrect aiming can also reduce the driver's ability to see the road properly. This could also lead to an accident. Headlight aim should be checked and adjusted:

- every 12 months
- when a headlamp bulb is replaced
- when there are special situations of vehicle rear loading, or
- when front end sheet metal is repaired or adjusted.

There are several methods of checking the aiming of headlights. These include the use of an aiming screen, a mechanical aimer and an optical aimer. The simplest method uses a screen set 25 feet (7.6 metre) in front of the vehicle and a perfectly level floor. With the car aligned perpendicular to the screen, the low beam and high beam can be checked. As shown in Fig.9.15 are the high-intensity area (hot-spot) reference lines. The low beams are adjusted so that the hot spot is below and to the right of the reference lines. The high beams are centered on the vertical and horizontal centrelines of the lamps.

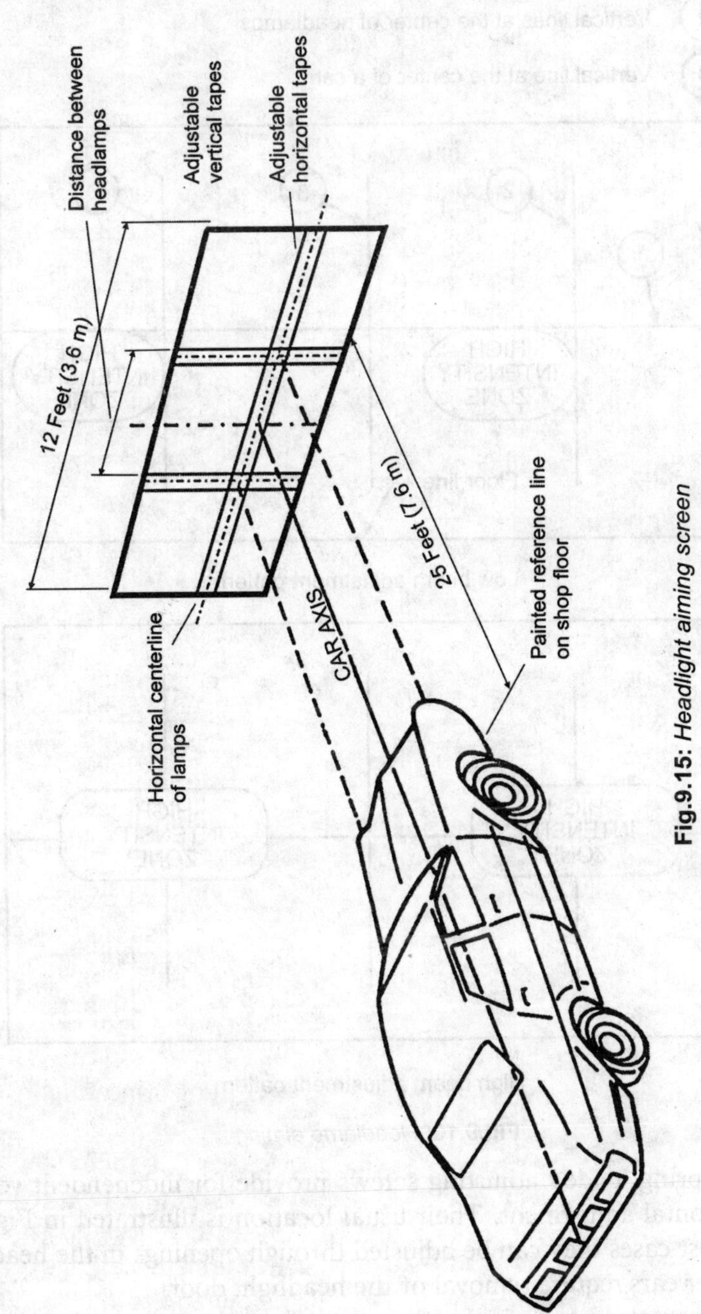


**Fig.9.15:** *Headlight aiming screen*

(1) Horizontal line at the center of headlamps

(2) Vertical lines at the center of headlamps

(3) Vertical line at the center of a car

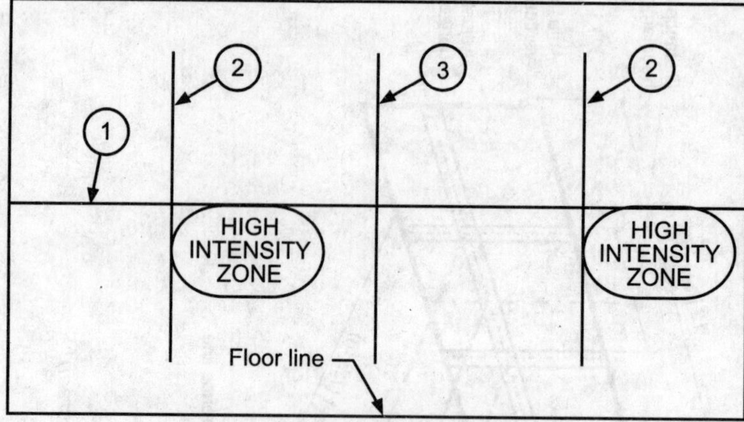


Low beam adjustment pattern

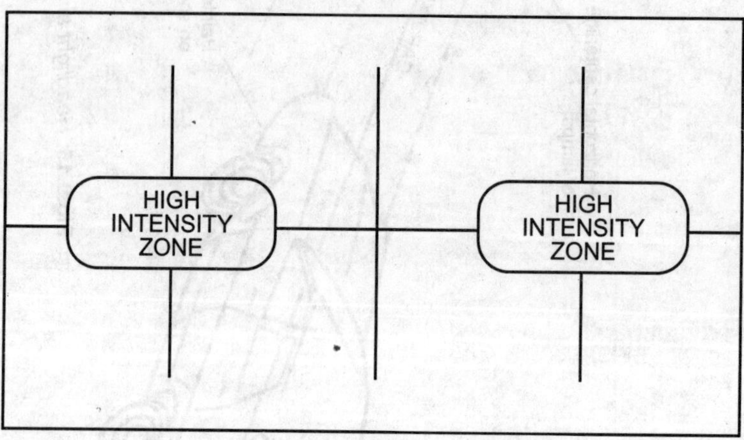


High beam adjustment pattern

**Fig.9.16:** *Headlamp aiming*

The spring loaded adjusting screws provide for independent vertical and horizontal adjustment. Their usual location is illustrated in Fig.9.16 and in most cases they can be adjusted through openings in the headlight door. Older cars require removal of the headlight door.

For accurate results in headlamp aiming, the process begins with vehicle preparation and ends with proper set up and use of equipment. Here are some vehicles preparations tips for headlamp aiming.

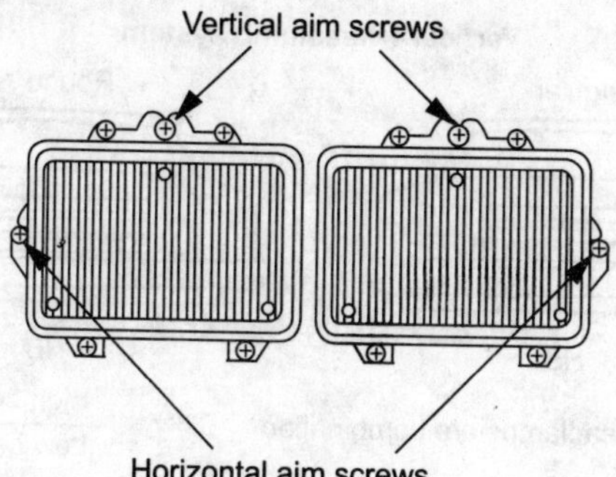


**Fig.9.17:** *Headlamp adjustment*

- Position the vehicle on a reasonably flat surface.
- The tyre pressure should be equalised to the specifications recommended by the manufacturer or to the pressure normally used by the owner.
- The vehicle should be loaded with an average weight. This includes (a) the driver, (b) fuel tank half to full, and (c) normal load carried in the trunk.
- Rock the vehicle sideways to equalise the springs.

## HEADLAMP ARRANGEMENT

The shape of the headlights may be round or rectangular. The headlights are arranged in three different ways.

The out-board headlights are numeral 2 lights (type 2) and the in-board are numeral 1 (type 1) headlights. The number (1 & 2) represents as follows:

1. One high beam filament
2. Two filaments: Low beam and high beam

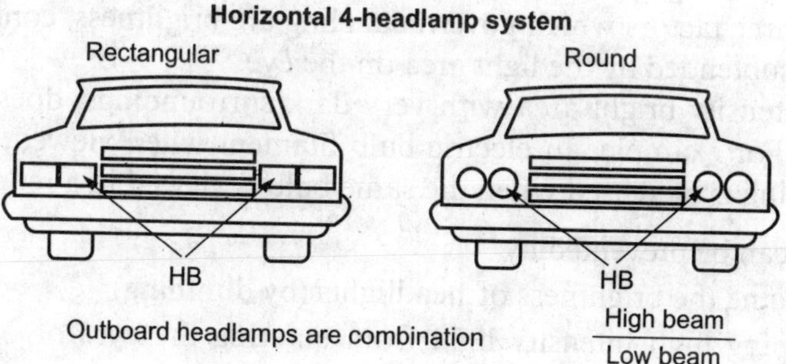

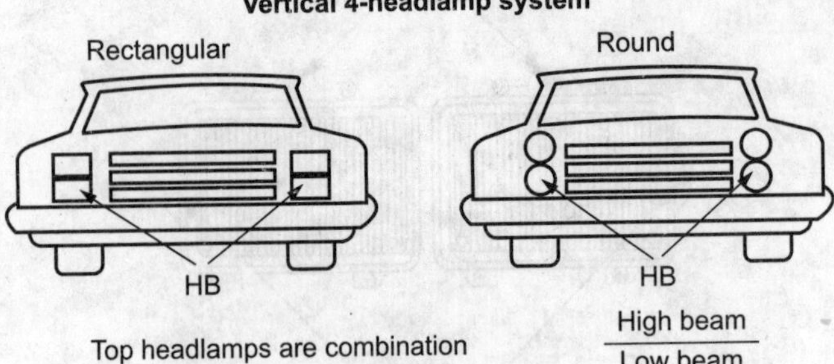


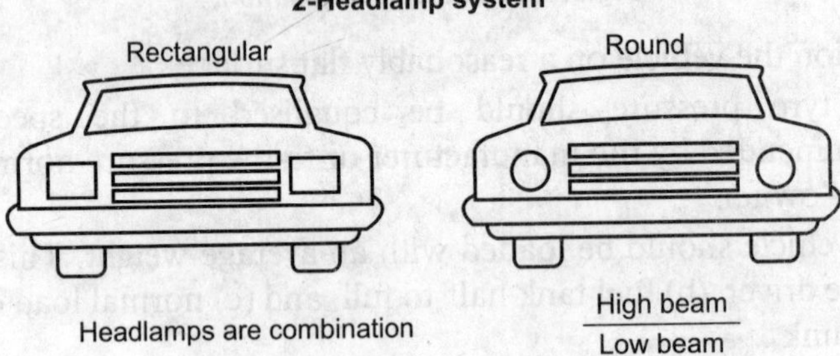


**Fig.9.18:** *Headlamp arrangement*

## HEADLIGHT DAZZLING

When two vehicles approach each other, the light intensity of the headlights falling on the eyes of the drivers will produce a sort of blindness temporarily. Such phenomenon is known as headlight dazzling. In other words, headlight dazzling is defined as the brightness which causes interference with vision. They blind oncoming traffic and are a major cause for accidents at night. Once the two vehicles pass each other, the effect of headlight dazzling will be reduced due to the improved field of vision.

The three factors which govern dazzling are brightness, contrast, and the angle subtended by the light area on the eye.

An intensity bright area with very dark surroundings, does not give dazzling (For example, an electric bulb filament when viewed at night). The dazzling is increased when the same bulb is viewed in a reflector.

Dazzling can be prevented by

- reducing the brightness of headlights (by dimming)
- stopping high intensity light from entering the eyes of on-coming driver
- reducing the contrast between the surrounding area and the headlights (by spreading lights to side)

## METHODS TO REDUCE DAZZLING

There are various methods to reduce dazzling effect. They are as follows:

### BULL'S EYE MASKING

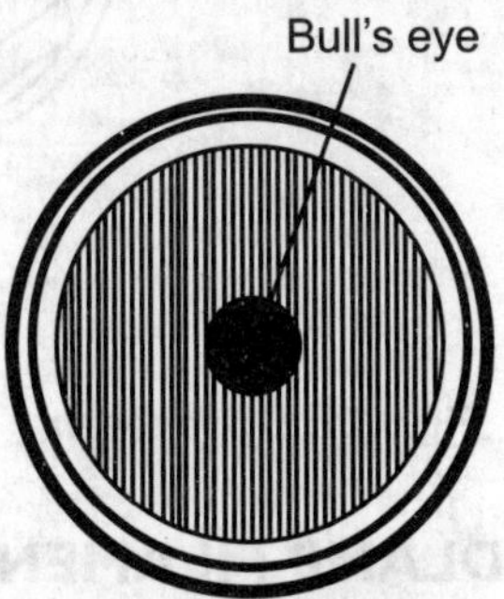


**Fig.9.19:** *Bull's eye*

The centre of the headlight assembly where the headlamp is fixed will induce dazzling effect when one sees it. If this is prevented to some extend, the dazzling effect is reduced. So at the centre of the headlight lens a small masking either by paint or by other means will help to reduce the dazzling effect.

### HALF MASKING OF HEADLIGHT LENS

The upper half of the headlight lens will be masked either by paint or by other means. This method will prevent the direct parallel rays emanating from the upper half of the headlight reflector falling on the opponent's eyes, thus reducing headlight dazzling effect.

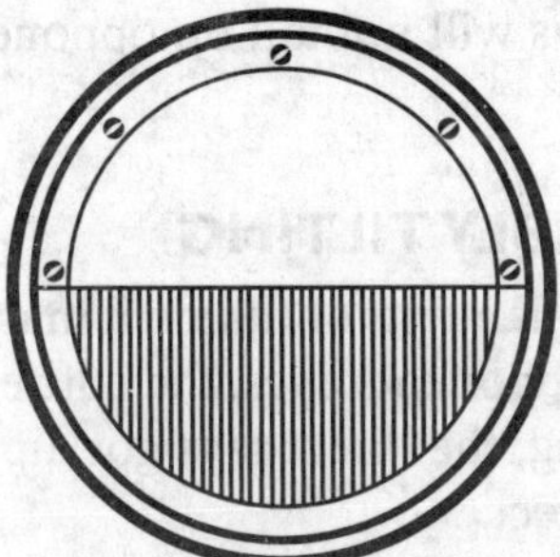

**Fig.9.20:** *Masking of headlight*

### PROVISION OF DIPPER BEAM

The headlight will be provided with a dipping beam. When the dipper beam is in operation, the direct parallel rays from the reflector will not fall on the opponent's eyes reducing the effect of dazzling.

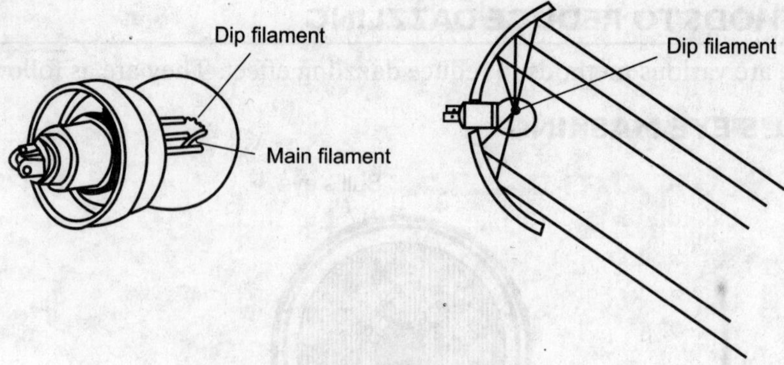


**Fig.9.21:** *Dipper beam*

## SHIELDING OF HEADLAMP FILAMENT

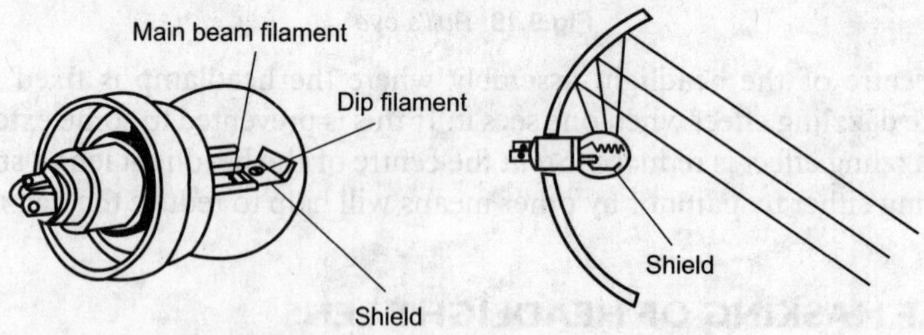


**Fig.9.22:** *Shielding of headlamp filament*

The dipper filament with a shield or cap can be used to reduce dazzling. The filament lower part will be provided with a shield or cap. This prevents the light rays from the filament falling on the lower half of the reflector so that the upper light rays will not fall on opponents eyes to get dazzling reduced.

## HEADLIGHT ASSEMBLY TILTING

Instead of providing two filaments or filament masking, in order to prevent the direct rays to fall on opponents eyes, the entire headlight assembly can be tilted. Due to this tilting, the direct rays got dipped towards the ground thus reducing dazzling effect.

## CONTRAST BETWEEN BRIGHT AND DARK AREAS

One of the reasons for headlight dazzle is the contrast between bright and dark areas. If the contrast is less then dazzling effect will be less. That is why dazzling effect is less in city drive since contrast between the headlights and its surroundings are less by the provision of street lights.

A side light provided in a vehicle will reduce the contrast between bright and dark areas. The side lights will give illumination to its sides and thus to the surrounding thus reducing the contrast between the headlights and the surrounding.

## AUTOMATIC ANTI-DAZZLING DEVICE

Automatic anti-dazzling device can be used to reduce dazzling effect. The vehicles are to be provided with light sensors. When the vehicles approach each other, the headlight rays falling on each vehicle sensor, will actuate an electronic circuit which will tilt the entire headlamp assemblies thus dipping the beam to reduce dazzling effect. This device is successful when all vehicles are fitted with light sensors.

## DIRECTIONAL WARNING LAMPS

When a driver intends to turn a vehicle to either side, a signal to other road users is necessary. This avoids confusion and accidents. Therefore, direction indicators (flashers) giving signals in the form of flash lamps (amber coloured) are provided on the vehicle. Flashers indicate other vehicles and pedestrians which direction the vehicle is going to be turned. These lights are usually operated by a lever located under the steering wheel. Most of them are of the self-cancelling type that is automatically switched off when the steering wheel is returned to straight-ahead position.

Three types of flasher units may be found in vehicles:

1. Thermal
2. Capacitor
3. Electronic

Of these, the electronic flasher is now virtually standard fitting on new and recent vehicles but many thermal flasher units are still in service. Capacitor types are of various designs and now of minor importance.

## THERMAL MAGNETIC FLASHER – HOT WIRE TYPE

It is based on hot wire principle. The thermal elongation of a thin resistance wire, under the influence of current passing through it, is made use for controlling the blinker contacts. The flasher unit consists of a steel core with a coil wound on it. It carries two armatures one on either side. The armature carries moving contact points. The armature on the right of the steel core is spring loaded. The armature on the left side of the steel core is connected to a fixed contact point through an actuating wire and a ballast resistor (Limiting resistance). The entire mechanism is housed inside a small cylindrical or rectangular container.

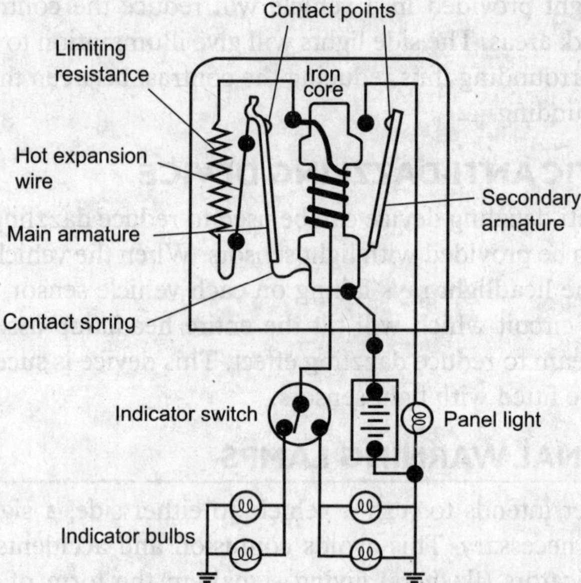


**Fig.9.23:** *Thermal magnetic flasher*

The working of the flasher unit is simple and takes the advantage of linear expansion of an actuating wire. When the switch on the dashboard is closed, current through the armature flows in the actuating wire. From the actuating wire, current comes to the ballast resistor. The flow of current is limited by the ballast resistor to a value, which will ensure that the lamp filaments do not glow. Due to the current flows in the actuating wire, it gets heated up and its length increases such that the contact point on it comes into contact with stationary contact point. This action closes the circuit and the lamps glow. Due to flow of current in the coil of the steel core, the armature is further attracted by the core and contacts become firm. Then the actuating wire gets short-circuited. It cools down and the contacts open. The cycle repeats itself. Numbers of flashes per minute are usually 60 to 120 (90±30).

When the main contacts are closed, the steel core attracts the armature which is on the right side towards it and contact point on it comes into contact with another fixed contact point to close the circuit of warning lamp on the dashboard.

## CAPACITOR-RELAY FLASHER UNIT

Capacitor relay type of flasher units are fitted to vehicles such as trucks, buses and heavy vehicles. The special features of these flasher units are relatively easy to design and generates constant flashing rate irrespective of number of bulbs they are required to operate.

Flasher units consist of an electromagnetic relay with the two separate but interconnected windings. These are identified in the diagram as winding A and B. Coil A consists of various turns of a fine gauge wire and is connected in series with a capacitor. Coil B consists of relatively few turns of a heavy gauge wire. This winding varies the front and rear indicator lamp currents.

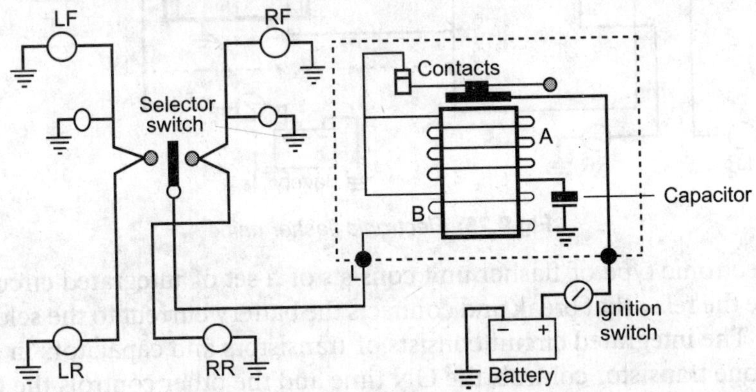


**Fig.9.24:** *Capacitor-relay flasher unit*

When the ignition switch is on and directional selector switch is selected to left or right-hand position, a current flow from battery negative via frame to the selector lamps and the directional switch to terminal L; via the winding B and the contacts to the terminal B; via the ignition switch to battery positive.

While the capacitor charging, the current flow through the coil A is in the opposite direction to the current flow in winding B. In other words, while the magnetic field generated by coil A is cancelled by the magnetic field generated by coil B. The net result is no magnetic field and the lamp contacts remain closed for a short period of time.

As the capacitor reaches the charged state, current flow via the coil A ceases. The time it takes to reach the charged state is determined by the capacitance of capacitor and the resistance of the coil A. The magnetic field of coil B is now capable to open the contacts and by this the lamps are extinguished. As the contacts open, the capacitor commences to discharge. This discharges current flow from the capacitor's negative plate, via frame to the selected lamps to terminal L then to coil B and coil A to the capacitor positive. The direction of current through the coil B and A are in the same. This establishes the continuous magnetic field for a period of time. At this stage the contacts are held open and the lights are remained in extinguished until the capacitor gets full discharged.

## ELECTRONIC FLASHER UNIT

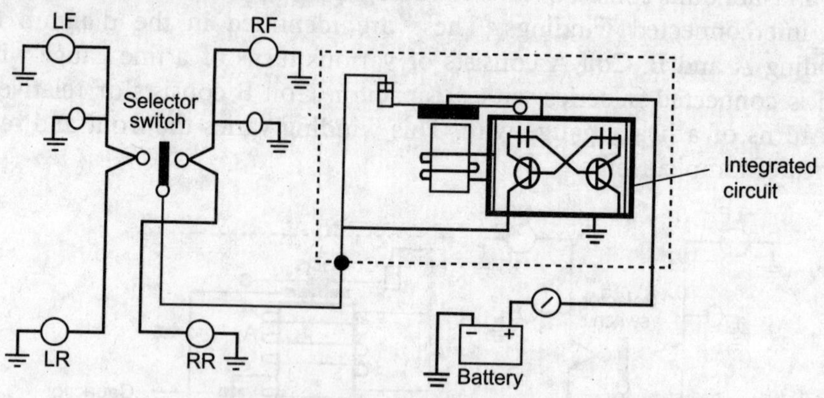


**Fig.9.25:** *Electronic flasher unit*

The electronic type of flasher unit consists of a set of integrated circuit to operate the relay that break and connects the battery current to the selected lamps. The integrated circuit consists of transistors and capacitors in such a way one transistor controls the ON time and the other controls the OFF time of the flasher unit.

When the ignition switch is on and selector switch is selected either left or right, the current from the battery negative flows through the frame to the selected bulbs and to the selector switch via contacts to the ignition switch to the battery positive. Instead of the relay, a power transistor can be used to do the switching action.

The electronic flashers are more preferred than the other type of flasher units because of its *special features*:

- They are more reliable and provides long life and excellent flashing stability.
- Additional features can be built into these units so that they are capable to perform more than one work at a time (for example, Hazard warning circuit).
- The cost of these units are not great compared with other types.

## LED HEADLIGHTS

Currently LED (Light-emitting diodes) headlamps are available on some vehicles. LEDs are semiconductors that change electrical energy into light. An individual LED does not produce enough light to serve as a headlamp. Ten to twenty LEDs are needed for ample forward lighting.

### Advantages of LED headlights

There are many reasons for using LEDs in headlights:

1. LEDs do not require a vacuum bulb or high voltage to work.

2. LED-based lighting sources require up to 40% less power than traditional lighting sources. This improves a vehicle's fuel economy.
3. LEDs provide a whiter light than xenon.
4. LEDs are mercury-free, making them environmental friendly.
5. Reliable. The average operating life of an LED is twice that of the vehicle itself. This means the headlamp may never need to be replaced.
6. LEDs are resistant to shock and vibration.
7. LEDs turn on quicker than ordinary bulbs. The turn on time is about 130 ms for the LEDs and 200 ms for bulbs. In time hours, LEDs have a typical rated life of over 50,000 hrs, compared with just a few thousand for incandescent lamps.
8. LED headlamps reduce dazzling (oncoming driver perception of glare)
9. LED headlamps are compact. This type of headlights is up to 55% thinner than other designs.

## Disadvantages

1. LEDs are sensitive to heat. The heat from the engine compartment and the lamp can cause failure.
2. LEDs require precise current control, which requires complicated electronic circuitry.

## REVIEW QUESTIONS

1. What is the approximate wattage of the headlamp?
2. With neat sketch explain any two types of headlamp.
3. Describe the construction of a sealed-beam headlight unit.
4. Write short note on LED lighting system in automobile.
5. What do you mean by headlight dazzling? How can it be prevented?
6. Write short note on lighting system of an automobile.
7. Why must headlights be aligned, or focused?
8. When should the (a) high, (b) low beam of headlights be used?
9. Why do we provide vacuum around the tungsten filament?
10. Why is it important that the headlamps are aimed correctly?
11. What is the use of Fog lamp?
12. Name the various lights in the vehicle.
13. Explain the construction of headlamp.
14. Describe headlamp beam setting and adjustments.
15. What is the advantage of Halogen lamp?

# UNIT 10 Accessories

## DASHBOARD INSTRUMENTS

Instruments are arranged on the instrument panel in front of the driver's seat to enable the driver to easily check the vehicle condition. The instrument panel tells the driver of detailed and present conditions of the vehicle by means of meters (gauges) and lights. It generally includes the following meters, gauges and indicator lights.

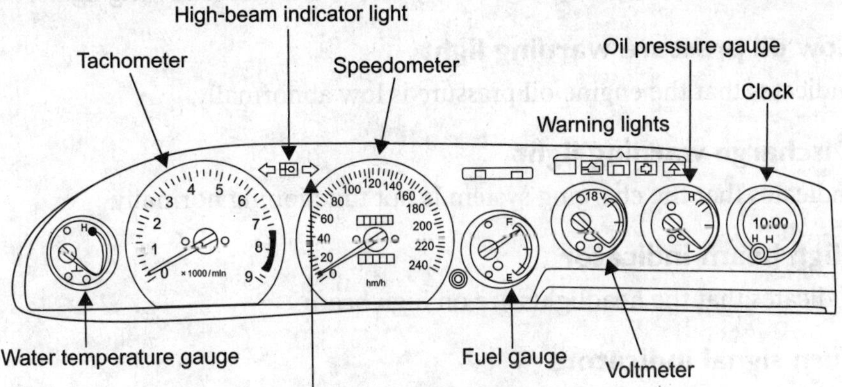


**Fig.10.1:** *Dashboard instruments*

## PANEL METERS

Meters may be of the analog type (with an indicator needle) or the digital type (with a numerical display). Here we will focus on explaining the analog meters.

### Speedometer

Speedometer indicates the vehicle speed. It also has odometer indicating the distance travelled, and the trip meter, which can be reset to 'zero' as desired.

### Tachometer

Indicates the engine speed in rpm.

### Voltmeter

Indicates the battery voltage or alternator output voltage.

### Oil pressure gauge

Indicates the engine oil pressure.

### Fuel gauge

Indicates the fuel level in the fuel tank.

### Water (engine coolant) temperature gauge

Indicates the engine coolant temperature.

## WARNING LIGHTS

Light display for level-over indication and for electrical parts functioning and abnormality indication generally consists of the following lights:

### Low oil pressure warning light

Indicates that the engine oil pressure is low abnormally.

### Discharge warning light

Indicates that the charging system is not functioning normally.

### High beam indicator

Indicates that the headlights are on high beam.

### Turn signal indicators

Indicate that the right or left turn signal lights are flashing.

### Hazard warning indicators

Indicate that both right and left turn signal lights are flashing.

### Low fuel level warning light

Indicates that the remaining fuel is insufficient.

### Brake system warning light

Indicates that the parking brake is applied or brake fluid is insufficient.

### Open door warning light (courtesy light)

Indicates that a door is not fully closed.

## SPEEDOMETER

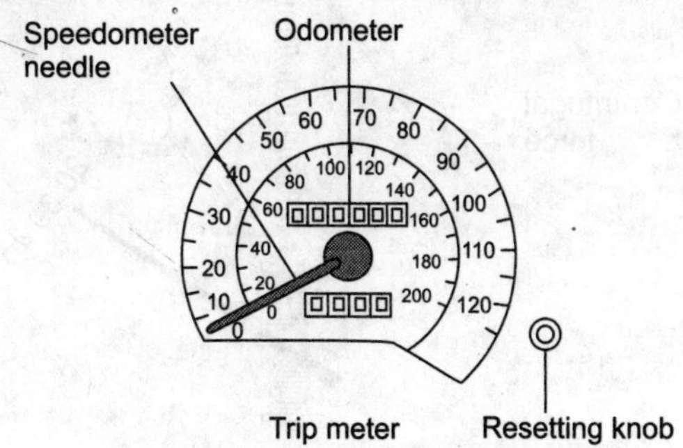


**Fig.10.2:** *Speedometer*

The speedometer indicates the speed at which the automobile is running. It is usually made in combination with the odometer that registers the total distance travelled. Many speedometers are further provided with a trip meter that can be reset to zero by the driver.

The speedometer is driven by a gear mounted on the rear end of transmission output shaft, the motion being transmitted to the speedometer by means of a flexible cable. The speed is indicated in kilometres (or miles) per hour.

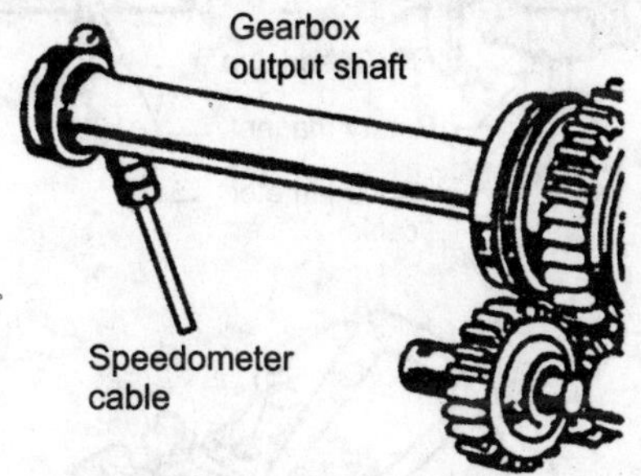


**Fig.10.3:** *Gear mounted speedometer*

There are many types of speedometers, such as the centrifugal type, magnetic type, and digital type. The magnetic type and digital type speedometers are most commonly used today.

### Centrifugal speedometer

As shown in Fig.10.4, the centrifugal type utilises the centrifugal force of two ball weights. When the shaft turns, the balls move outward and raises up the toothed tube, which turns the pinion and pointer to indicate the speed.

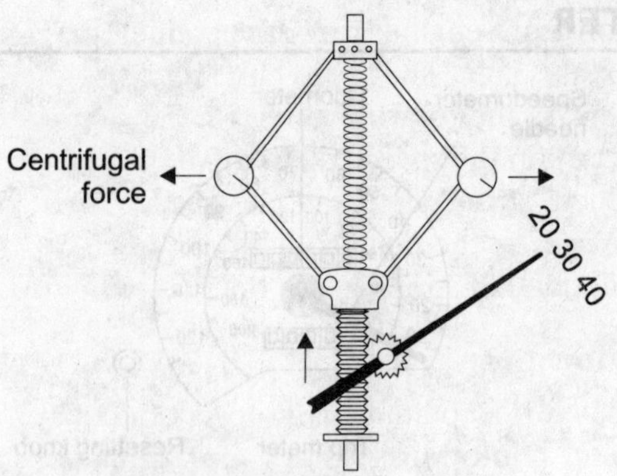


**Fig.10.4:** *Centrifugal speedometer*

## Magnetic speedometer

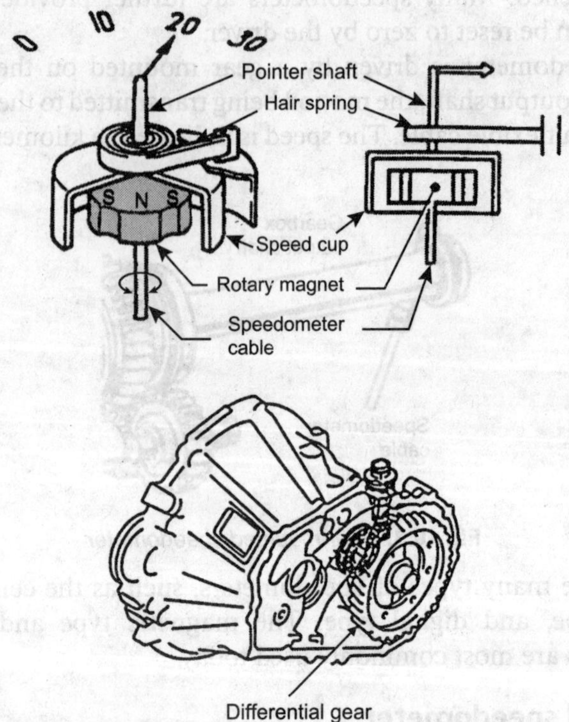


**Fig.10.5:** *Magnetic speedometer*

A magnetic speedometer uses a magnet to indicate the vehicle speed based on wheel revolutions. The revolutions are transmitted to the magnetic speedometer by a worm gear from the differential gear of the transaxle (on front-engine front-drive vehicles) or from the output shaft of the transmission (on front-engine, rear-drive vehicles). The rotary magnet is driven by a flexible cable. The speed cup fixed on the pointer shaft rotates with the pointer. As the rotary magnet rotates, a vortex magnetic field is generated to act upon the speed cup to drive the pointer.

As the vehicle speed increases, the speed cup torque increases and indicates the speed by the balance with the hair spring force. When the vehicle stopped, the hair spring keeps the pointer at the zero position.

### Digital speedometer

A small slotted wheel is rotated by the speedometer cable. A light-emitting diode (LED) and phototransistor are positioned so that they face each other, separated by the slotted wheel. As the wheel rotates, light from the LED reaches the phototransistor each time a slot in the wheel lines up with the LED. The phototransistor produces a voltage pulse each time a slot passes the LED. The microcomputer receives these voltage pulses from the sensor. At the same time, the internal clock provides a time reference. The microcomputer counts the number of voltage pulses received in a certain period of time and calculates the vehicle speed in miles per hour or kilometres per hour. The number calculated is converted to the voltages necessary to operate the digital display.

## ODOMETER

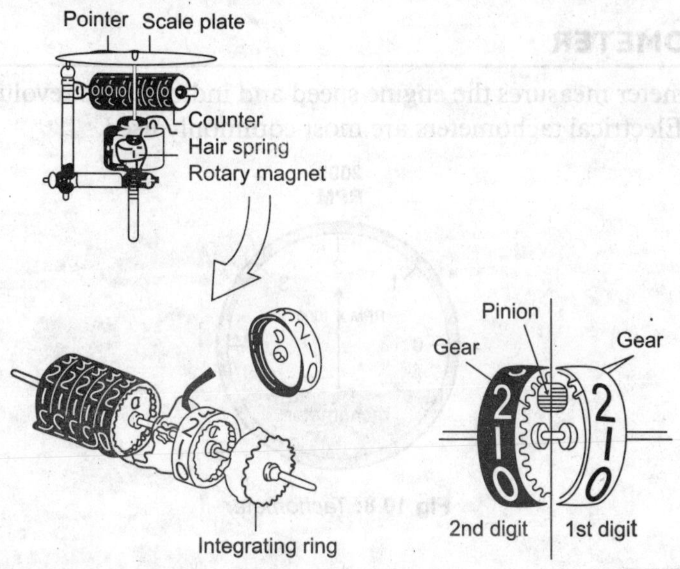


**Fig.10.6:** *Odometer*

The odometer registers the total distance travelled after the automobile comes off the assembly line, and has the capacity to register up to 99999.9

kilometers. A special gearing provided at the speedometer shaft drives the integrating ring for odometer indication.

When the first digit makes one rotation and the figure changes from 9 to 0, two teeth on the first digit wheel rotate the second digit wheel by two teeth via the pinion.

## TRIP METER

Some of the recent cars are equipped with two odometers, one being a small capacity type that can be reset to zero. This short distance odometer is usually called a trip meter. The number of digits is less and the indication can easily be reset to zero by pressing the trip meter knob.

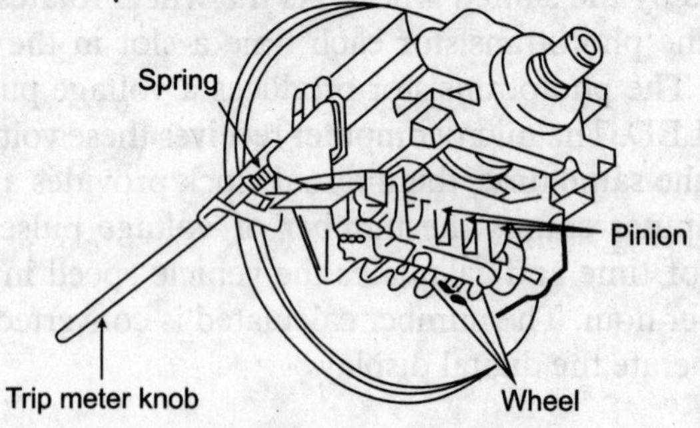


**Fig.10.7:** *Trip meter*

## TACHOMETER

A tachometer measures the engine speed and indicates the revolutions per minute. Electrical tachometers are most commonly used.

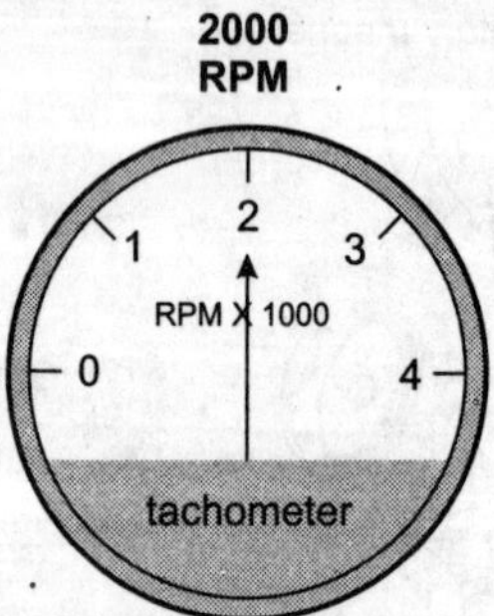


**Fig.10.8:** *Tachometer*

## AMMETER

The ammeter indicates the amount of current discharging from the battery to the various electric parts except the starter or the amount of current charging into the battery from the generator. The ammeter is made up of a moving core mounting the pointer, and an electromagnet. The pointer deflects in accordance with the amount of current flowing through the electromagnetic coil.

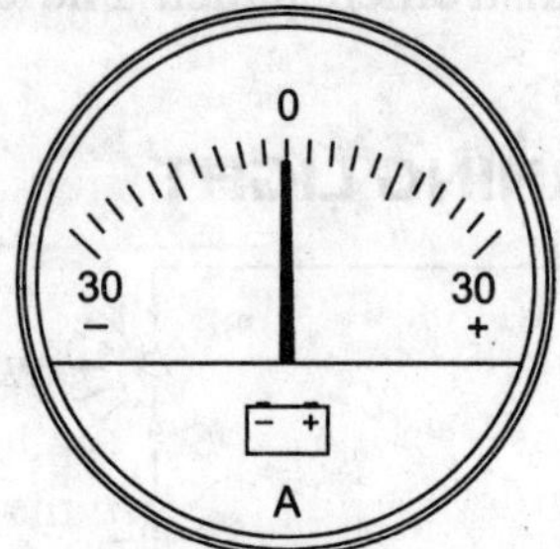


Fig.10.9: *Ammeter*

Instead of the ammeter, some automobiles are equipped with a charge warning light, which lights to warn the driver when the generator is not charging the battery properly. This type is simpler in construction but has the disadvantages of not knowing whether the battery is being charged or discharged at low speed when the lamp is not lighted and there is no way of knowing how much the battery is being charged.

## VOLTMETER

A voltmeter generally indicates the voltage across the battery terminals and helps to check the charging voltage and estimates the charged battery's capacity. Moving magnet voltmeters are most commonly used.

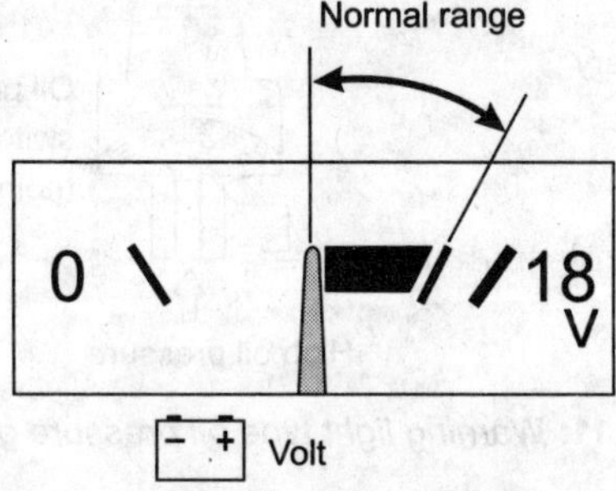


Fig.10.10: *Voltmeter*

## ENGINE AND VEHICLE MONITORING SYSTEM

Engine and vehicle monitoring system has some indicators and gauges to convey the information about the engine and the vehicle like engine coolant temperature, battery charging, fuel level in the tank, lubricant oil pressure and the vehicle speed etc., to the driver during vehicle running so that the drive can take a proper action to avoid the hazards that can cause to the engine or to the vehicle.

In general the engine monitoring system using lamps as indicators carries three lamps on the instrument panel. The lamps used are monitoring the following systems.

## OIL PRESSURE WARNING LIGHT

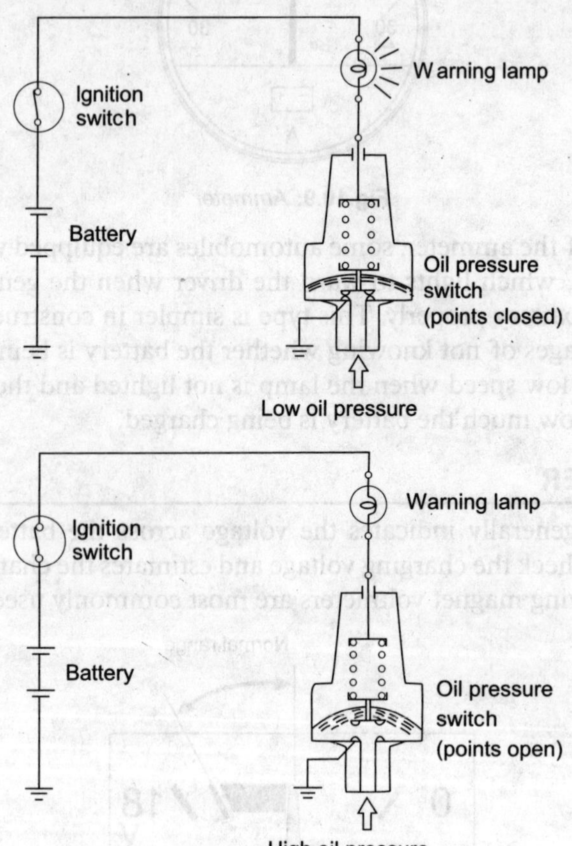


**Fig.10.11:** *Warning light type oil pressure gauge*

The engine lubricant system is attached with a warning lamp which lights whenever the engine oil pressure falls below a specified value. Under normal operating condition when the engine is at rest, ignition switch is turned on the lamp lights. When the engine is running and oil pump build up the oil pressure in the system to a predetermined value, the lamp is switched off by the oil pressure opening a set of conductors in the pressure switch.

## ALTERNATOR CHARGE WARNING LIGHT

This lamp lights when the ignition switch is turned on with the engine attached or when the alternator fails to charge with the engine running.

## COOLANT TEMPERATURE WARNING LIGHT

The temperature warning light is wired through the ignition switch and it is connected to a thermo switch screwed into some point in the coolant chamber of the engine block. The contact points which are made of bimetal strip in the thermo switch are normally open and does not allow the current to flow as the ignition switch is turned on.

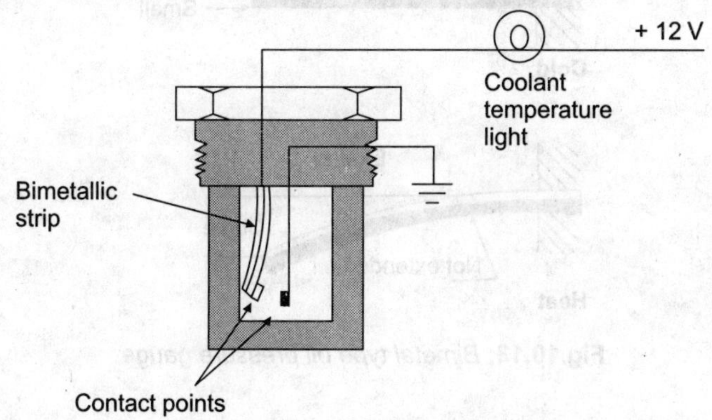


**Fig.10.12:** *Coolant temperature warning light*

When the temperature is gradually increase and reaches a critical point, the bimetal gets bend and thus the points get contact each other and allows the current to flow thus makes the warning light to glow.

## GAUGES

These are the gauges which are used in the place of warning lights. The gauges indicate the gradual reading of the particular indications like charging rate, pressure level, quantity of fuel or temperature range. Although the readings may not be accurate but more informative compare with the warning lights.

These gauges works on basis of either one of two effects of an electric current, which are explained below:

### Magnetic effect

Whenever the current flows in a conductor it will produce a magnetic field around that conductor.

### Thermal effect

The electric current will heat any conductor through it passes. The coil surrounded a bimetal strip is used in this case. Whenever the heat is supplied to bimetal strip it gets bend due to difference in the coefficient of expansion of metals.

The bimetal element is made up of two different metals with different heat expansion coefficients welded together so that when heated, the bimetal element will not remain straight but will curve toward the side with the smaller heat expansion coefficient. This property has been utilized in the bimetal type pressure gauge.

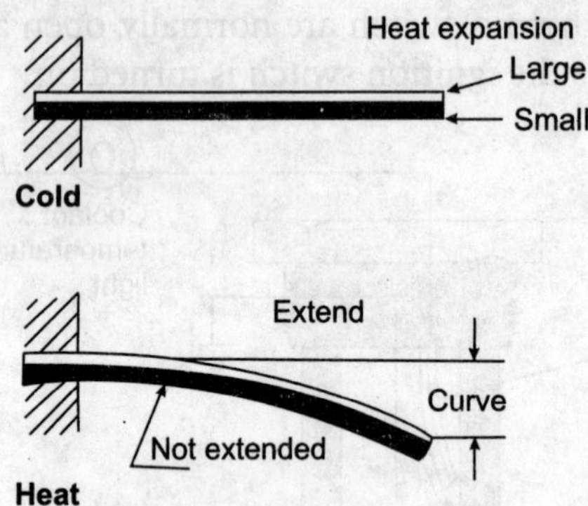


**Fig.10.13:** *Bimetal type oil pressure gauge*

## FUEL GAUGE

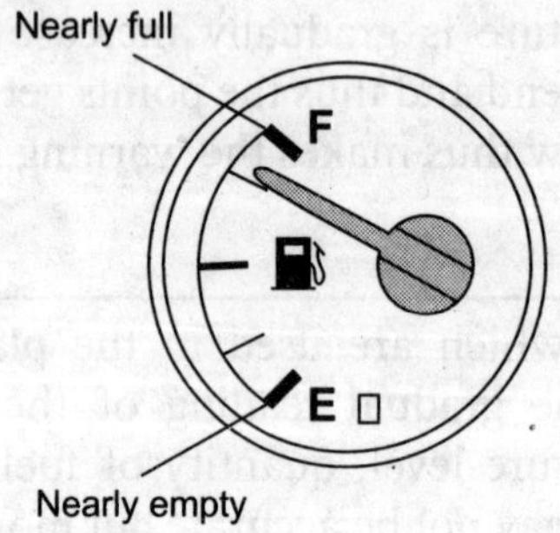


**Fig.10.14:** *Fuel gauge*

## MAGNETIC TYPE

The diagram below shown is a magnetic type of fuel gauge circuit, in which there are two units.

### Tank unit

It consists of a variable resistance with a sliding contact whose position is determined by a float on the fuel. The one end of the resistance is connected to the earth in tank unit. The other end is connected to dash unit using a single wire. The resistance in the tank unit controls the current flow in the coils of the dash unit according to the level of fuel.

## Dash board unit

There are two electromagnets set at 90° each other. Between electromagnets there is a soft iron core upon which a pointer is fixed. The first coil is called as limiting coil and other is called as operating coil. The operating coil is attached with the extended pole to give increased magnetic field.

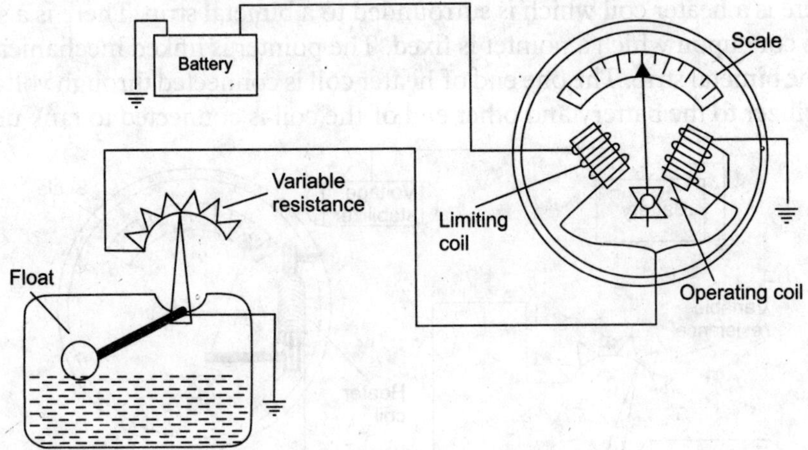


**Fig.10.15:** *Magnetic type fuel gauge*

## Operation

When the fuel tank is empty the float brings the variable resister contact points such that there will be direct contact and no resistance. At this stage the whole current will pass from battery through the ignition switch, limiting coil and the tank unit. There will be no current through the operating coil. The magnetic power of limiting coil is more and operating coil is less so the needle shows *Empty position*.

When the fuel tank is full the float brings the variable resistance contact point to such that full resistance is in series with the tank unit. Now the current cannot flow through the tank unit because of resistance and all the current will move from battery through ignition switch, limiting coil and the operating coil. Since the operating coil gets the current it will convert into a powerful magnet and attracts the needle towards it. Thus the pointer shows full position.

## THERMAL TYPE

The diagram below shown is a magnetic type of fuel gauge circuit, in which there are two units

## Tank unit

It consists of a variable resistance with a sliding contact whose position is determined by a float on the fuel. The one end of the resistance is connected to the earth in tank unit. The other end is connected to dash unit using a single wire. The resistance in the tank unit controls the current flow in the coils of the dash unit according to the level of fuel.

## Dash board unit

There is a heater coil which is surrounded to a bimetal strip. There is a soft iron core upon which a pointer is fixed. The pointer is linked mechanically to the bimetal strip. The one end of heater coil is connected through voltage stabilizer to the battery and other end of the coil is connected to tank unit.

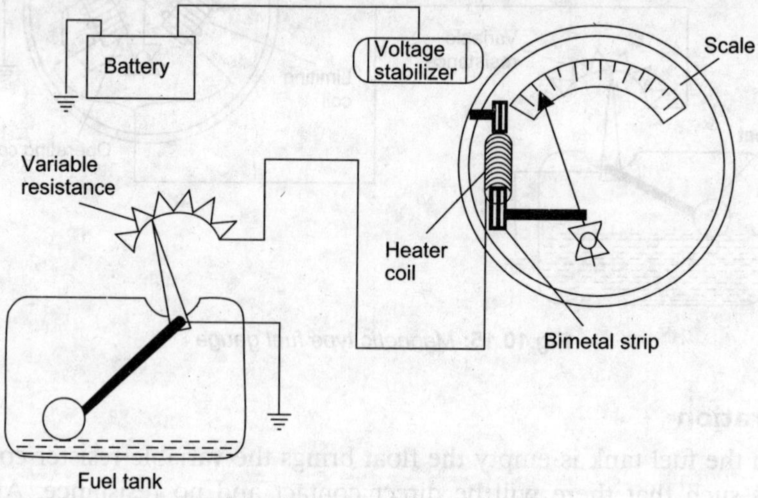


**Fig.10.16:** *Thermal type fuel gauge*

## Operation

When the fuel tank is empty the float brings the variable resister contact points such that the resistance is series with the tank unit. At this stage the whole current cannot pass in the circuit so there is no current in heater coil. The bimetal strip in the coil will be straight and the pointer shows *empty position*.

When the fuel tank is full the float brings the variable resistance contact point such that there in no resistance in the circuit. Now the current can flow from battery through the voltage stabilizer then through the heater coil to the tank unit. As the heater coil gets the current it heats up the bimetal strip and the strip will be bent to pull the linkage of the pointer. Thus the pointer shows *full position.*

## OIL PRESSURE GAUGE

The oil pressure gauge indicates the oil pressure in the engine for easy detection of a problem in the lubrication system. There are two types of oil gauges: the Burdon tube type and the electric type. Pressure indication may be by meter or by warning light.

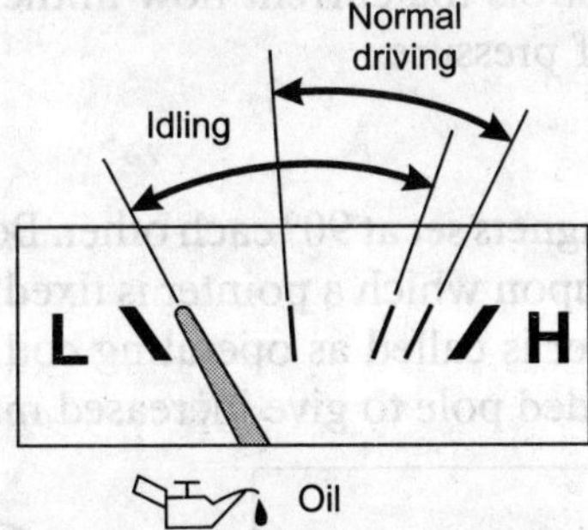


**Fig.10.17:** *Oil pressure gauge*

## BURDON TUBE TYPE

As shown in Fig.10.18, the Burdon tube (curved tube) when subjected to pressure tends to straighten out. This motion is transmitted by gears to the pointer to perform the function of a gauge. This type is practically no longer in use today.

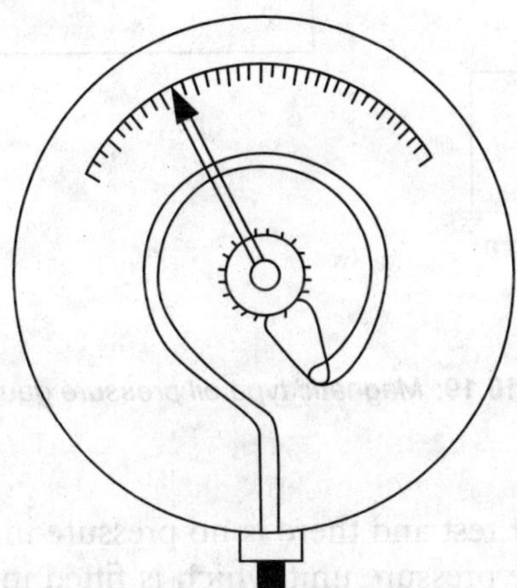

**Fig.10.18:** *Burdon tube oil pressure gauge*

## MAGNETIC TYPE

The Fig.10.19 is a magnetic type of fuel gauge circuit, in which there are two units.

### Pressure unit

It consists of a variable resistance with a sliding contact whose position is determined by a diaphragm which operates by the oil pressure. The one end of the resistance is connected to the earth in pressure unit. The other end is connected to dash board unit using a single wire. The resistance in the pressure unit controls the current flow in the coils of the dash unit according to the level of pressure.

### Dash board unit

There are two electromagnets set at 90° each other. Between electromagnets there is a soft iron core upon which a pointer is fixed. The first coil is called as limiting coil and other is called as operating coil. The operating coil is attached with the extended pole to give increased magnetic field.

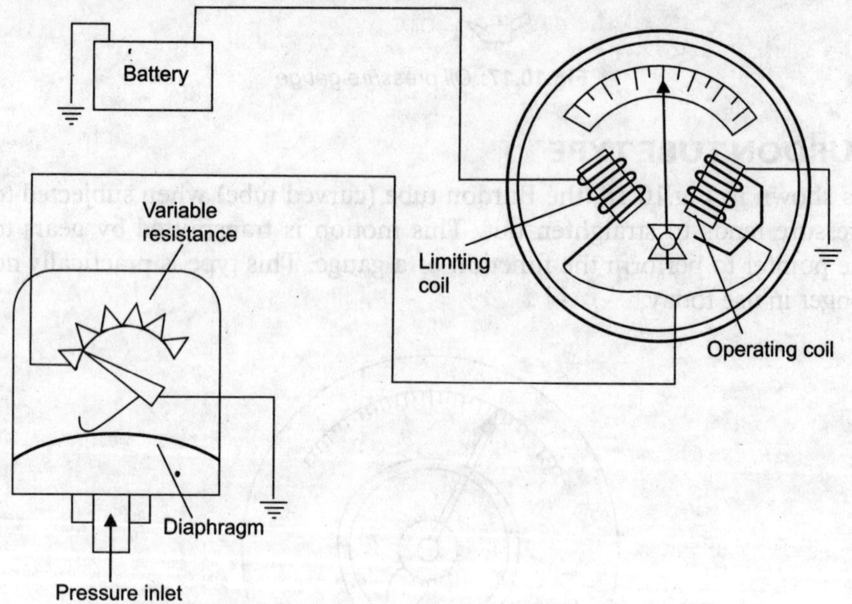


**Fig.10.19:** *Magnetic type oil pressure gauge*

### Operation

When the engine is at rest and there is no pressure in the lubricating line. The diaphragm in the pressure unit which is fitted in the lubrication line will be straight and brings the variable resister contact points such that there will be direct contact and no resistance. At this stage the whole current will pass from battery through the ignition switch, limiting coil and the pressure unit. There will be no current through the operating coil. The magnetic power of limiting coil is more and operating coil is less so the needle shows '*low pressure position*'.

When the pressure in the lubricating line is developed high the diaphragm of pressure unit brings the variable resistance contact point to such that full resistance is in series with the pressure unit. Now the current cannot flow through the pressure unit because of resistance and all the current will move from battery through ignition switch, limiting coil and the operating coil. Since the operating coil gets the current it becomes a powerful magnet and attracts the needle towards it. Thus the pointer shows 'h*igh pressure position*'.

## THERMAL TYPE

### *Low oil pressure/No oil pressure*

The bimetal element in the sender is provided with contact points, and moves the pointer of the gauge in accordance with the current flowing in the receiver heat wire.

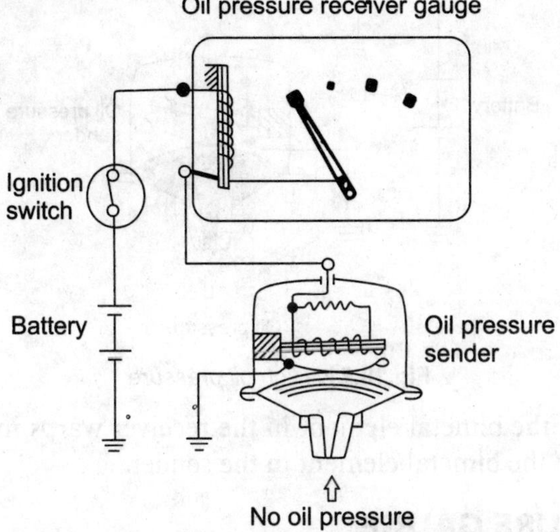


**Fig.10.20:** *Low oil pressure/No oil pressure*

When the oil pressure is zero, the points are open and no current flows when the ignition switch is turned on. Therefore, the pointer remains at zero. When there is low oil pressure, the diaphragm pushes the point, establishing a light contact. A current then flows through the heat wires in the receiver and sender.

Because the point contact pressure is low, the points are opened by warping of the bimetal element due to a slight current. Since the sender side points reopened by a current flowing even for a very short time, the temperature of the bimetal element in the receiver gauge does not rise, so the warping of the bimetal element is small. As a result, the pointer deflection is slight.

## Oil pressure high

As the oil pressure rises, the diaphragm pushes the points strongly, raising the bimetal element. So the current must flow for a long time because the points are not opened unless the bimetal element bends upward enough to counteract this. Since the current flows for a long time until opening of the sender side points, the temperature of the bimetal element in the receiver rises, increasing its amount of warpage. As a result, the pointer deflects greatly.

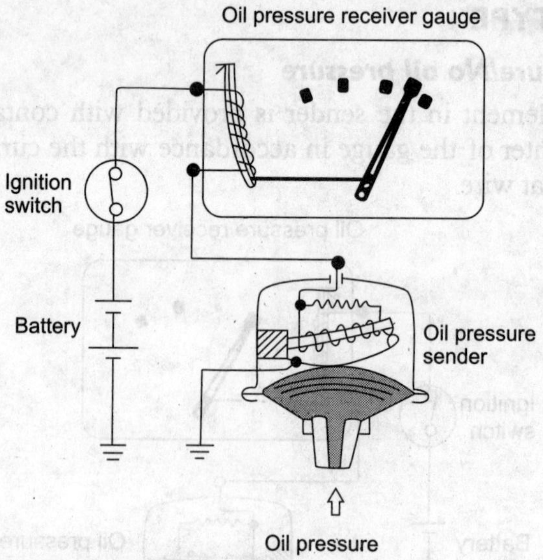


**Fig.10.21:** *High oil pressure*

Therefore, the bimetal element in the receiver warps in proportion to the warpage of the bimetal element in the sender.

## TEMPERATURE GAUGE

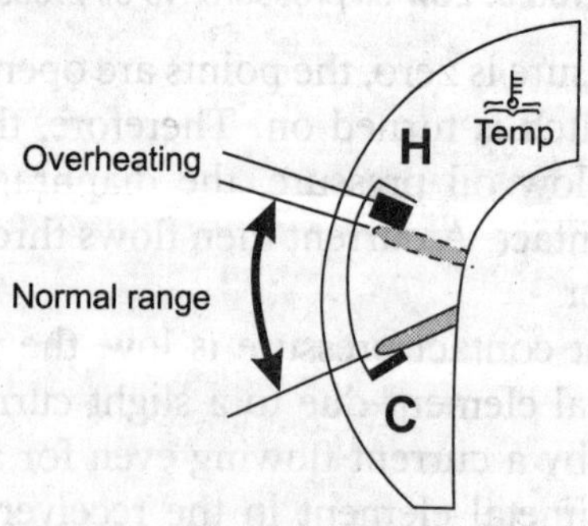


**Fig.10.22:** *Temperature gauge*

## MAGNETIC TYPE

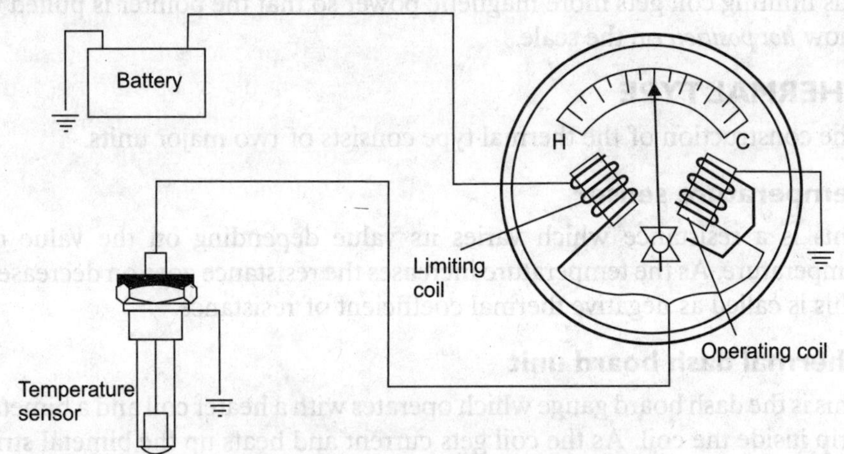


**Fig.10.23:** *Magnetic type temperature gauge*

The construction of the magnetic type of consists of two major units:

### Temperature sensor

This is a resistance which varies its value depending on the value of temperature. As the temperature increases the resistance goes on decreases. This is called as negative thermal coefficient of resistance.

### Magnetic dash board unit

This is the dash board gauge which operates by two electromagnets situated at 90° with each other. A soft iron core is kept between two coils and a pointer is fixed to it. One of the coils is called as limiting coil and other is operating coil. The operating coil is added with extended pole to give increased magnetic power.

### Operation

When the engine is at rest or just started, the temperature of coolant in the cooling system is still cool only. At this stage the temperature sensor is cold and acts as a resistance thus the current can pass from battery, through the ignition switch, limiting coil and the operating coil. Thus operating coil gets magnetic power so that the pointer is pulled by the operating coil to show *cold position* on the scale.

When engine is running at high speeds the coolant gets heat. As the temperature of coolant increases the temperature sensor goes on loses its resistance power and becomes conductor. At critical temperature the thermal sensor acts as a complete conductor and full current is flowing from battery through ignition switch, limiting coil and the thermal sensor.

Thus no current is passing through operating coil in the gauge. Because of this limiting coil gets more magnetic power so that the pointer is pulled to show *hot position* on the scale.

## THERMAL TYPE

The construction of the thermal type consists of two major units.

### Temperature sensor

This is a resistance which varies its value depending on the value of temperature. As the temperature increases the resistance goes on decreases. This is called as negative thermal coefficient of resistance.

### Thermal dash board unit

This is the dash board gauge which operates with a heater coil and a bimetal strip inside the coil. As the coil gets current and heats up the bimetal strip and bends which operates pointer through the linkage on scale.

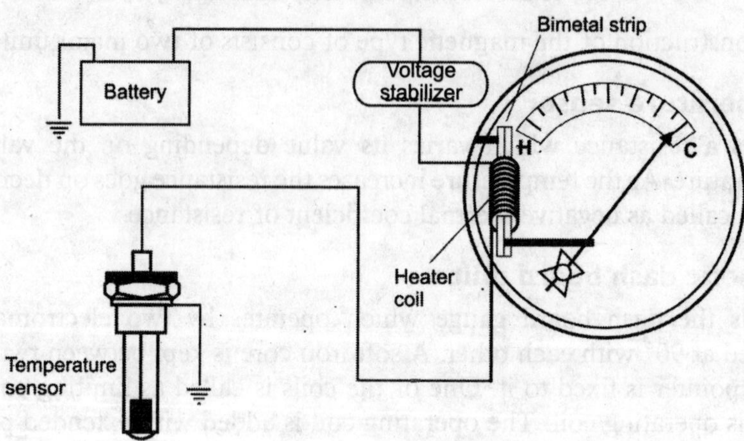


**Fig.10.24:** *Thermal type temperature gauge*

### Operation

When the engine is at rest or just started, the temperature of coolant in the cooling system is still cool only. At this stage the temperature sensor is cold and acts as a resistance thus no current can be passed from battery, through the ignition switch, dashboard gauge and the thermal sensor. Thus the heater coil in the gauge also not heated and the bimetal strip is straight and the pointer show cold position on the scale.

When engine is running at high speeds the coolant also gets heat. As the temperature of coolant increases the temperature sensor goes on loses its resistance power and becomes conductor. At critical temperature the thermal sensor acts as a complete conductor and full current is flowing from battery through ignition switch, gauge and the thermal sensor. Thus the heater coil in the gauge gets current and heats up. Because of this heat the bimetal strip bends. Thus the pointer is pulled to show hot position on the scale.

## HORN

The horn is an electromagnetic vibrating device, which is operated by the driver. It provides a warning sound to other road users. The horns are usually wired directly to battery from the fuse panel. The majority of automobiles utilise a horn relay. With a relay, the horn button on the steering wheel or column completes a circuit to ground which closes a relay, and the heavy current flow required by the horn then travels from the relay to the horn.

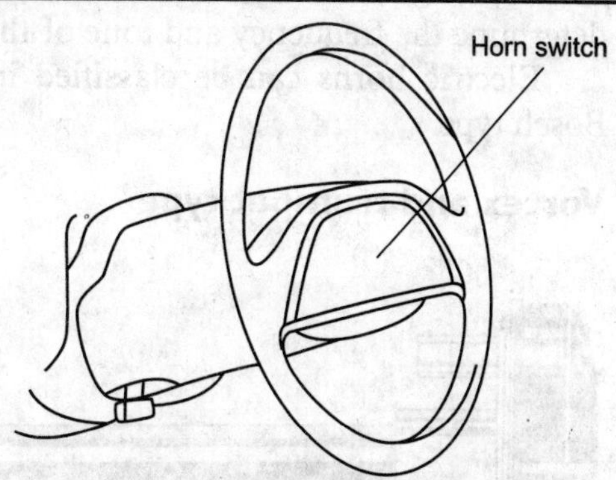


**Fig.10.25:** *Horn switch*

Horns are manufactured in several different tones or frequencies ranging between 1800 and 3550 Hz. Car manufacturers can select various horn tones for a particular car sound. The horns can be roughly classified into the electric type, air type, and vacuum type.

## ELECTRIC HORN

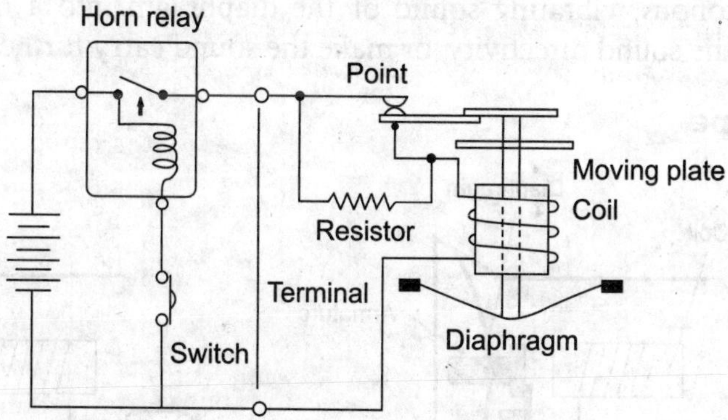


**Fig.10.26:** *Electric horn*

Electric horn operates by vibrating a metal diaphragm with an electromagnet.

It consists of coil winding, armature, diaphragm, and contact points. When the horn button is depressed, current starts flowing through the field coil produces a magnetic field that pulls the armature with the diaphragm and opens the contact points due to which circuit is broken. So that current cannot flow. The magnetic field in the coil collapses and the diaphragm is released. The contact points are closed then. The cycle again starts which is repeated again and again resulting in continuous vibration of the diaphragm which produces sound. The design and shape of the horn determine the frequency and tone of the sound.

Electric horns can be classified into vortex and trumpet type, and Bosch type.

## Vortex and trumpet type

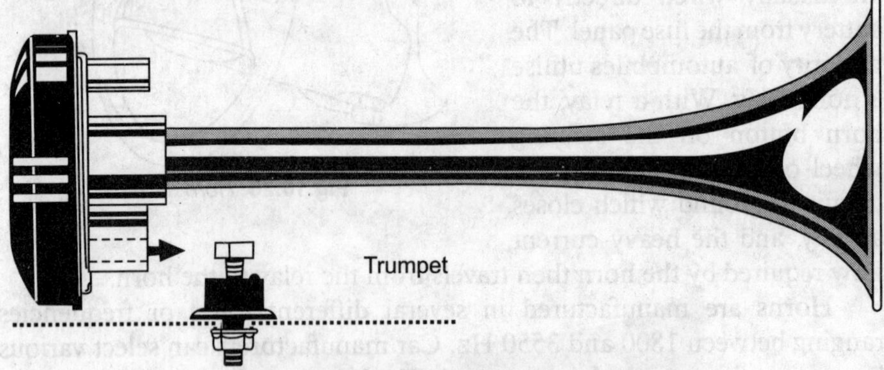


**Fig.10.27:** *Trumpet type horn*

In addition to the sound producing device, the horn is provided with vortex or trumpet type sound resonating devices. The purpose of this is to change the monotonous, vibrating sound of the diaphragm into a more richer tone, give the sound directivity, or make the sound carry further.

## Bosch type

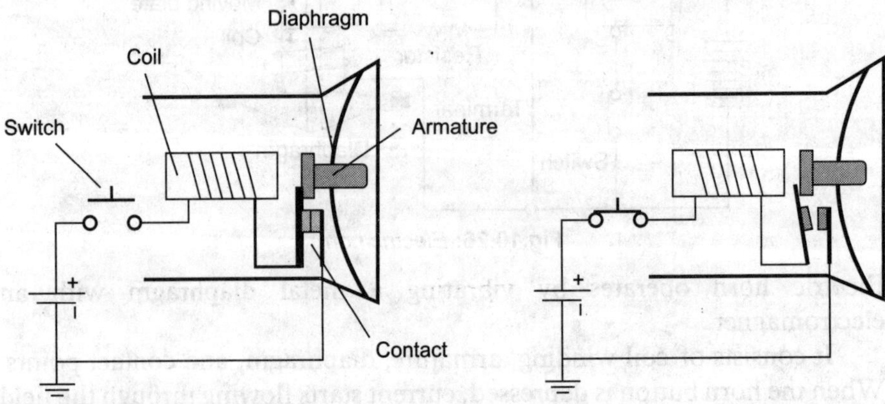

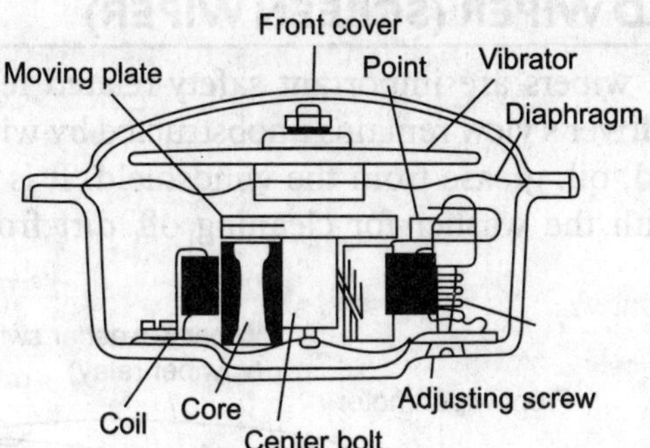


**Fig.10.28:** *Bosch type horn*

As shown in Fig.10.28, a vibrator (with diaphragm) is used to resonate with the vibrating sound. This type is mainly used in motorcycles and midget cars.

## AIR HORN

Air horns are used by large trucks and other big motor vehicles. These horns operate on compressed air from the air reservoir that vibrates the diaphragm to produce the warning sound.

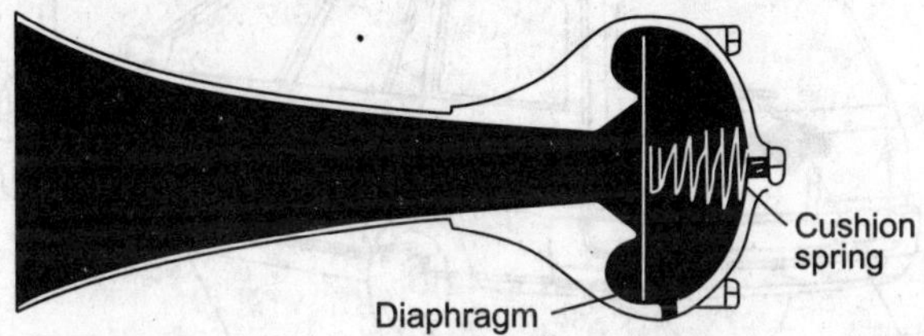


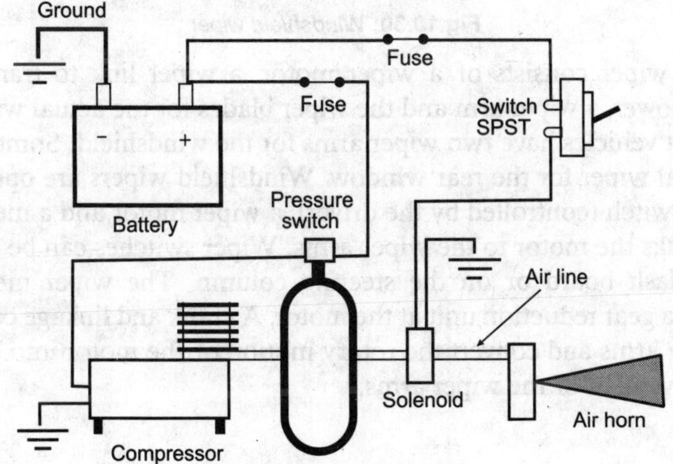


**Fig.10.29:** *Air horn*

## WINDSHIELD WIPER (SCREEN WIPER)

The windshield wipers are important safety-related features installed to ensure that the driver's view remains unobstructed by wiping off rainwater, snow, bugs, mud, oil, grease from the windshield. It is frequently used in combination with the washer for cleaning off dirt from the windshield glass.

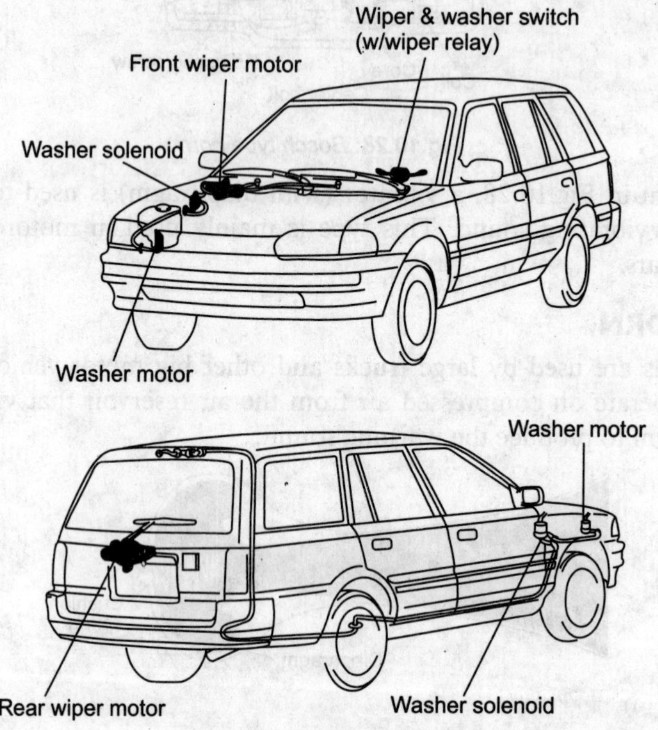


**Fig.10.30:** *Windshield wiper*

The wiper consists of a wiper motor, a wiper link to transmit the driving power, a wiper arm and the wiper blades for the actual wiping.

Most vehicles have two wiper arms for the windshield. Some have an additional wiper for the rear window. Windshield wipers are operated by a wiper switch (controlled by the driver), a wiper motor and a mechanism which links the motor to the wiper arms. Wiper switches can be mounted on the dash board or on the steering column. The wiper mechanism includes a gear reduction unit at the motor. A crank and linkage connect to the wiper arms and convert the rotary motion of the motor into back and forth movement of the wiper arms.

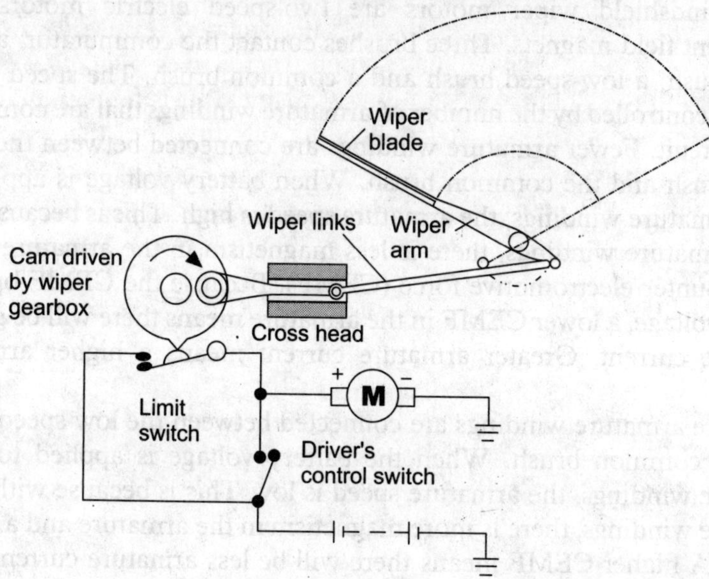


**Fig.10.31:** *Windshield wiper mechanism*

## Wiper motor

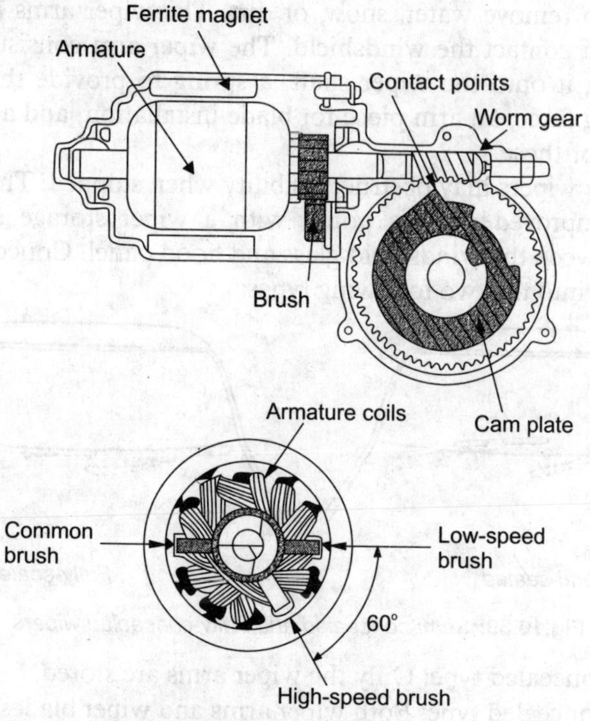


**Fig.10.32:** *Windshield wiper motor*

Most windshield wiper motors are two-speed electric motors with permanent field magnets. Three brushes contact the commutator: a high-speed brush, a low-speed brush and a common brush. The speed of the motor is controlled by the number of armature windings that are connected in the circuit. Fewer armature windings are connected between the high-speed brush and the common brush. When battery voltage is applied to fewer armature windings, the armature speed is high. This is because with fewer armature windings, there is less magnetism in the armature and a lower counter electromotive force (CEMF). Because the CEMF opposes battery voltage, a lower CEMF in the armature means there will be greater armature current. Greater armature current means a higher armature speed.

More armature windings are connected between the low-speed brush and the common brush. When the battery voltage is applied to more armature windings, the armature speed is low. This is because with more armature windings, there is more magnetism in the armature and a higher CEMF. A higher CEMF means there will be less armature current. This results in a lower armature speed.

## Wiper arm

Wiper arms are mechanical arms, which sweep back and forth across the windshield to remove water, snow, or dirt. The wiper arms have rubber blades, which contact the windshield. The wiper arm consists of a head for mounting it onto the wiper shaft, a spring to provide the necessary blade holding force, an arm piece for blade installation, and a retainer for retaining all of these.

Ordinary wipers may obstruct visibility when stopped. The concealed wiper has improved on this point, with a wiper storage space being provided between the windshield glass and hood panel. Concealed wipers are classified into the two following types:

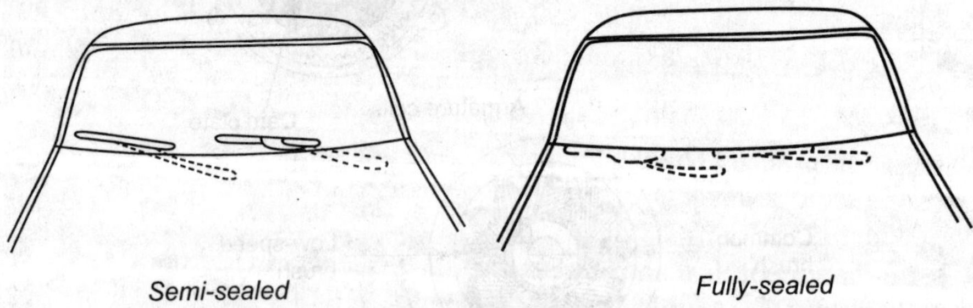


**Fig.10.33:** *Semi-concealed and Fully-concealed wipers*

1. Semi-concealed type: Only the wiper arms are stored.
2. Fully-concealed type: Both wiper arms and wiper blades are stored.

### Wiper link

The wiper link converts the rotary motion of wiper motor to the oscillating motion of the wiper shaft.

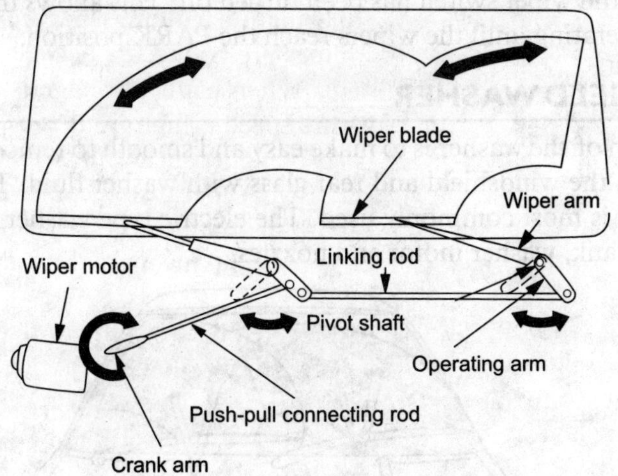


**Fig.10.34:** *Wiper link*

### Wiper blade

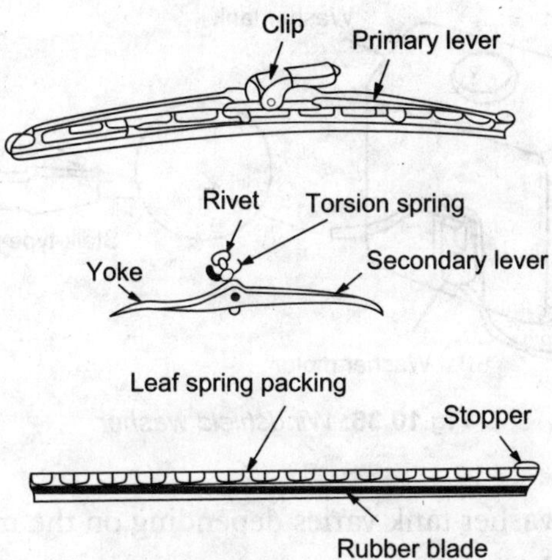


**Fig.10.35:** *Wiper blade*

The wiper blade consists of a rubber blade for wiping the glass surface, a combination of leaf spring packing and several levers, and clips for installing the blade on the wiper arm.

### Park contacts

When windshield wipers are turned off, they assume a rest position at the lower moulding edge of the windshield. This rest position is called the PARK position. A set of switch contacts, called park contacts, is located inside the motor assembly. The park contacts supply current to the motor after the wiper switch has been turned off. This allows the motor to continue operating until the wipers reach the PARK position.

## WINDSHIELD WASHER

The function of the washer is to make easy and smooth to remove dust and insects from the windshield and rear glass with washer fluid. The electric type washer is most commonly used. The electric type washer consists of the washer tank, washer motor and nozzles.

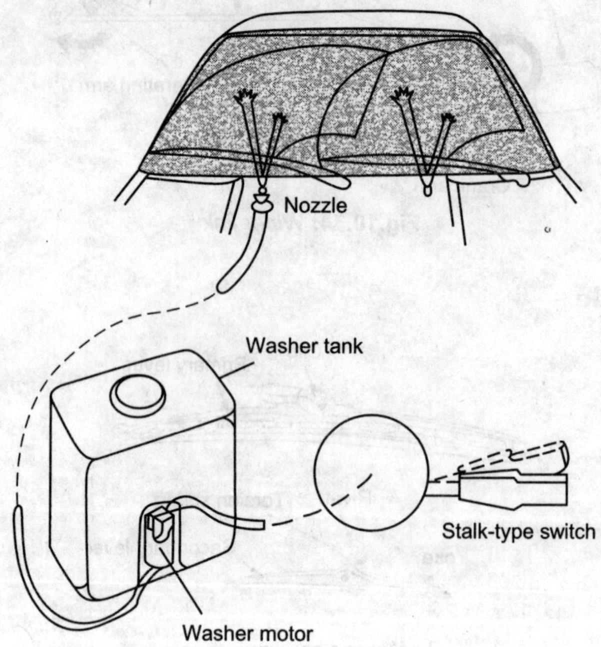


**Fig.10.36:** *Windshield washer*

### Washer tank

The shape of the washer tank varies depending on the mounting position and available space.

### Washer motor (Pump)

The washer motor drives the pump, discharging washer fluid from the tank. There are several types of pump are used:

1. Gear type
2. Squeeze type
3. Centrifugal type

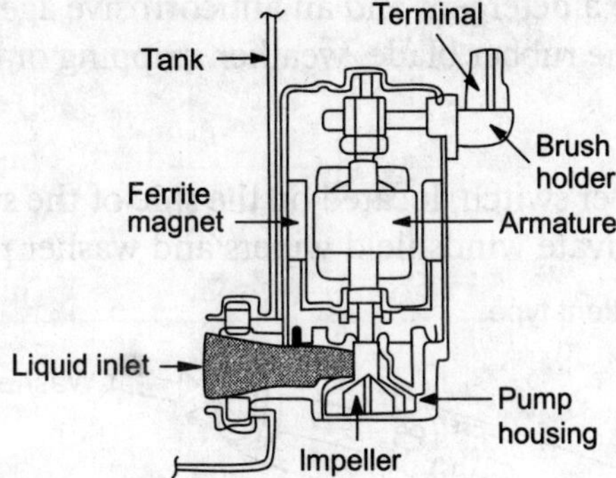


**Fig.10.37:** *Washer pump*

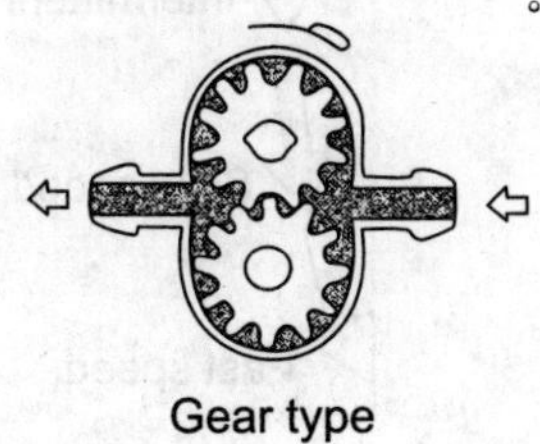


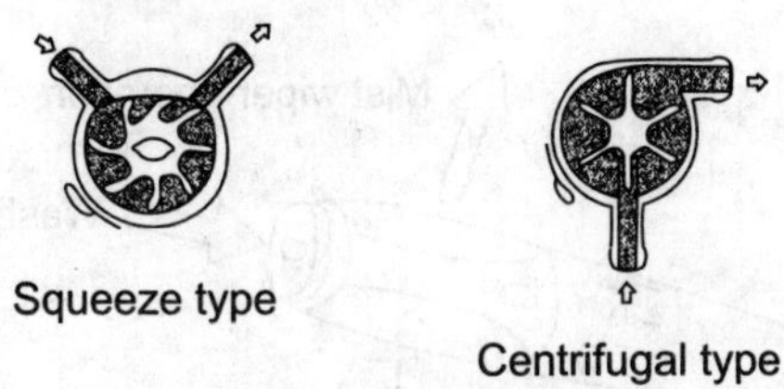


**Fig.10.38:** *Types of washer pump*

The centrifugal type is widely used because of its durability due to the use of fewer contacting parts. Since the centrifugal type is only capable of sending the washer fluid to the washers, but is not capable of drawing the fluid up from the tank, it is installed below the washer tank.

## Nozzles

The nozzle (Fig.10.36) is made of a copper, aluminium or resin pipe with one or two orifices. At present, resin nozzles fitted with adjustable orifices are most commonly used.

The orifice is 0.8 – 1.0 mm in diameter and one or two in number. The normal type has uniform discharge from each orifice without spreading.

## Washer fluid

Washer fluid consists of anti-freeze (isopropyl alcohol, ethylene glycol or methanol) added to a detergent and an anticorrosive agent. The fluid used must not damage the rubber blade, weather stripping or body paint.

## Wiper switch

The windshield wiper switch, located on the side of the steering column, is driver's input to activate windshield wipers and washer pump.

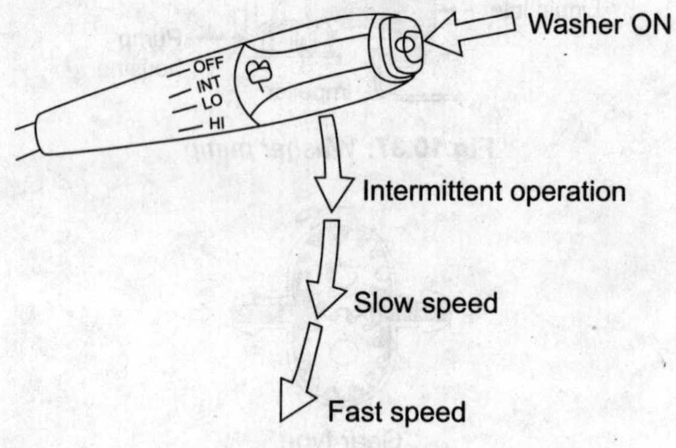


Mist type

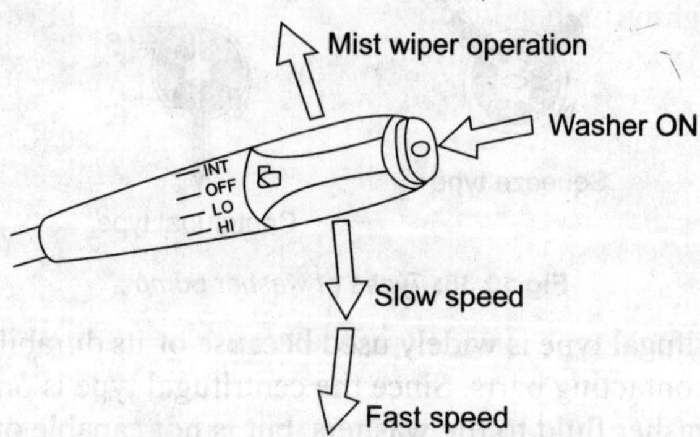


**Fig.10.39:** *Windshield wiper switch*

Today's windshield wipers can include an intermittent or delay function which causes the wiper to swipe the windshield periodically, rather than continuously. The driver can choose the length of time between swipes by adjusting a knob on the windshield wiper lever. Prior to the invention of intermittent wipers in the 1960's, cars usually had high and low speed continuous wipers (mist type), so the switch would allow the driver to vary the speed at which the wiper motor moves the windshield wipers only in this way. Intermittent wipers function by varying resistance in the wiper circuit. More resistance means a longer delay time.

## POWER WINDOW

Power windows are electric windows which can be raised and lowered at the touch of a button rather than the traditional handle which has to be turned manually for raising and lowering car windows. In hand-crank window, turning the winding handle oscillates the two parallel links which move the window carrier up and down squarely and smoothly.

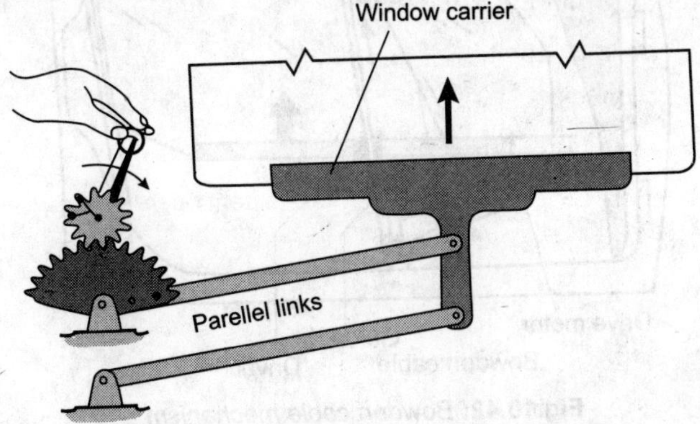


**Fig.10.40:** *Hand-crank window mechanism*

Power windows are now common on most vehicles. A car with power window is slightly higher in price as compared to the car with normal windows.

### Mechanism

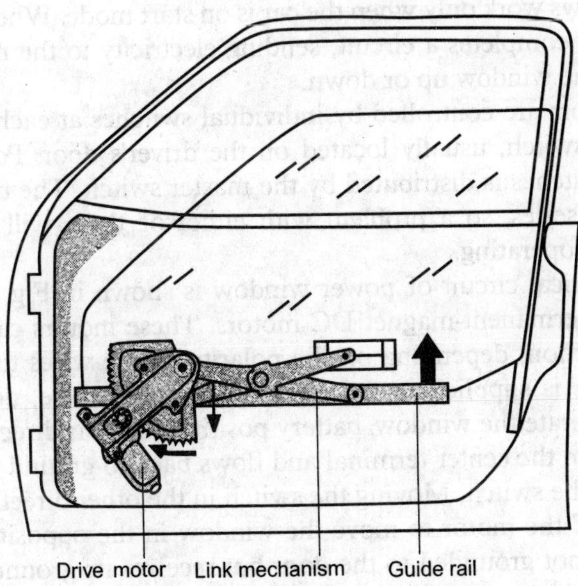


**Fig.10.41:** *Scissor-acting lever mechanism*

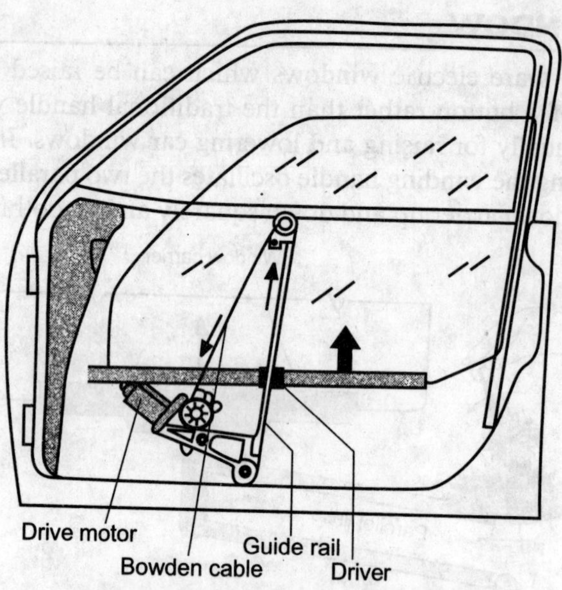


**Fig.10.42:** *Bowden cable mechanism*

Force is transmitted from the motor to the window glass will vary by manufacturer and vehicle model. They will usually comprise a gearbox and scissor-acting lever mechanism. Alternatively a flexible rack or cable assembly can be used as shown in Figs.10.41 and 10.42.

## Working of power windows

Power windows work only when the car is on start mode. When the switch is activated it completes a circuit, sending electricity to the motor which then moves the window up or down.

The motors are controlled by individual switches at each window or by a master switch, usually located on the driver's door. Power for the individual switches is distributed by the master switch. The two switches are wired in series, so a problem with either of them will prevent the window from operating.

The electrical circuit of power window is shown in Fig.10.43. Most systems use permanent-magnet DC motors. These motors can be driven in either direction, depending on the polarity of the wires to the motor. Battery power is supplied to the center of a double-pole, double-throw switch. To operate the window, battery positive power is directed through the motor from the center terminal and flows back to ground through the other side of the switch. Moving the switch in the other direction reverses the polarity of the motor to move the window in the opposite direction. The motor is not grounded to the door but receives its ground through a connection at the master switch.

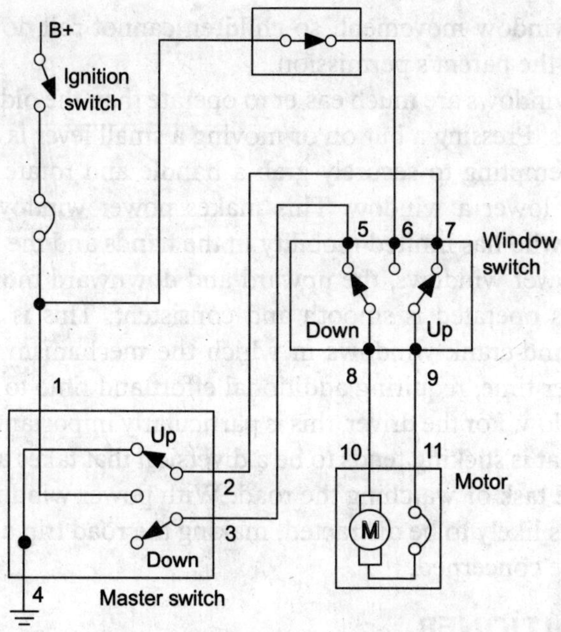


**Fig.10.43:** *Circuit of power window*

## Features

Many power windows have a variety of added features such as one-touch opening, lockout switches, and courtesy power on.

Single touch switches prevent accidental closing by having the user pull up on the switch to close the window or come with sensors that automatically reverse operation if an obstruction is detected in the window path.

A power window lockout switch allows the driver to prevent passengers, often children, from operating the windows. Some vehicles allow window operation from outside the car using a remote.

Many cars are also equipped with a feature called "courtesy power on." This feature allows the windows to continue operating for a brief interval after the ignition key is removed so that they can be adjusted without reinserting the key.

The motor drive is usually through a self-locking worm reduction gear. This ensures that the window cannot move unintentionally or be forced open when the motor is not energised.

## Advantages

The advantages provided by these power windows are:

1. Power windows ensure passengers safety. Driver has complete control

of the window movement, so children cannot roll down a window without the parent's permission.

2. Power windows are much easier to operate than the older hand-crank windows. Pressing a button or moving a small lever is a simpler task than attempting to securely grab a handle and rotate it in order to raise or lower a window. This makes power windows a must for anyone who has limited mobility in the hands and the lower arms.
3. With power windows, the upward and downward movement as the device is operated is smooth and consistent. This is in contrast to some hand-crank windows in which the mechanism may begin to stick over time, requiring additional effort and time to open or close the window. For the driver, this is particularly important, since a hand crank that is sticking tends to be a diversion that takes attention away from the task of watching the road. With power windows, the driver is far less likely to be distracted, making the road trip a little safer for everyone concerned.

## AIR-CONDITIONER

'Air conditioner' is the general term for the equipment that maintains the air in a room at a comfortable temperature and humidity. When the room temperature is high, the heat is taken away so as to lower the temperature (cooling) and conversely, when the room temperature is low, heat is supplied so as to raise the temperature (heating). In addition, moisture is added to or removed from the air so as to attain a comfort level of humidity.

## COMFORT ZONE

Most people feel comfortable in a relatively narrow temperature zone of about 70 to 80°F. When the effect of humidity is taken into consideration, this range becomes a little wider. Humidity is simply the amount of moisture in the air. When the air has absorbed as much moisture as it can, the relative humidity is said to be 100%. The actual amount of humidity varies in accordance with the temperature, because warm air holds more moisture than cold air. One can feel just as cool when the temperature is 80°F and the relative humidity is 30% as when the temperature is 70°F and the relative humidity is 90%.

The key to whether you feel hot or cold is the ability of your body perspiration to evaporate quickly and easily. If the air is dry, perspiration evaporates quickly and the heat leaving the body makes it feel cool. But when the moisture content of the air is high, perspiration cannot evaporate as rapidly, so less body heat is removed and you feel warmer.

Most automotive air conditioning and heating systems are integrated into a "climate control" system that controls the temperature, humidity and air circulation by cooling the air inside the passenger compartment when it's hot outside and heating it when the outside air is cold.

## COMPONENTS OF AUTOMOBILE AIRCONDITIONER

The air conditioner for an automobile generally consists of a heating unit (heater), a cooling unit (refrigeration system) with a moisture remover, and a ventilator.

### HEATER

A device that heats the air in the car or fresh outside air drawn into the passenger compartment of the vehicle.

There are various types of heater, including the hot water heater, combustion heater and exhaust heater, but ordinarily the hot water heater is used.

### PRINCIPLE

In the hot water heater system, engine coolant is circulated through the heater core to make the heater core hot. Then a blower is used to blow cold air through the hot heater core to heat the air.

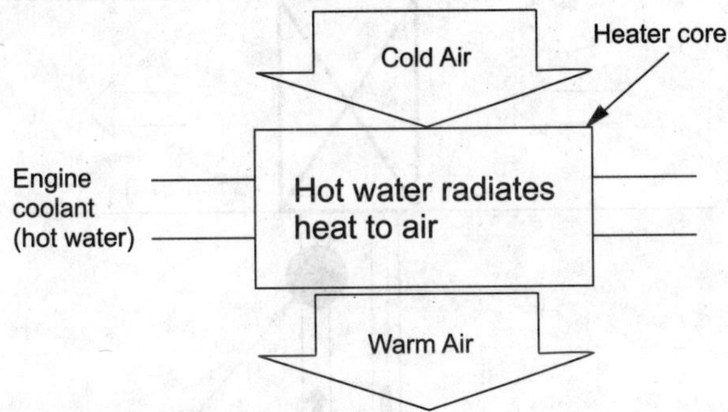


**Fig.10.44:** *Principle of heater*

### TYPES

There are two types of hot water heater, differing in the system used to adjust the temperature. One is the air-mix type and the other is the water flow control type.

#### Air-mix type

This type uses an air-mix control damper which changes the air temperature by adjusting the proportion of cold air allowed to pass through the heater core and the proportion of air which does not pass through the heater core.

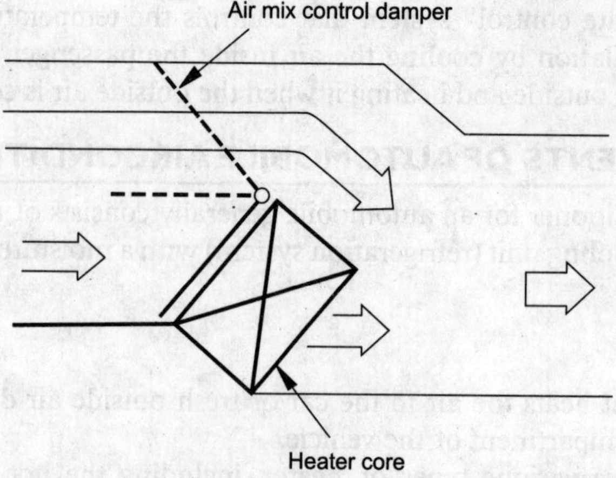


**Fig.10.45:** *Air-mix type*

Recently, this type of heater system is used in many models.

### Water-flow control type

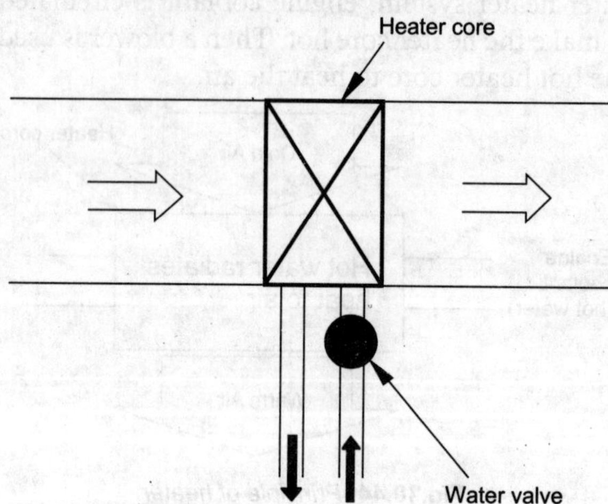


**Fig.10.46:** *Water flow control type*

This type controls the temperature by adjusting the amount of water flowing through the heater core with a water valve, thus changing the temperature of the heater core itself and adjusting the temperature of the air passing through the heater core. This type of heater system is used for the rear heater in vehicles.

## COMPONENTS OF HOT WATER HEATER

The hot water heater which is ordinarily used in an automobile consists of the components shown in the Fig.10.47.

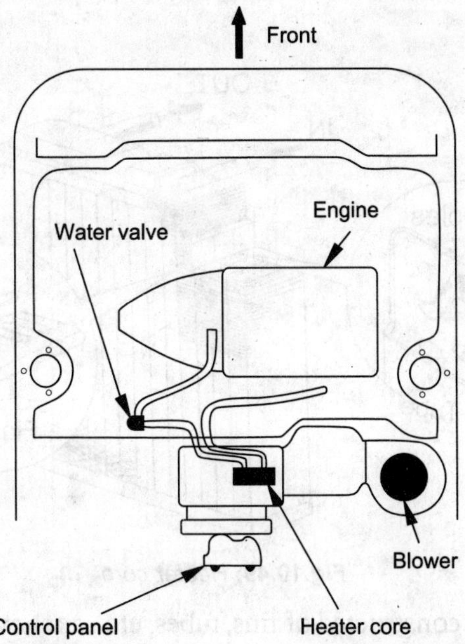


**Fig.10.47:** *Components of a heater*

## Water valve

The water valve is installed in the engine's coolant circuit and controls the amount of engine coolant that is allowed to enter the heater core. The driver operates the water valve by moving the temperature control lever on the control panel.

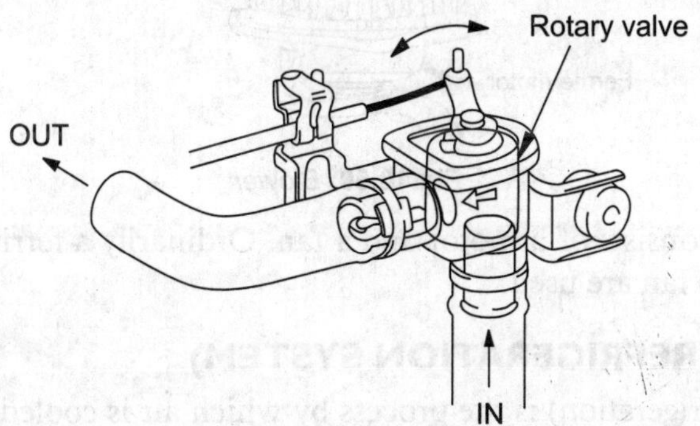


**Fig.10.48:** *Water valve*

**Note:** Some recent models do not have a water valve. In such models, the engine coolant constantly flows through the heater core. The heater outlet temperature is regulated by the driver, who can open and close the dampers located in the heater core housing.

### Heater core

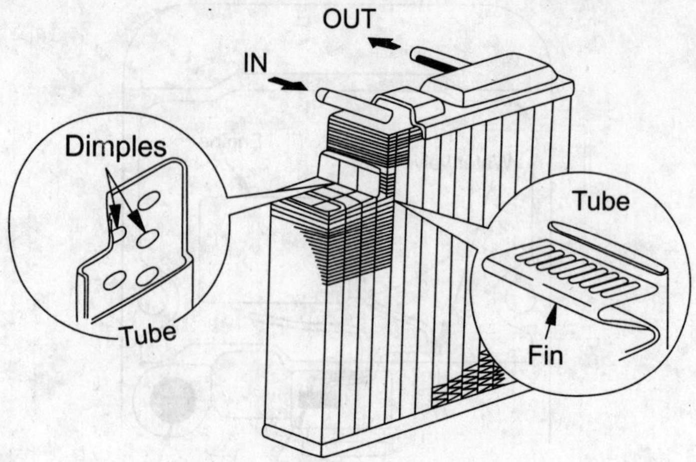


**Fig.10.49:** *Heater core*

The heater core is constructed of fins, tubes, etc., as shown in the Fig.10.49. Recently, a type with a large number of dimples added to the tubes in order to improve the heat output performance of the heater core is used in some models.

### Blower

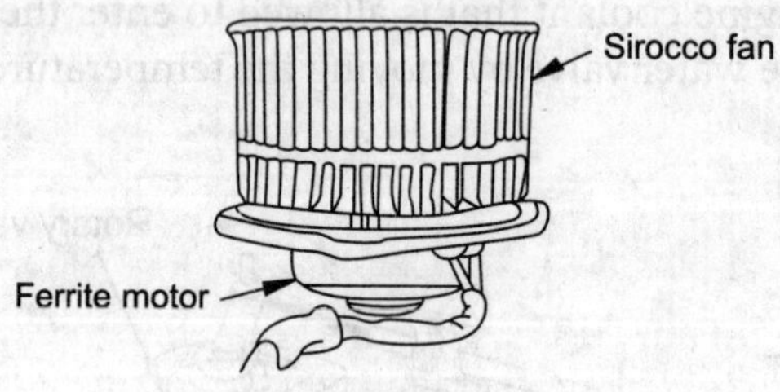


**Fig.10.50:** *Blower*

The blower consists of a motor and a fan. Ordinarily a ferrite type motor and a sirocco fan are used.

## COOLER (REFRIGERATION SYSTEM)

Cooling (Refrigeration) is the process by which air is cooled, cleaned and dehumidified before entering, or re-entering, the passenger compartment. Basically, an air conditioning system removes heat from the passenger compartment by absorbing it and carrying it outside, where it is released into the atmosphere.

The basic components of an automobile refrigeration system consist of the compressor, condenser, receiver/dryer, expansion valve, and evaporator. In addition, there are other devices and functions which help bring out the system's full potential, such as anti-frosting, engine stall prevention, engine idle-up, etc.

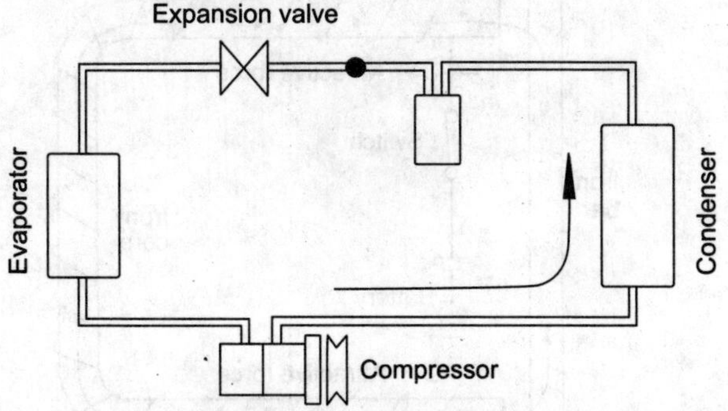


**Fig.10.51:** *Components of refrigeration system*

## COMPRESSOR

After being converted into low-temperature, low-pressure gas, the refrigerant is compressed by the compressor, and changed into high-temperature, high-pressure gas. It is then sent to the condenser.

Compressors are classified as follows:

### Crank type

In the reciprocating compressor, the rotation of the compressor crankshaft is changed to the reciprocating movement of a piston.

### Swash plate type

A number of paired pistons are set on the swash plate at intervals of 72 degrees for a 10-cylinder compressor or at intervals of 120 degrees for a 6-cylinder compressor. When one side of a piston is in a compression stroke, the other is in a suction stroke.

### Through-vane type

Each vane of the through-vane compressor forms an interval component with its opposite. There are two of such vane pairs, each mounted at right angles to the other in slots in the rotor. As the rotor turns, the vanes shift in the radial direction as their ends slide across the inside surface of the cylinder.

## MAGNETIC CLUTCH

A magnetic clutch is used to connect and disconnect the compressor from the engine. Its main components are the stator, rotor, and pressure plate.

## Principle

When current is fed to the coil in the Fig.10.52, magnetic force is generated in the iron core and attracts the iron bar.

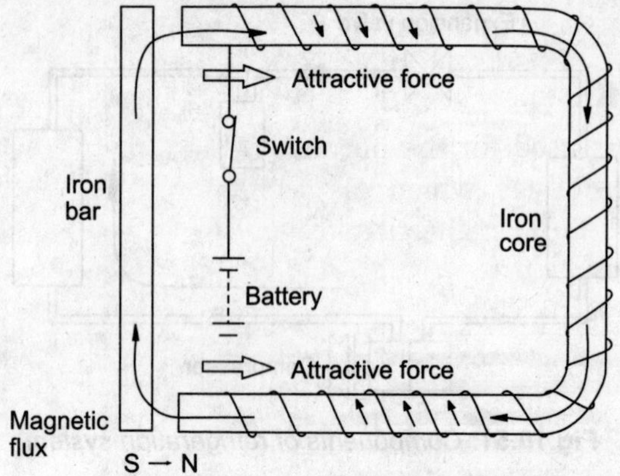


**Fig.10.52:** *Principle of magnetic clutch*

## Construction

A magnetic clutch consists of a stator, a rotor with pulley, and a pressure plate to engage the drive pulley and compressor magnetically.

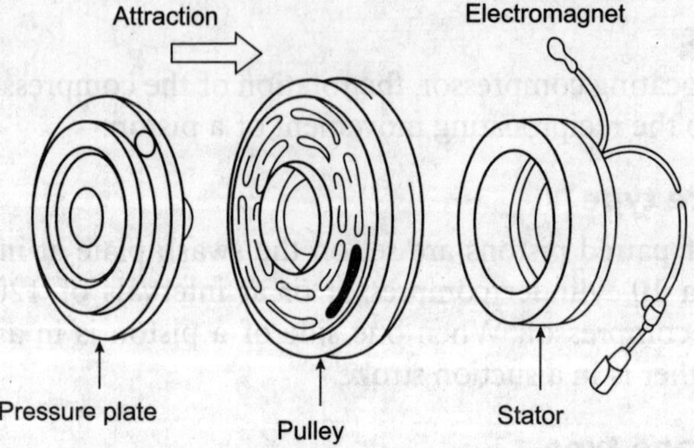


**Fig.10.53:** *Components of a magnetic clutch*

The stator is fixed on the compressor housing, and the pressure plate is attached to the compressor shaft.

## Operation

Whenever the engine is operating, the pulley is rotating since it is connected to the crankshaft by a drive belt, but the compressor does not operate until the clutch is energised. When the air-conditioning system is on, the amplifier supplies current to the stator coil. Then the electromagnet attracts the pressure plate and pulls the plate against the friction facings on the pulley. The friction between the facing and plate causes the clutch assembly to rotate as a unit, driving the compressor.

## CONDENSER

The condenser is used for the purpose of cooling and robbing the heat from the refrigerant gas, which has been compressed by the compressor into high-temperature, high-pressure gas, so as to change this gas into liquid refrigerant.

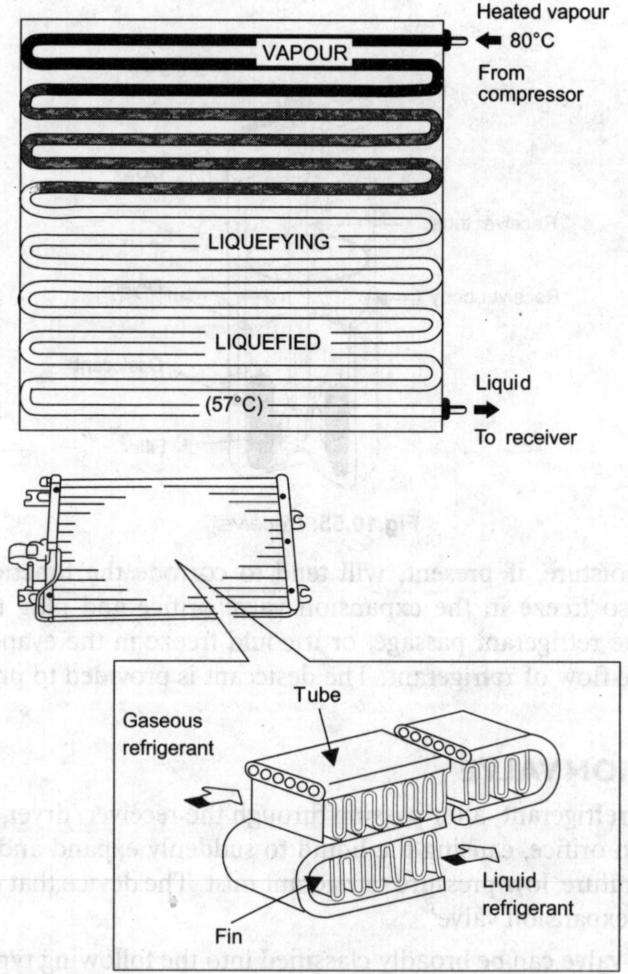


**Fig.10.54:** *Condenser*

The condenser is installed at the front end of the vehicle to enable forcible cooling by the air drawn in by the engine radiator fan and the air stream created by the car when running.

Recently, there are models which are equipped with an electric fan exclusively for the condenser.

## RECEIVER/DRYER

A filter and desiccant sealed in a container, called the receiver and dryer, has been provided for the purpose of (1) temporarily storing the refrigerant liquefied by the condenser so as to allow a supply conforming with the cooling load; and (2) removing the dirt and moisture that could be injurious to the refrigeration system if allowed to get in. A sight glass is provided on the top of the receiver/dryer to allow the flow of the refrigerant to be viewed.

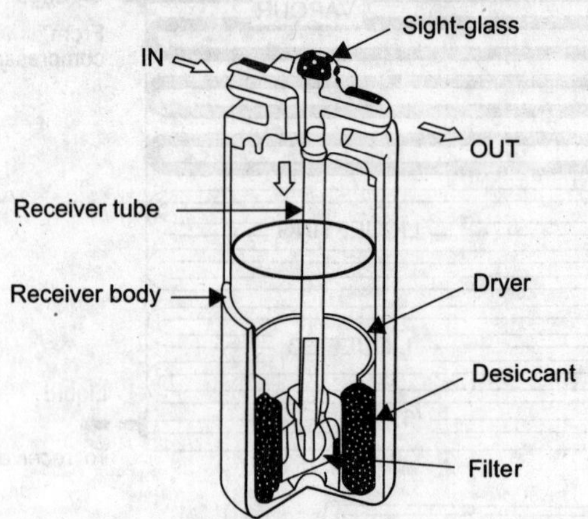


**Fig.10.55:** *Receiver*

The moisture, if present, will tend to corrode the functional parts. It could also freeze in the expansion valve orifice and plug the orifice, blocking the refrigerant passage, or it could freeze in the evaporator and obstruct the flow of refrigerant. The desiccant is provided to prevent such problems.

## EXPANSION VALVE

The liquid refrigerant, after passing through the receiver/dryer, is injected out from an orifice, causing the liquid to suddenly expand and become a low-temperature, low-pressure refrigerant mist. The device that does this is called the 'expansion valve'.

Expansion valve can be broadly classified into the following types:

- constant-pressure expansion valve
- thermal type expansion valve

A thermal type expansion valve is used in the refrigeration system of most vehicles.

## Function

1. Injects the high-temperature, high-pressure liquid refrigerant out from a small orifice (after it has passed through the receiver/dryer), causing the liquid to suddenly expand and atomise into low-temperature, low-pressure mist.
2. Regulates the amount of refrigerant supplied to the evaporator, based on the cooling load, in order to provide maximum cooling effectiveness at all times. As a result, the refrigerant constantly turns into a gaseous state at the evaporator outlet, regardless of changes in the cooling load or compressor rpm.

## Operation

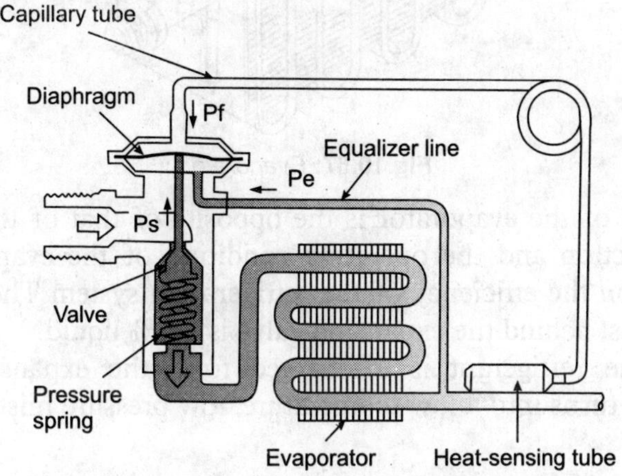


**Fig.10.56:** *Operation of expansion valve*

The amount of refrigerant flowing into the expansion valve after being liquefied in the condenser is determined by the vertical movement of the valve, depending on the difference between the vapour pressure $P_f$ inside the heat-sensing tube and the sum of pressures $P_s$ and $P_e$, where $P_s$ is the hold-down pressure of the pressure spring and $P_e$ the vapour pressure inside the evaporator.

When the cooling load is large, the refrigerant gas temperature at the evaporator outlet will be high. Consequently, the temperature and pressure within the heat-sensing tube will also be high so the valve will be pressed down, causing a larger amount of refrigerant to circulate. Conversely when the cooling load is small, the reverse action takes place, causing a smaller amount of refrigerant to circulate.

There is an inner equalising type and an external equalising type of thermal expansion valve, depending on where the vapour pressure within the evaporator is measured, but both are identical in principle.

## EVAPORATOR

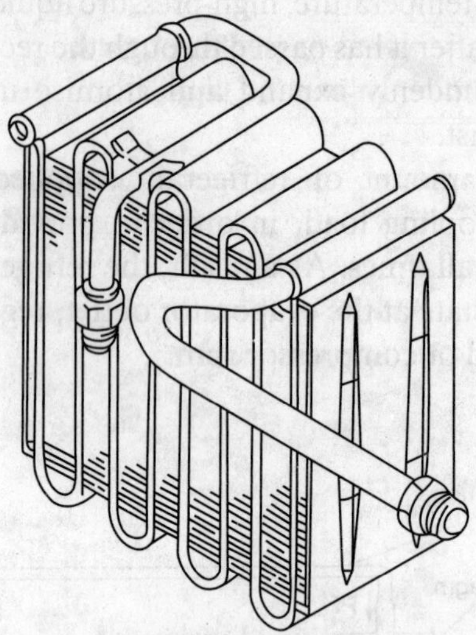

**Fig.10.57:** *Evaporator*

The purpose of the evaporator is the opposite of that of the condenser. The construction and the operating condition of the evaporator has a great effect on the efficiency of the refrigeration system The state of the refrigerant just behind the expansion valve is 100% liquid.

Once the refrigerant is discharged from the expansion valve, it immediately turns into a low-temperature, low pressure mist and starts to vapourise.

The evaporator is made of aluminium. There are three types:

- Plate fin type
- Serpentine fin type
- Drawn cup type

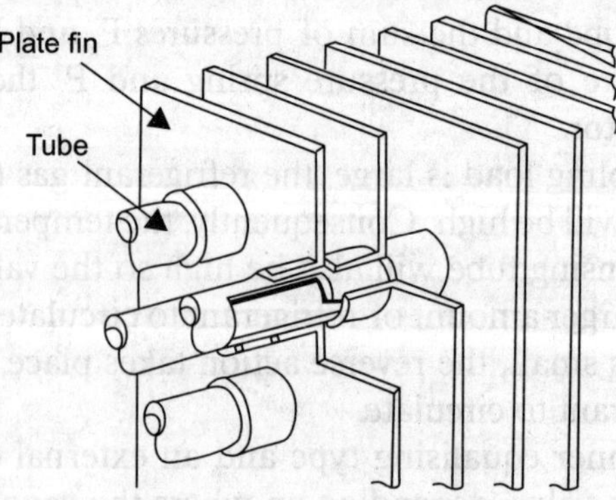


**Fig.10.58:** *Plate fin type*

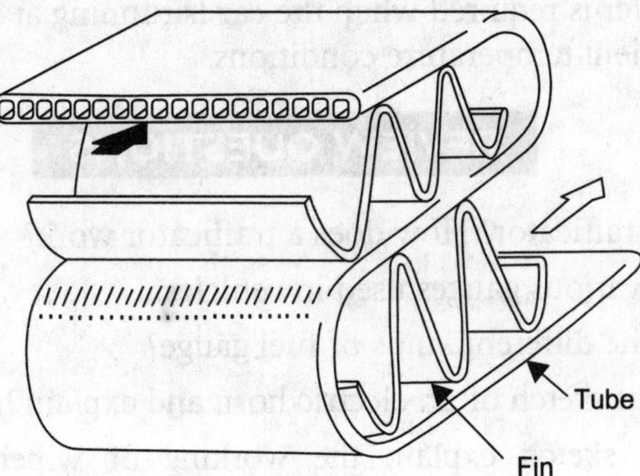


**Fig.10.59:** *Serpentine fin type*

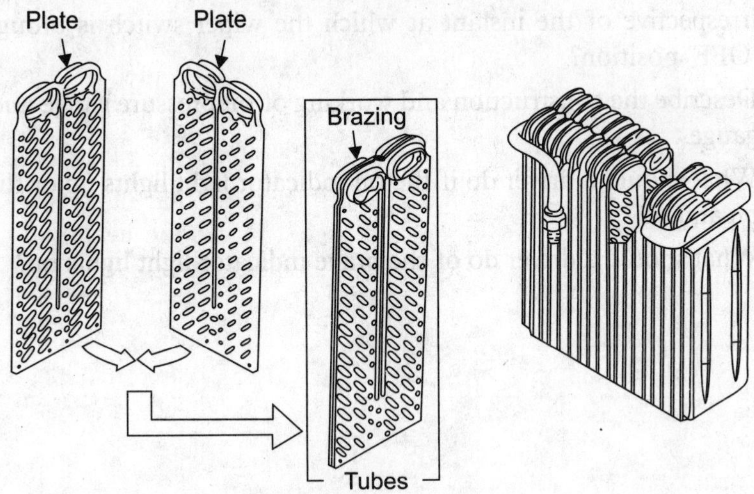


**Fig.10.60:** *Drawn cup type*

## CAPACITY

In comparison to domestic air-conditioning, very large amount of air-conditioning capacity is required for a car. This is due to the fact that car body is made of metal and there is large glass area in the passenger compartment. Typically, the capacity of a car air-conditioner is 1 to 4 tons. At high speeds, the refrigerating system consumes as much as 8 H.P with a unit capacity of 3 tons approximately.

## FUEL ECONOMY

Because the cooling capacity of the automobile air-conditioning is quite high, the engine driving compressor may absorb 1 to 10 H.P and the fuel economy of the car is affected. A loss in economy level of the order of 1 to 1.5 km/litre can occur due to use of air-conditioning. The maximum compressor power is required when the car is running at maximum speed under high ambient temperature conditions.

## REVIEW QUESTIONS

1. What is a trafficator? How does a trafficator work?
2. Name the various gauges used in vehicles.
3. What are the different kinds of fuel gauge?
4. Draw a neat sketch of an electric horn and explain how it works.
5. With neat sketch explain the working of wiper system of an automobile.
6. How is it made sure that wiper arm stops only at extreme position irrespective of the instant at which the wiper switch is brought to 'OFF' position?
7. Describe the construction and working of oil pressure gauge and fuel gauge.
8. What should a driver do if the oil indicator light lights up or flickers on and off?
9. What should a driver do of the brake indicator light lights up?

# UNIT 11 Electronic Engine Management

Government has setup stringent norms for controlling pollution and to meet international emission standards. As a result the recent trend is to shift from conventional mechanical carbureted systems to electronic engine management system and fuel injection.

With the implementation of electronic engine management system (EEMS) it is possible to achieve good engine response, easy starting, better derivability, excellent comfort and good idling apart from good fuel efficiency and better reliability.

The system can be divided into three distinct elements with three operational phases. The three system elements are:

1. Input sensors
2. Electronic Control Unit (A microcomputer)
3. Output actuators

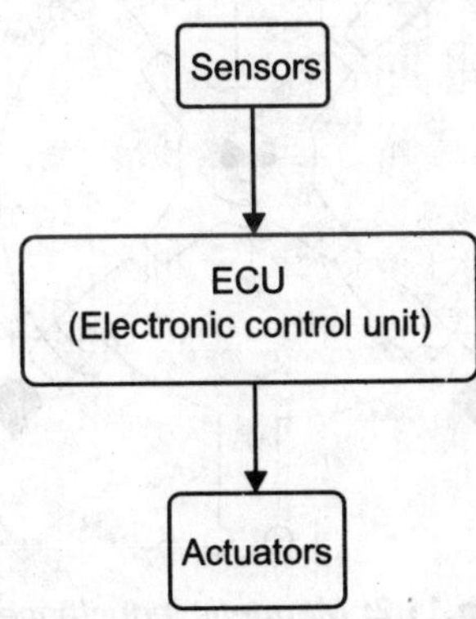


**Fig.11.1:** *Elements of engine management system*

Modern engine controls are capable of managing fuel delivery, idle speed control, electronic spark advance, and emission systems with extraordinary speed and accuracy.

In an electronic control system, these functions are managed in three phases.

- The input phase of electronic control allows the electronic control unit (ECU) to monitor engine operating conditions, utilising information from the input sensors.
- The process phase of electronic control requires the ECU to use this input information to make operating decisions about the fuel and spark advance systems.
- The output phase of electronic control requires the ECU to control the output actuators, the fuel injectors, and igniter to achieve the desired fuel metering and spark timing.

## SENSOR PRINCIPLE

### INDUCTIVE

Magnetic inductance sensors consist of a coil of wire around an iron core plus a permanent magnet. When a toothed rotor made from a magnetic material is rotated past the coil and magnet, the magnetic lines of force move and cut through the coil. The lines of force cutting through the coil will produce the voltage output.

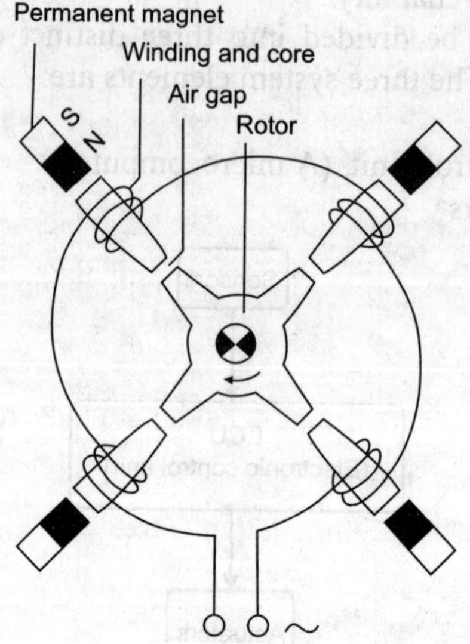


**Fig.11.2:** *Magnetic inductance*

This type of sensor is commonly used as a wheel speed sensor on ABS equipped vehicles.

### HALL GENERATOR

A Hall-generator consists of a permanent magnet, a Hall IC, and several thin metal plates called 'shutter blades'.

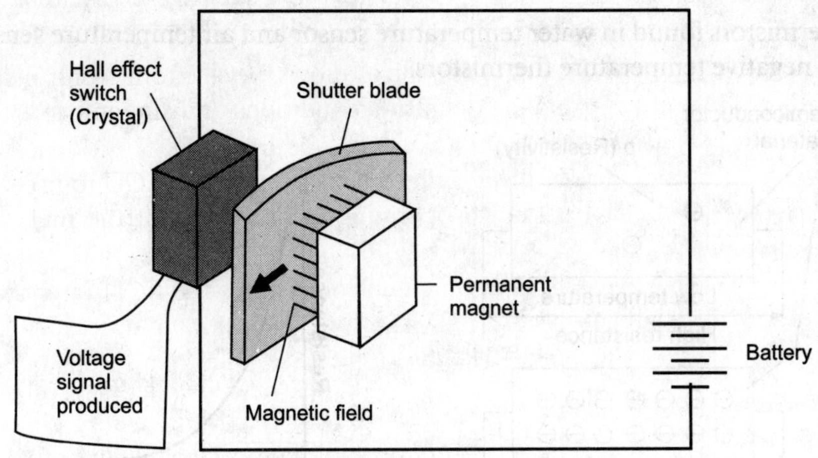


**Fig.11.3:** *Hall effect*

Since the Hall IC is placed next to a permanent magnet, the Hall IC has a magnetic field around it. As long as a magnetic field is present around the device, no electrical signal is produced.

Hall generator consists of shutter blades which are attached to a round disk. Each time a shutter blade passes between the permanent magnet and the Hall IC, the shutter blade prevents the magnetic field from reaching the Hall IC. When there is no magnetic field around the switch, the Hall IC produces a voltage signal that is sent to the ignition module.

## HOT WIRE

A wire or film resistance mesh is kept in the intake manifold and current is supplied to it through the ECU. A wire or film resistance is heated above air temperature. The incoming air flow cools the resistor. The voltage is then increased to maintain the temperature. The increase in voltage will be proportional to air flow. The voltage increase is a true measure of the quantity of air and also of its temperature. Depending upon the airflow sensed by this device the ECU decides the duration for which the fuel injector should stay open to achieve the desired air-fuel ratio.

## THERMISTORS

Thermistors are variable resistors whose resistance changes with temperature. Thermistors can have either a negative temperature coefficient or a positive temperature coefficient. A thermistor with a negative temperature coefficient will decrease in resistance as the temperature is increased. On the other hand, a thermistor with a positive temperature coefficient will increase in resistance as the temperature is increased.

Thermistors found in water temperature sensor and air temperature sensor are negative temperature thermistors.

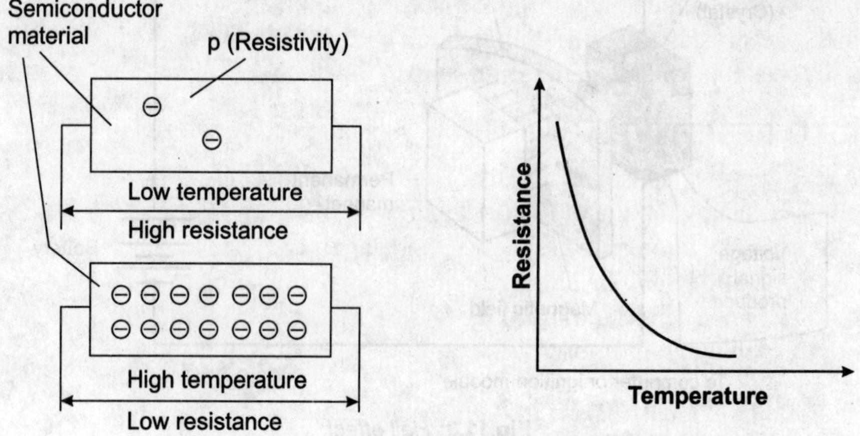


**Fig.11.4:** *Thermistor*

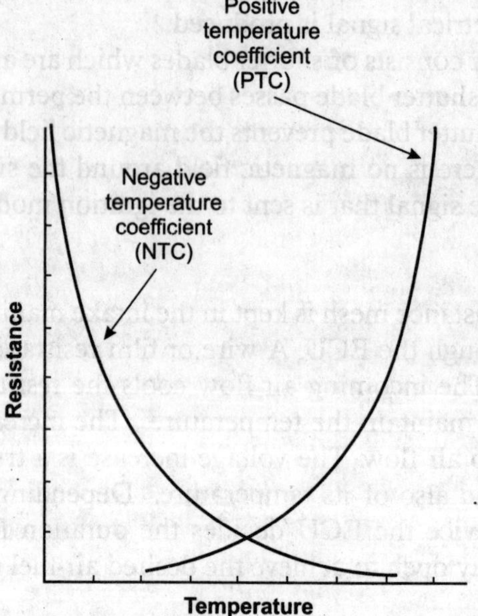


**Fig.11.5:** *PTC and NTC*

When either the air temperature or the coolant temperature is low, the respective thermistor's resistance increases and the computer receives a high voltage signal at the respective sensor wire. Conversely, a high temperature results in a low voltage signal due to the lower resistance of the thermistor.

## PIEZO ELECTRIC

Piezo-electricity is generated by pressure on certain crystals such as quartz which will develop a potential difference (voltage) on the crystal face. When the crystal flexes or vibrates, an AC voltage is produced. Knock sensor uses this voltage generating sensor.

## PIEZO RESISTIVE

A piezo-resistive sensor is a resistor circuit constructed on a thin silicon wafer. Physically flexing or distorting the wafer a small amount changes its resistance. The change in resistance value is converted to a voltage signal that is sent to the ECU. This type of sensor is usually used as a pressure sensing device such as a manifold pressure sensor. This sensor is also used to sense turbocharger boost.

## SENSORS IN ENGINE MANAGEMENT

In order to provide the correct amount of fuel for every operating condition, the engine electronic control unit (ECU) needs input from a number of sensors.

| Sensor | Function |
|---|---|
| Air flow sensor | Detects intake air volume as a voltage ratio using a potentiometer. |
| Manifold pressure sensor | Measures changes in intake manifold pressure (vacuum) |
| Speed sensor | Measures the speed of a component, and the change in speed of a component. |
| Throttle position sensor | Detects engine load and idle condition according to throttle valve opening. |
| Oxygen sensor | Detects amount of residual oxygen in exhaust gas. |
| Temperature sensor | Detects engine coolant temperature.<br>Detects intake air temperature. |
| Detonation sensor | Detects engine knocking. |

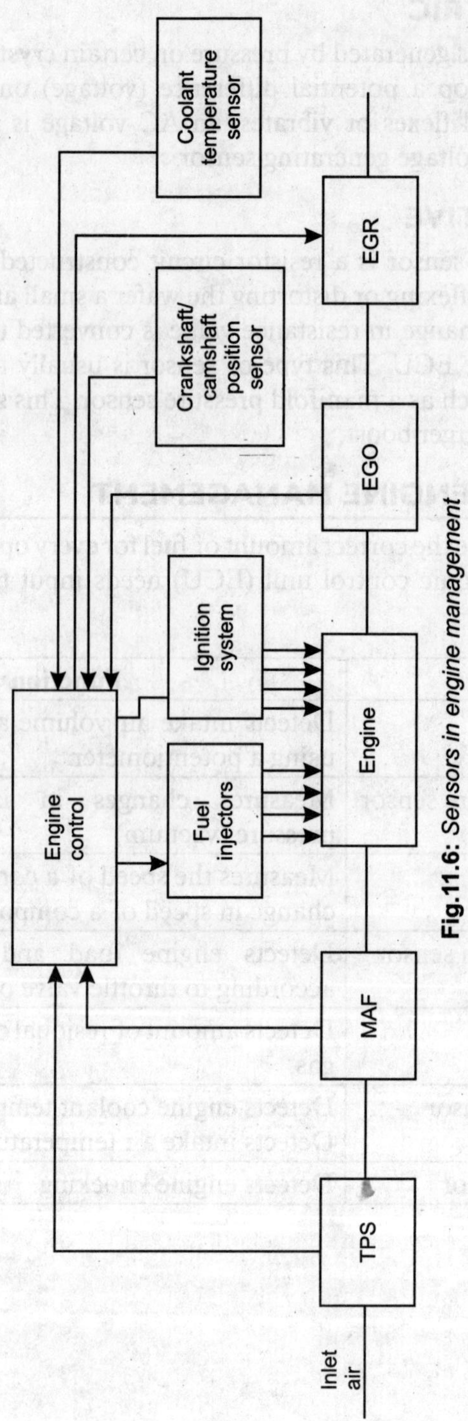


**Fig.11.6:** *Sensors in engine management*

## MASS AIR-FLOW SENSOR

Mass Airflow Sensor (MAF) monitors the mass of air entering the engine. The mass airflow sensors convert the amount of air drawn into the engine into a voltage signal. The ECU needs to know intake air volume to calculate engine load. This is necessary to determine how much fuel to inject, when to ignite the cylinder, and when to shift the gearbox. The airflow sensor is located directly in the intake air stream, between the air cleaner and throttle body where it can measure incoming air.

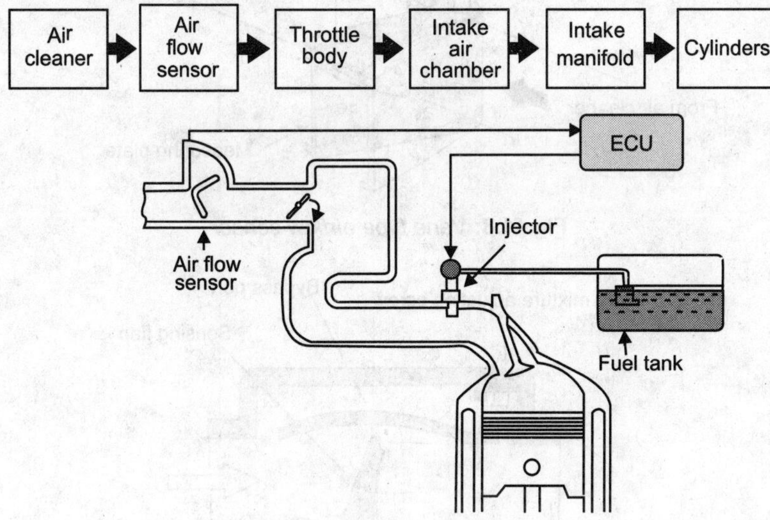


**Fig.11.7:** *Mass airflow sensor*

Measuring the airflow is done by an air flap, as with the mechanical system or by the more recent hot wire or hot film sensors.

## VANE (FLAP) TYPE AIR FLOW SENSOR

The air flow sensor measures the volume of air passing through the chamber and into the engine. Movement of the sensor flap varies a voltage, which is fed to the ECU as one of its inputs.

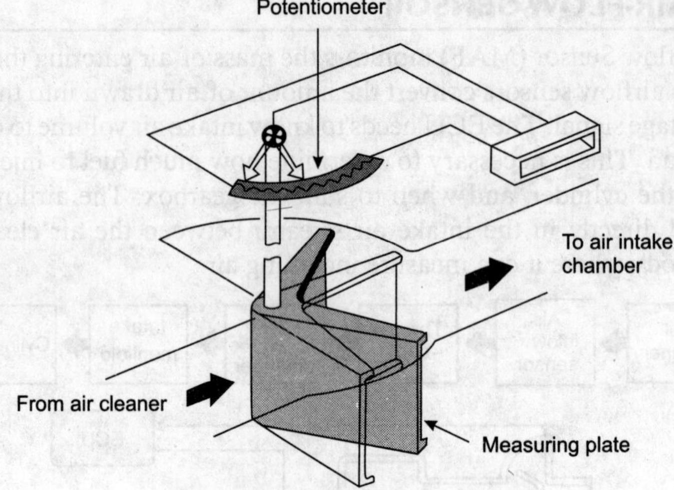


**Fig.11.8:** *Vane type airflow sensor*

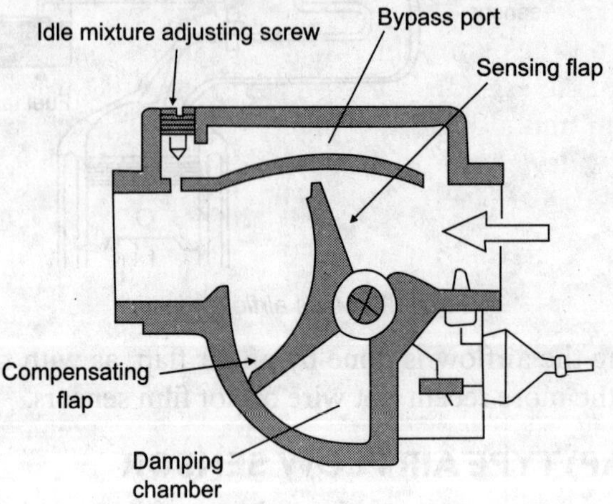


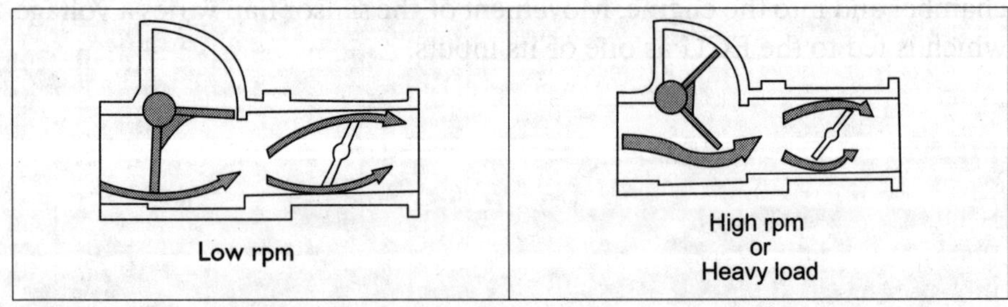


**Fig.11.9:** *Operation of airflow sensor*

## HOT WIRE SENSOR

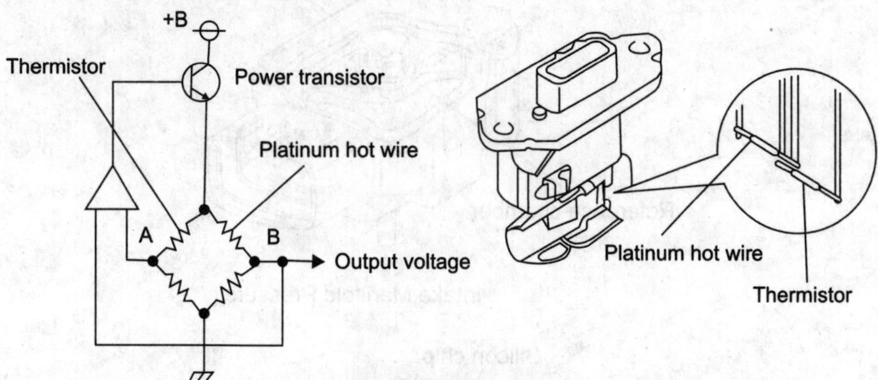


**Fig.11.10:** *Hot wire sensor*

A wire or film resistance mesh is kept in the intake manifold and current is supplied to it through the ECU. A wire or film resistance is heated above air temperature. The incoming air flow cools the resistor. The voltage is then increased to maintain the temperature. The increase in voltage will be proportional to air flow. The voltage increase is a true measure of the quantity of air and also of its temperature. Depending upon the airflow sensed by this device the ECU decides the duration for which the fuel injector should stay open to achieve the desired air-fuel ratio.

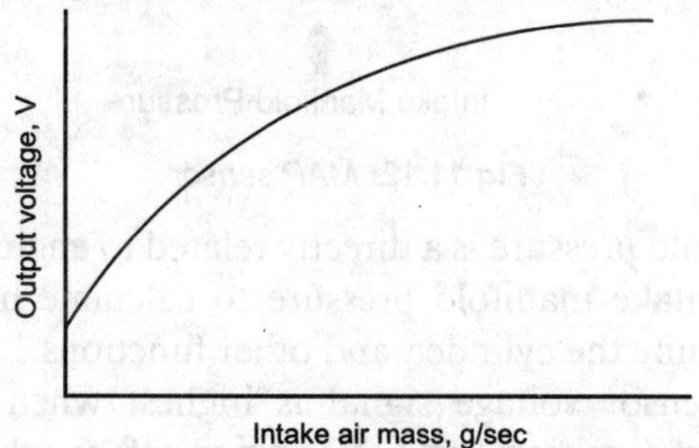


**Fig.11.11:** *Increase in voltage is proportional to airflow*

## MANIFOLD PRESSURE SENSOR

Changes in engine speed and load cause changes in intake manifold pressure. The sensor measures these pressure changes and converts them into an electrical signal. It is called a manifold absolute pressure (MAP) sensor (sometimes referred to as vacuum sensor).

This sensor which may be incorporated into the ECU itself, is connected to the manifold by a small vacuum pipe. The depression in the manifold is converted into an electrical signal for use as an input.

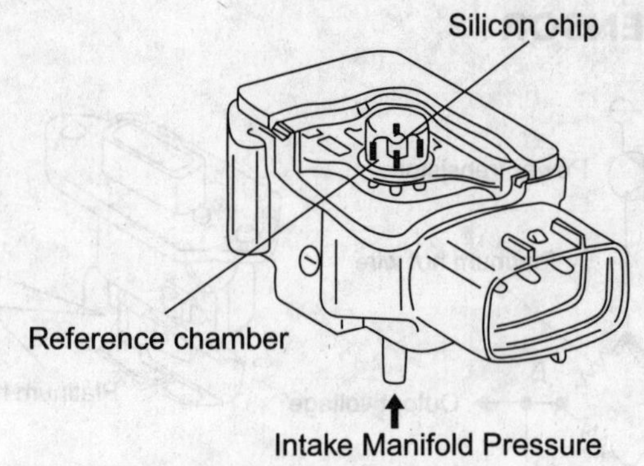


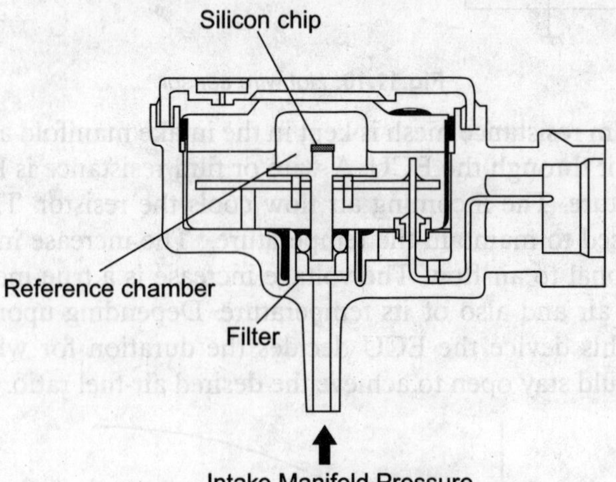


**Fig.11.12:** *MAP sensor*

Intake manifold pressure is a directly related to engine load. The ECU needs to know intake manifold pressure to calculate how much fuel to inject, when to ignite the cylinder, and other functions.

The MAP sensor voltage signal is highest when intake manifold pressure is highest (ignition key ON, engine off or when the throttle is suddenly opened). The MAP sensor voltage signal is lowest when intake manifold pressure is lowest on deceleration with throttle closed.

## Operation

In the MAP sensor there is a silicon chip mounted inside a reference chamber. On one side of the chip is a reference pressure. This reference pressure is either a perfect vacuum or a calibrated pressure, depending on the application. On the other side is the pressure to be measured. The silicon chip changes its resistance with the changes in pressure. When the silicon chip flexes with the change in pressure, the electrical resistance of the chip changes. This change in resistance alters the voltage signal. The ECU interprets the voltage signal as pressure and any change in the voltage signal means there is a change in pressure.

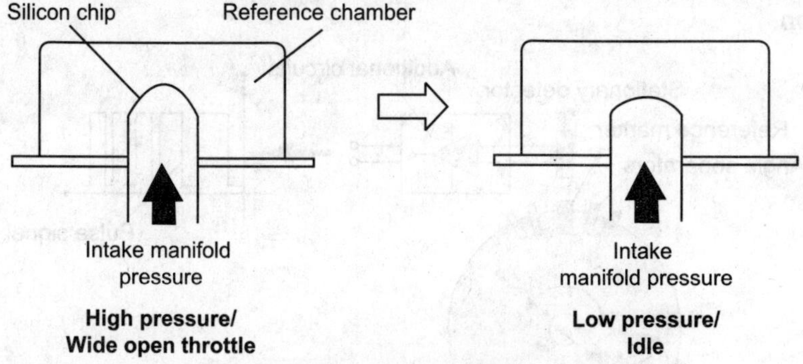


**Fig.11.13:** *Operation of MAP*

The silicon chip flexes as pressure changes. The amount the silicon chip flexes determines the output voltage signal.

## POSITION/SPEED SENSORS

Position sensors and speed sensors are similar devices that operate based on pulse detection and counting. These provide information to the ECU about the position of a component, the speed of a component, and the change in speed of a component. The following sensors provide this data:

- Camshaft position sensor (also called G sensor)
- Crankshaft position sensor (also called NE sensor)
- Vehicle speed sensor

The camshaft position sensor, crankshaft position sensor, and one type of vehicle speed sensor are of the pick-up coil type sensor.

## COMPONENTS OF POSITION/SPEED SENSORS

The fundamental components of these sensors include a toothed disc mounted on the shaft, a stationary detector and electronic circuitry (Fig.11.14). The disc has teeth or marks distributed evenly around the circumference. When the shaft spins, teeth or marks pass by the stationary detector and generate a series of electrical pulses. The frequency of these pulses is detected and converted to the corresponding shaft rotation speed. The relative angular position can be obtained by counting the number of pulses. Additional reference markers are required to sense the absolute position of the crank or camshaft. The Crank position sensor is usually mounted in the crankcase over the toothed wheel on the crankshaft. The

Camshaft position sensor is usually mounted in line with the toothed wheel on the end of the camshaft. It is usually hidden under the timing chain cover on the front of the engine. Some sensors have two sets of discs with a constant phase shift in order to determine the direction of the rotary motion.

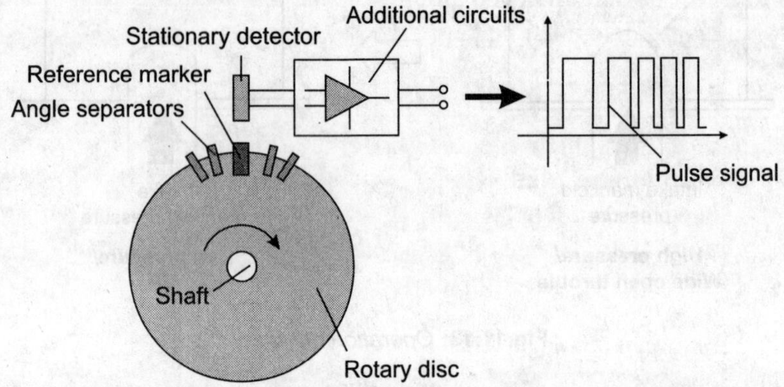


**Fig.11.14:** *Components of general position/speed sensors*

## POSITION SENSORS

The crank angle sensor provides accurate information of crankshaft speed and position. It is also known as crank angle sensor. The crank angle sensor signal is used by the ECU to calculate correct ignition timing, correct fuel injector timing and control of engine speed. The crankshaft position sensor may be mounted externally on the crankcase wall or it may be inside the housing of the ignition distributor.

Camshaft position sensor provides additional information for ignition and fuel injection timing. The CAM sensor or camshaft position sensor's role is to signal the ECU the camshaft position. The crank and cam sensor operates in synchronization with each other. The CAM sensor is frequently used in determining which injector to fire in a sequential system and for the coil on-plug ignition systems coil firing event.

To determine which cylinder is in its power stroke, while the vehicle's computer (for example) monitors the rotating position of the camshaft which is relative to the crankshaft position using a camshaft position (CMP) sensor. It will use this information to adjust the spark timing and the operation of the fuel injectors.

## TYPES OF POSITION SENSORS

Generally, there are four types of crankshaft or camshaft position sensors: magnetic pick-up coils, Hall-effect sensors, magneto-resistive element (MRE) sensors, and optical sensors. The first three utilise the metal teeth of a wheel to generate a series of electric pulses, based on changes in the magnetic field, as the teeth come into the proximity of the sensor. Optical position sensors use a light-emitting diode (LED) light source and a photodiode to detect optical marks on the shaft or slots in a disk as they move past the detector. Optical sensors have a good degree of accuracy and are suitable for both high-speed and low-speed applications, but the optical components and disc must be kept clean to ensure an accurate reading.

## VEHICLE SPEED SENSOR

The purpose of the vehicle speed sensor is to inform the ECU how fast the car is going down the road. The sensor can be mounted on the gearbox or in the speedometer. The sensor shown here is driven by a gear on the gearbox output shaft. Speed sensing is used by the ECU in some cars to limit top speed and also to improve fuel economy and drivability.

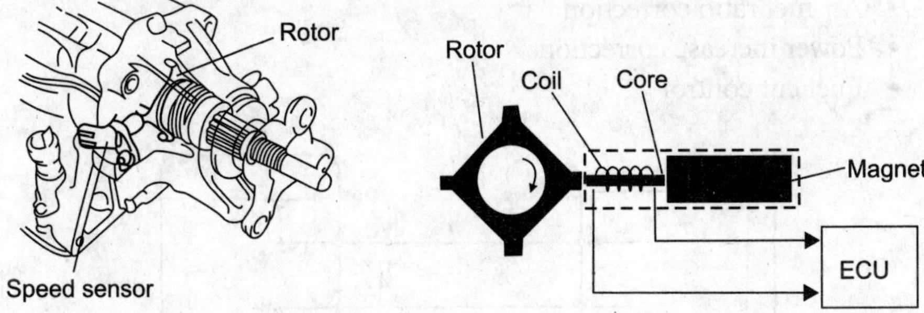


**Fig.11.15:** *Vehicle speed sensor mounted on the gearbox output shaft*

## THROTTLE POSITION SENSOR

The throttle position sensor indicates position of the throttle valve. The throttle position sensor is mounted on the throttle body and converts the throttle valve angle into an electrical signal. As the throttle opens, the signal voltage increases.

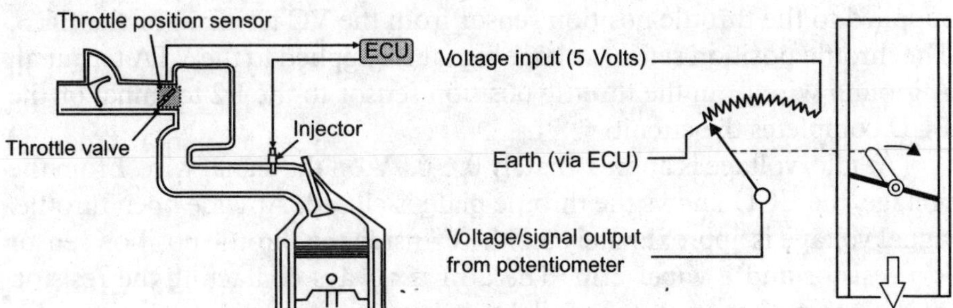


**Fig.11.16:** *Throttle position sensor*

## Principle

A wiper arm inside the sensor is mechanically connected to a moving part, such as a valve or vane. As the part moves, the wiper arm also moves. The wiper arm is also in contact with a resistor. As the wiper arm moves on the resistor, the signal voltage output changes. At the point of contact the available voltage is the signal voltage and this indicates position. The closer the wiper gets to VC voltage (Fig.11.17), the higher the signal voltage output. From this voltage, the ECU is able to determine the position of a component.

## Operation

The ECU uses throttle valve position to decide the following:

- Engine mode: idle, part throttle, wide open throttle
- Switch off AC and emission controls at wide open throttle (WOT)
- Air-fuel ratio correction
- Power increase correction
- Fuel cut control

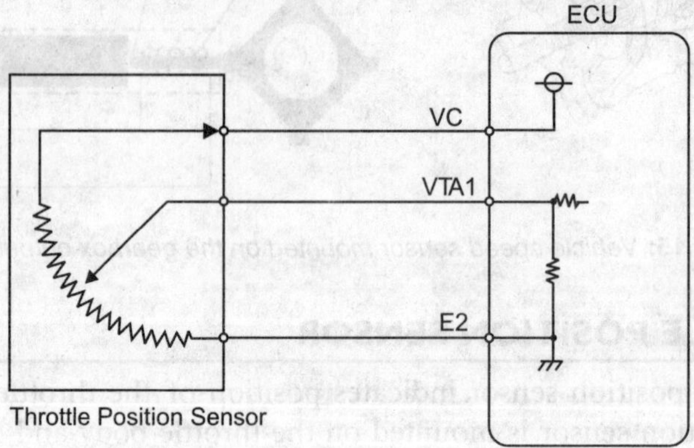


**Fig.11.17:** *Throttle position sensor circuit*

The basic throttle position sensor requires three wires. Five volts are supplied to the throttle position sensor from the VC terminal of the ECU. The throttle position sensor voltage signal is supplied to the VTA terminal. A ground wire from the throttle position sensor to the E2 terminal of the ECU completes the circuit.

At idle, voltage is approximately 0.6-0.9V on the signal wire. From the voltage, the ECU knows the throttle plate is closed. At wide open throttle, signal voltage is approximately 3.5-4.7V. Inside the throttle position sensor is a resistor and a wiper arm. The arm is always contacting the resistor. At the point of contact, the available voltage is the signal voltage and this indicates throttle valve position. At idle, the resistance between the VC and VTA terminal is high, therefore, the available voltage is approximately 0.6-0.9V. As the contact arm moves closer the VC terminal, resistance decreases and the voltage signal increases to 3.5-4.7V.

## OXYGEN SENSOR (LAMBDA SENSOR)

The oxygen oensor (also known as Lambda sensor) is located in the exhaust manifold and measures the amount of unburned oxygen in your engine's exhaust. Based on the amount of oxygen, it sends a signal to your engine computer which then adjusts the air-fuel mixture for optimum engine performance and emission control.

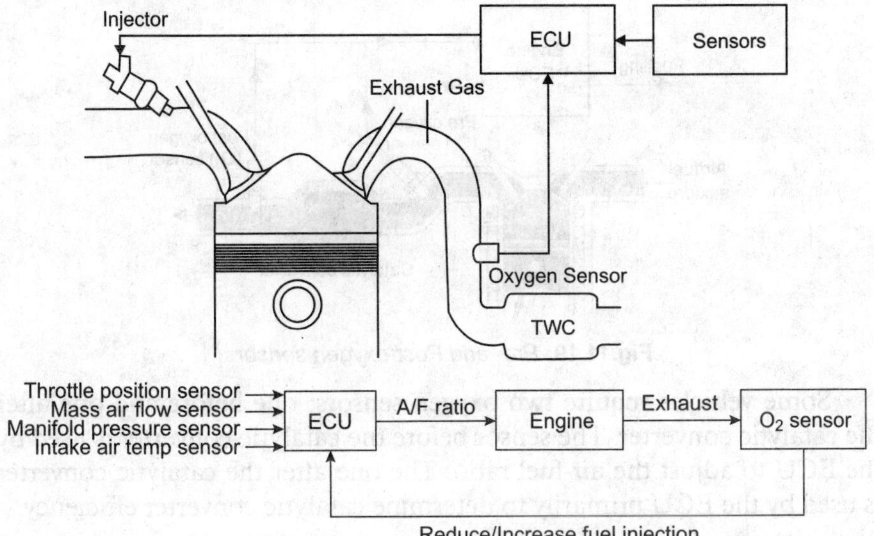


**Fig.11.18:** *Operation of oxygen sensor*

The idle, low speed and power circuits supply varying air-fuel ratios to meet different operating conditions of an engine. These systems depend on mechanical controls which often are not as accurate as the emission laws require. Excessive rich mixture may cause excessive HC and CO in the exhaust gases. When mixture is too lean, excessive NOx may appear in exhaust gas and the engine may stall. Modern electronic systems control the air-fuel ratio of the mixture flowing from the carburetor or electronic fuel injection system to engine cylinders. These systems monitor the amount of oxygen in the exhaust gases. They use an oxygen sensor, also called a lambda sensor which is installed in the exhaust pipe. It detects left-over oxygen in the exhaust gas and sends the data to the control unit. If the exhaust gas is low in oxygen, the air-fuel mixture is too rich. The fuel is using up most of the oxygen in the air fuel mixture. If the exhaust gas is too high in oxygen, the air-fuel mixture is too lean. Depending on the amount of oxygen in exhaust gas, the ECU adjusts the amount of fuel injected by the injectors accordingly.

There are different types of oxygen sensors, but two of the more common types are:

- the narrow range Zirconia oxygen sensor, the oldest type, simply called the oxygen sensor
- wide range oxygen sensor, the newest style, called the Air/Fuel ratio sensor
- Also used on very limited models in the early 90s, was the Titania oxygen sensor.

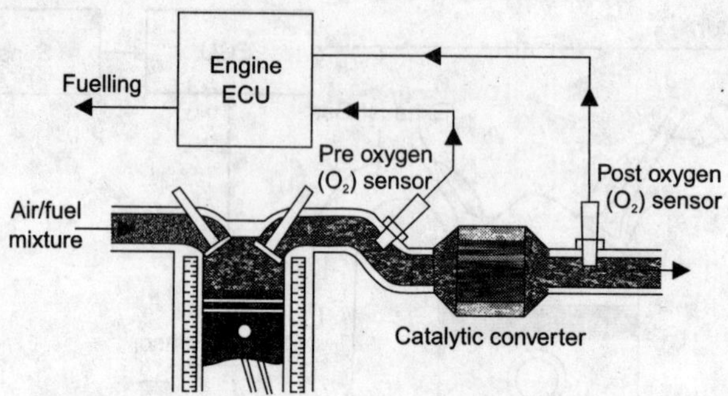


**Fig.11.19:** *Pre- and Post-oxygen sensors*

Some vehicles require two oxygen sensors: one before and one after the catalytic converter. The sensor before the catalytic converter is used by the ECU to adjust the air-fuel ratio. The one after the catalytic converter is used by the ECU primarily to determine catalytic converter efficiency.

## ZIRCONIA OXYGEN SENSOR

This type of oxygen sensor has been in service the longest time. It is made of zirconia (zirconium dioxide), platinum electrodes, and a heater. The oxygen sensor generates a voltage signal based on the amount of oxygen in the exhaust compared to the atmospheric oxygen. The zirconia element has one side exposed to the exhaust stream, the other side open to the atmosphere. Each side has a platinum electrode attached to zirconium dioxide element. The platinum electrodes conduct the voltage generated.

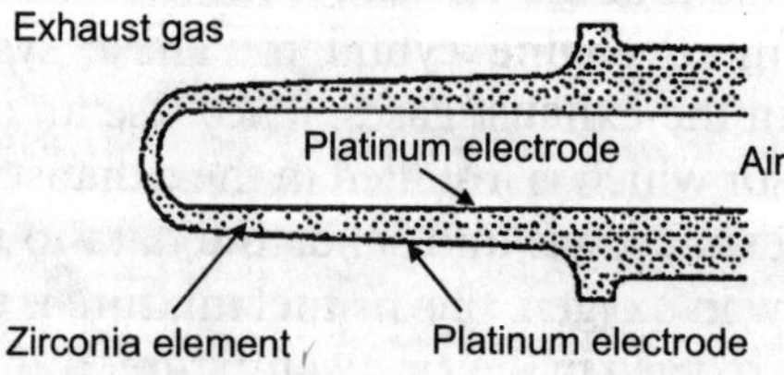


**Fig.11.20:** *Zirconia oxygen sensor*

## Operation

When exhaust oxygen content is high, oxygen sensor voltage output is low. When exhaust oxygen content is low, oxygen sensor voltage output is high. The greater the difference in oxygen content between the exhaust stream and atmosphere, the higher the voltage signal.

| Exhaust Oxygen Content | Oxygen sensor output | Air/Fuel ratio |
|---|---|---|
| Low | High, >0.45 V | Rich |
| High | Low, <0.45 V | Lean |

From the oxygen content, the ECU can determine mixture strength (rich or lean) and adjusts the fuel mixture accordingly. A rich mixture consumes nearly all the oxygen, so the voltage signal is high (>0.45 V). A lean mixture has more available oxygen after combustion than a rich mixture, so the voltage signal is low (<0.45 V). At the stoichiometric air-fuel ratio (14.7:1), oxygen sensor voltage output is approximately 0.45 V.

## Voltage range

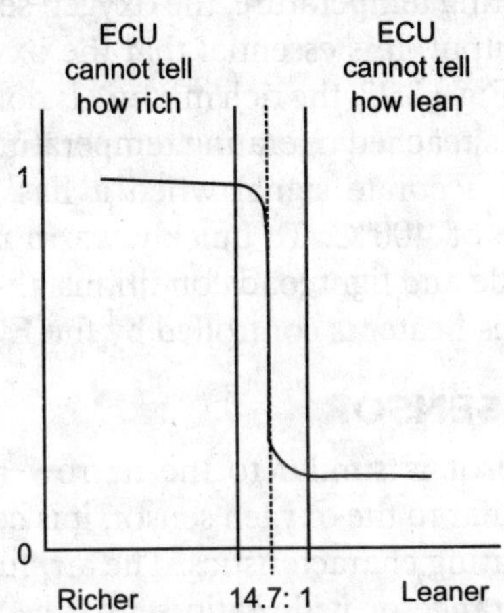


**Fig.11.21:** *Oxygen sensor voltage range*

The oxygen sensor is unable to detect how rich or lean the air-fuel mixture is. Small changes in the air-fuel ratio from the stochiometric point radically change the voltage signal. This type of oxygen sensor is sometimes referred to as a *narrow range sensor* because it cannot detect the small changes in the exhaust oxygen content produced by changes in the air-fuel mixture. The ECU will continuously add and subtract fuel producing a rich/lean cycle.

**Sensor output Vs Temperature**

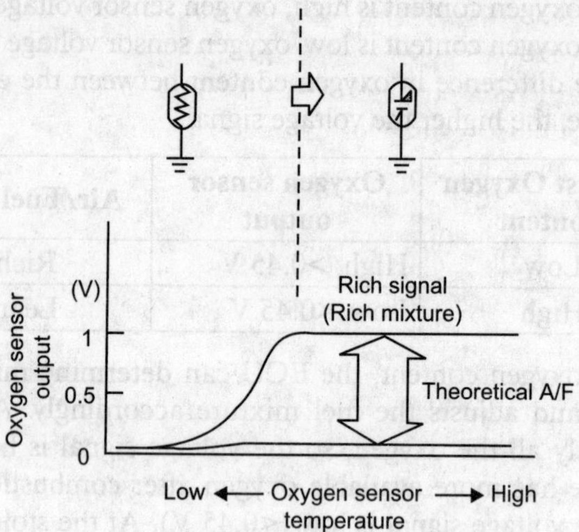


**Fig.11.22:** *Oxygen sensor Vs Temperature*

When cold, the oxygen sensor acts as a resistor until it reaches operating temperature. At operating temperature, the oxygen sensor acts as a battery. For accurate signal output, it is essential that the oxygen sensor is kept at high temperatures. In Fig.11.22, the rich mixture is not accurately measured until oxygen sensor has reached operating temperature. The oxygen sensor will only generate an accurate signal when it has reached a minimum operating temperature of 400°C. To quickly warm up the oxygen sensor and to keep it hot at idle and light load conditions, the oxygen sensor has a heater built into it. This heater is controlled by the ECU.

## AIR/FUEL RATIO SENSOR

The air/fuel ratio sensor is similar to the narrow range oxygen sensor. Though it appears similar to the oxygen sensor, it is constructed differently and has different operating characteristics. The air/fuel ratio sensor is also referred to as a wide range or wide ratio sensor because of its ability to detect air/fuel ratios over a wide range.

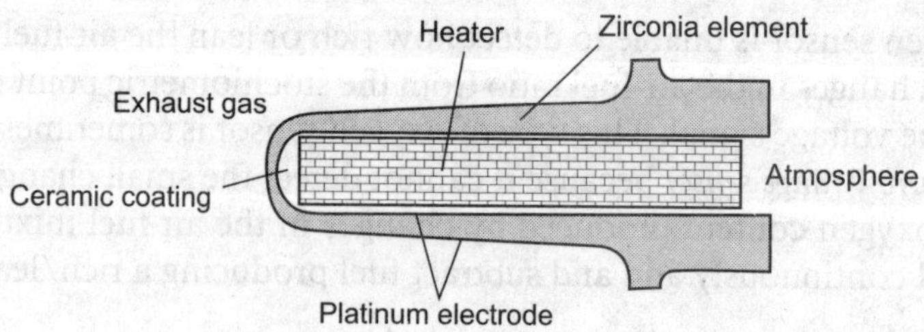


**Fig.11.23:** *Air/Fuel ratio sensor*

Also, the voltage signal is proportional to the change in the air/fuel mixture. This allows the ECU to more accurately judge the exact air/fuel ratio under a wide variety of conditions and quickly adjust the amount of fuel to the stoichiometric point. This type of rapid correction is not possible with the narrow range oxygen sensor. With an air/fuel sensor, the ECU does not follow a rich lean cycle.

The advantage of using the air/fuel sensor is that the ECU can more accurately meter the fuel for reducing emissions. To accomplish this, the air/fuel sensor:

- operates at approximately 650°C, much hotter than the oxygen sensor 400°C.
- changes its current (amperage) output in relation to the amount of oxygen in the exhaust stream.

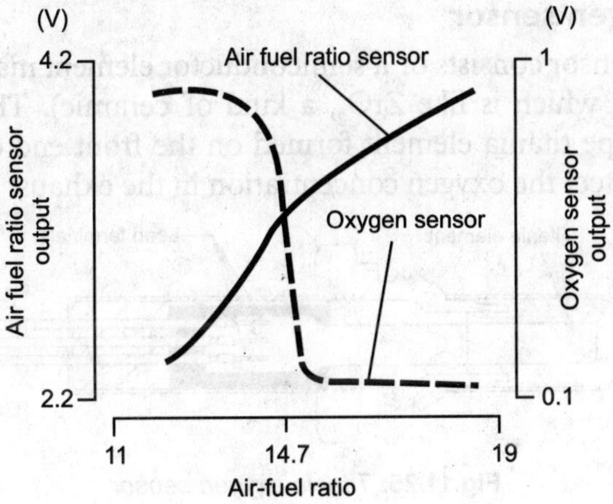


**Fig.11.24:** *Air/fuel ratio Vs Sensor output*

## Operation

A detection circuit in the ECU detects the change and strength of current flow and puts out a voltage signal relatively proportional to exhaust oxygen content. The air/fuel sensor voltage output is the opposite of what happens in the narrow range oxygen sensor. Voltage output through the detection circuit increases as the mixture gets leaner. The air/fuel sensor is designed so that at stoichiometry, there is no current flow and the voltage put out by the detection circuit is 3.3 Volts. A rich mixture, which leaves very little oxygen in the exhaust stream, produces a negative current flow. The detection circuit will produce a voltage below 3.3 Volts. A lean mixture, which has more oxygen in the exhaust stream, produces a positive current flow. The detection circuit will now produce a voltage signal above 3.3 Volts.

| Exhaust oxygen content | Current flow direction | Voltage signal | Air/Fuel mixture |
|---|---|---|---|
| Low | Negative | < 3.3 V | Rich |
| Stoichiometric | 0 | 3.3 V | 14.7:1 |
| High | Positive | > 3.3 V | Lean |

The detection circuit is always measuring the direction and how much current is being produced. The result is that the ECU knows exactly how rich or lean the mixture is and can adjust the fuel mixture much faster than the oxygen sensor based fuel control system. Therefore, there is no cycling that is normal for a narrow range oxygen sensor system. Instead, air/fuel sensor output is more even and usually around 3.3 Volts.

## Titania oxygen sensor

This oxygen sensor consists of a semiconductor element made of titanium dioxide ($TiO_2$, which is like $ZrO_2$, a kind of ceramic). This sensor uses a thick film type titania element formed on the front end of a laminated substrate to detect the oxygen concentration in the exhaust gas.

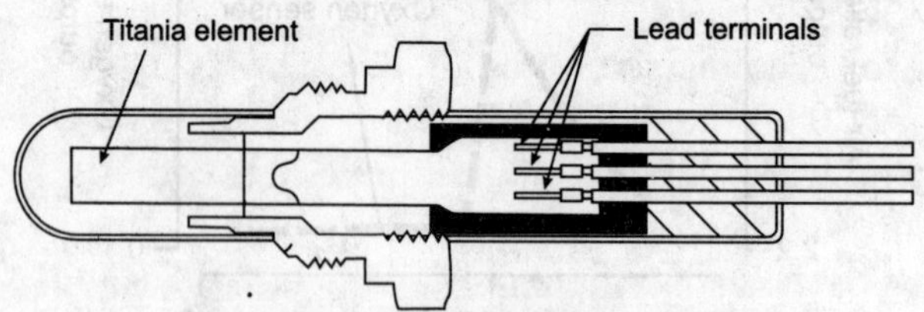


**Fig.11.25:** *Titania oxygen sensor*

## Operation

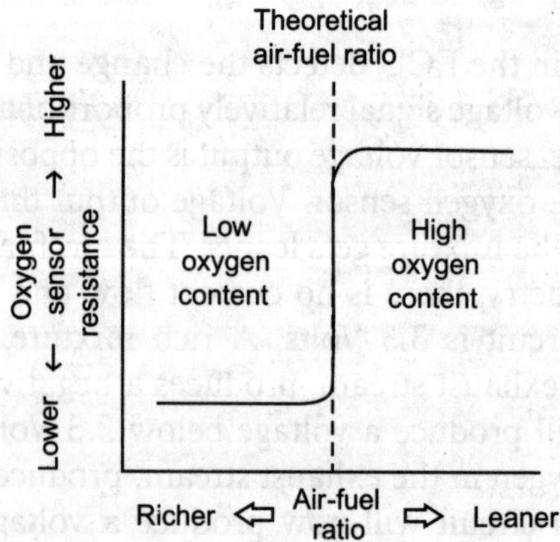


**Fig.11.26:** *Sensor resistance Vs Air-fuel ratio*

The properties of titania are such that its resistance changes in accordance with the oxygen concentration of the exhaust gas. This resistance changes abruptly at the boundary between a lean and a rich theoretical air/fuel ratio, as shown in the Fig.11.26. The resistance of titania also changes greatly in response to changes in temperature. A heater is, thus built into the laminated substrate to keep the temperature of the element constant.

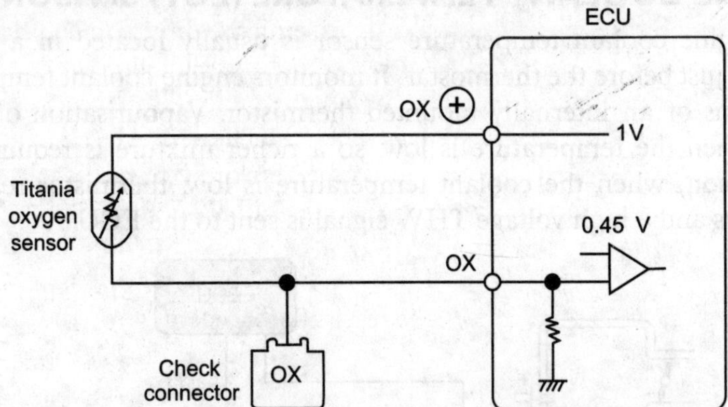


**Fig.11.27:** *Titania sensor circuit*

This sensor is connected to the ECU as shown in the circuit diagram (Fig.11.24). A 1.0 V potential is supplied at all times to the OX positive (+) terminal by the ECU. The ECU has a built-in comparator that compares the voltage drop at the OX terminal (due to the change in resistance of the titania) to a reference voltage (0.45 V). If the result shows that the OX voltage is greater than 0.45 V (oxygen sensor resistance is low), the ECU judges that the air-fuel ratio is rich. If the OX voltage is lower than 0.45 V (oxygen sensor resistance high), it judges that the air-fuel ratio is lean.

## TEMPERATURE SENSOR

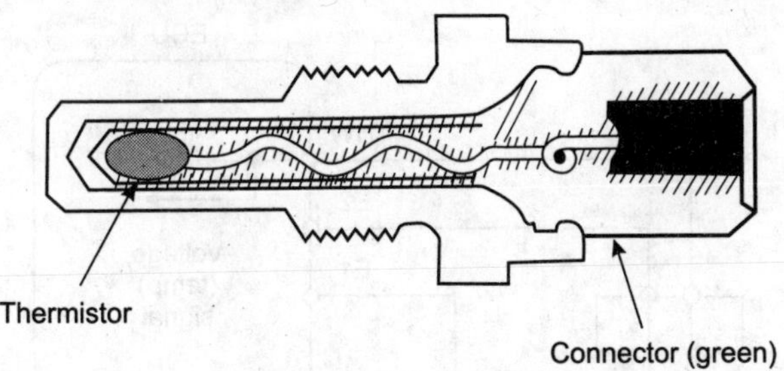


**Fig.11.28:** *Temperature sensor*

The ECU needs to adjust a variety of systems based on temperatures. It is critical for proper operation of these systems that the engine reach operating temperature and the temperature is accurately signalled to the ECU. For example, for the proper amount of fuel to be injected the ECU must know the correct engine temperature. Temperature sensors measure engine coolant temperature (ECT), intake air temperature (IAT) and exhaust recirculation gases (EGR).

## ENGINE COOLANT TEMPERATURE (ECT) SENSOR

The engine coolant temperature sensor is usually located in a coolant passage just before the thermostat. It monitors engine coolant temperature by means of an internally mounted thermistor. Vapourisation of fuel is poor when the temperature is low, so a richer mixture is required. For this reason, when the coolant temperature is low, thermistor resistance increases and a high voltage THW signal is sent to the ECU.

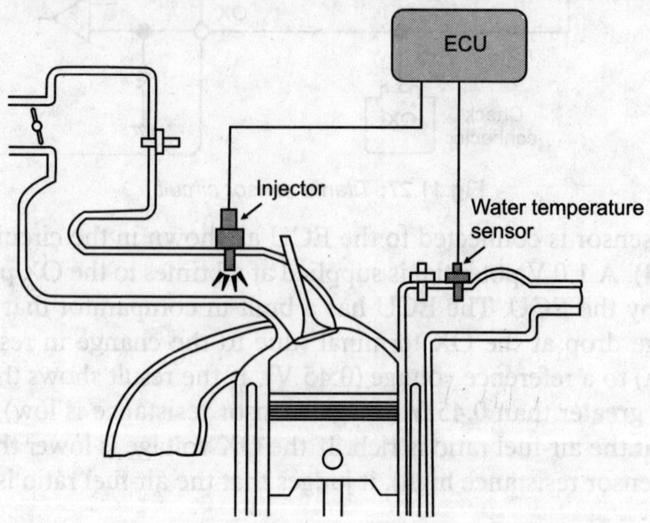


**Fig.11.29:** *Engine coolant (water) temperature sensor*

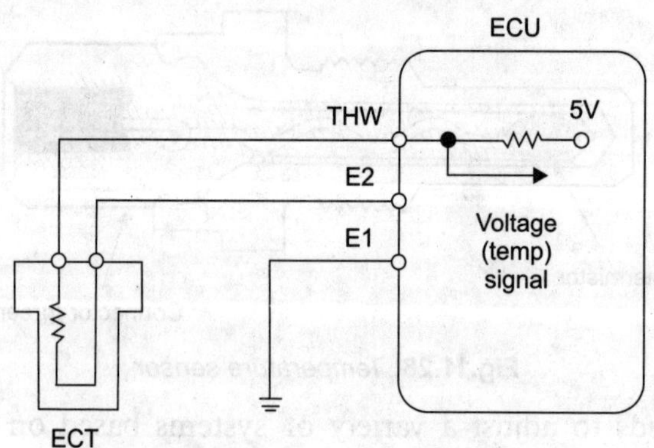


**Fig.11.30:** *Circuit of temperature sensor*

Based on this signal, the ECU increases fuel injection volume to improve drivability during cold engine operation. Conversely, when the coolant temperature is high, a low voltage THW signal is sent to the ECU, which decreases fuel injection volume.

## INTAKE AIR TEMPERATURE (IAT) SENSOR

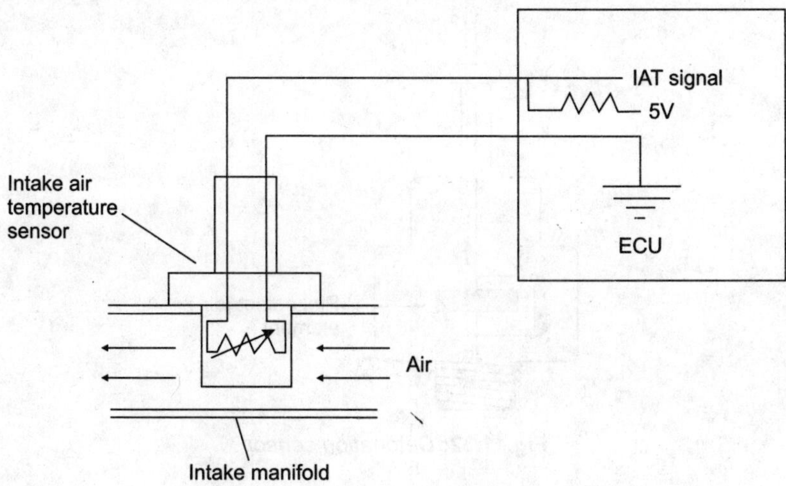


**Fig.11.31:** *Intake air temperature sensor*

The intake air temperature (IAT) sensor detects the temperature of the incoming air stream. The IAT can be located on the air cleaner or on an intake manifold. Just like the coolant temp sensor, the intake air temp sensor is a variable resistor. This particular sensor has a resistance of 355 ohms at 20° C and just 475 ohms at 70° C. It sure on a hot day the car intake air temperature would be 70°C.

It detects ambient temperature on a cold start and intake air temperature as the engine heats up the incoming air.

> One strategy the ECU uses to determine a cold engine start is by comparing the ECT and IAT signals. If both are within 8°C of each other, the ECU assumes it is a cold start.

## DETONATION SENSOR

A knocking combustion process leads to an extremely sharp rise in cylinder pressure and produces pressure oscillation. Depending on its intensity the piston and crank components may be overloaded or engine may overheat and lead to serious mechanical damage. The knock sensor feeds the information about engine knocking to an ECU which retards the ignition timing to reduce the knock detected.

The knock sensor is like a microphone listening for the sounds of knocking (detonation). It is designed to separate out the special noise which means that knocking is occurring. Many electronic fuel injection (EFI) engines run ignition timing very close to knocking, meaning that this sensor is vital if engine damage isn't to occur on a bad batch of fuel or on a very hot day.

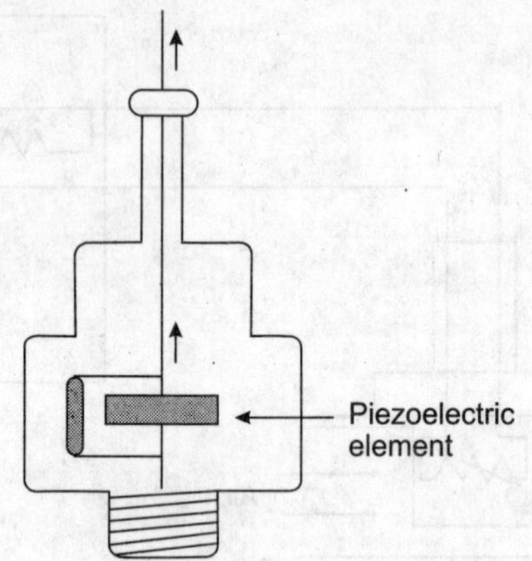


**Fig.11.32:** *Detonation sensor*

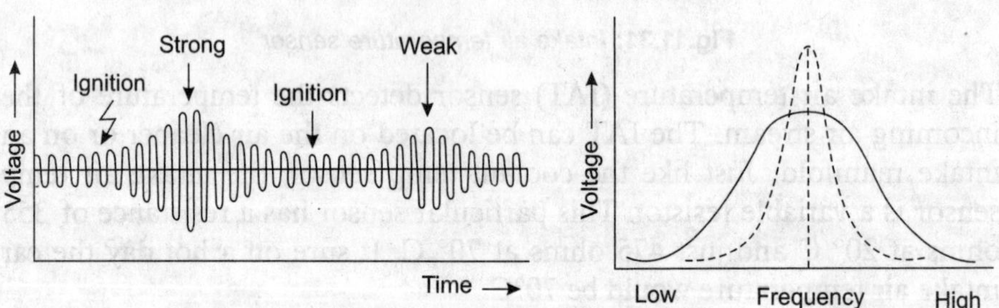


**Fig.11.33:** *Knock signal*

The knock sensor detects engine knock and sends a voltage signal to the ECU. The ECU uses the knock sensor signal to control ignition timing. Engine knock occurs within a specific frequency range. The knock sensor, threaded into the engine block, cylinder head or intake manifold, is tuned to detect that frequency.

Inside the knock sensor is a piezoelectric element. Piezoelectric elements generate a voltage when pressure or a vibration is applied to them. The piezoelectric element in the knock sensor is tuned to the engine knock frequency.

The vibrations from engine knocking vibrate the piezoelectric element generating a voltage. The voltage output from the knock sensor is highest at this time.

## ELECTRONIC CONTROL UNIT (A MICROCOMPUTER)

A microcomputer is a device which receives information, processes it, and makes decisions based on a set of program instructions. The microcomputer exercises control over the output actuators to carry out these instructions.

With the ability to process information so rapidly, the modern ECU is capable of carrying out its programmed instructions with extreme accuracy. Engine management can address virtually every condition the engine will encounter so that for any engine condition, the ECU will deliver optimum fuel and spark.

## ACTUATORS

The computer uses the sensor data to control different systems on a vehicle through the use of actuators. An actuator is an electromechanical device such as a relay, solenoid or motor. Actuators can adjust engine idle speed, change suspension height or regulate the fuel metered into the engine.

Actuators that could be fitted to an engine management system are stepper motor, solenoid, etc.

### STEPPER MOTOR

Stepper motors are commonly used to enable the ECU to control idle speed. In most fuel injection systems, the stepper motor controls an idle air bypass built into the throttle body.

Essentially, stepper motors are digital actuators; in other words, they are either on or off. They move in fixed increments in both directions, and can have over 120 steps of motion.

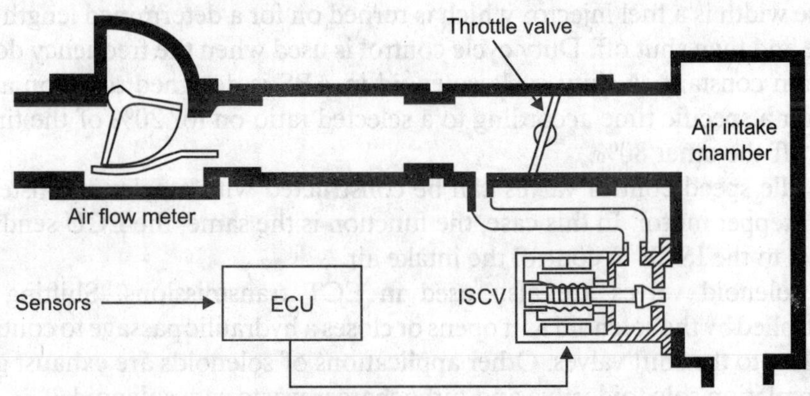


**Fig.11.34:** *Stepper motor*

In the air intake chamber a stepper motor is built into the idle speed control valve (ISCV) where it rotates a valve shaft either in or out. This in turn increases or decreases the clearance between the valve and the valve seat, thereby regulating the amount of air allowed to pass through. The

ISCV stepper motor allows 125 possible valve opening positions.

## SOLENOIDS

Like stepper motors, solenoids are digital actuators. One terminal is attached to battery voltage while the other is attached to the computer which opens and closes the ground circuit as needed. When energised, the solenoid may extend a plunger or armature to control functions such as vacuum flow to various emission-related systems or fuel injection. Most actuators are solenoids.

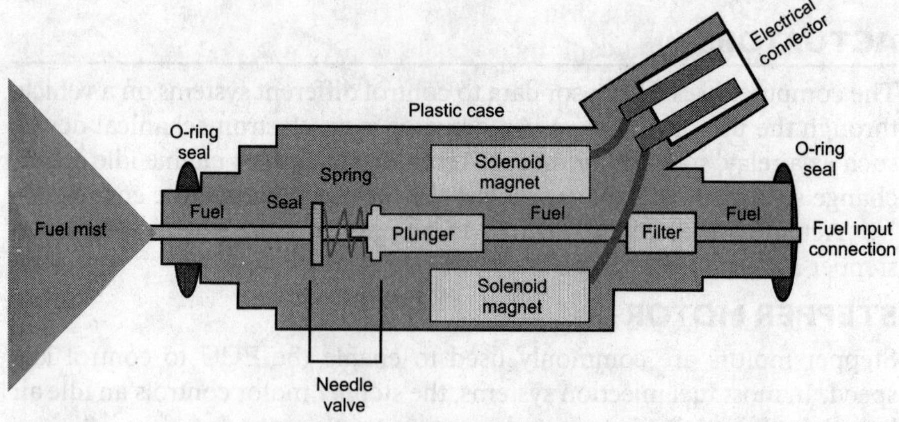


**Fig.11.35:** *Fuel injector solenoid*

Solenoids are controlled two ways: pulse width or duty cycle. Pulse width control is used when the frequency is not consistent. An example of pulse width is a fuel injector which is turned on for a determined length of time and then shut off. Duty cycle control is used when the frequency does remain constant. A duty cycle solenoid in ABS is designed to be on and off for a specific time according to a selected ratio-on for 20% of the time and off the other 80%.

Idle speed control valves can be constructed with a solenoid instead of a stepper motor. In this case, the function is the same; the ECU sends a signal to the ISCV to control the intake air.

Solenoid valves are also used in ECT transmissions. Shifting is controlled by the solenoid as it opens or closes a hydraulic passage to control oil flow to the shift valves. Other applications of solenoids are exhaust gas recirculation solenoid valve and turbocharger waste gate solenoid

## FEEDBACK CARBURETOR

Feedback carburetors were used in Chevy celebrity and some Toyota cars. An electronic feedback carburetor is a computer controlled carburetor, which provides better combustion by improved control of the air-fuel mixture, resulting in improved emission control. The feedback carburetor was introduced following the development of the 3-way catalytic converter. A three-way converter not only oxidizes hydrocarbon (HC) and carbon monoxide (CO), but also chemically reduces oxides of nitrogen (NOx). However, for the 3-way catalyst to work efficiently, the air-fuel mixture must be maintained very close to a 14.7:1 ratio. If the air-fuel mixture is too lean, NOx is not converted efficiently. If the mixture is too rich, HC and CO do not oxidize efficiently.

## Working

An oxygen sensor senses the amount of oxygen present in the exhaust stream. This signal sends to the computer ECU. The ECU then controls the electric bleed control valve (EBCV) to open or close to adjust air-fuel ratio in accordance with existing driving conditions.

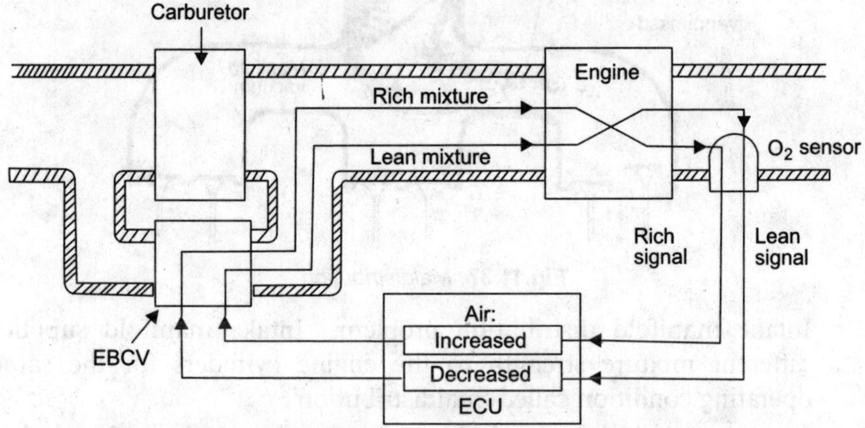


**Fig.11.36:** *Feedback carbuetor*

When the solenoid is on, an EBCV opens. This causes a lean mixture. When the solenoid is off, the mixture becomes rich as the EBCV closes. By maintaining optimal air-fuel mixture according to the engine operating conditions, this carburetor reduces HC, CO and NOx. This also improves drivability and fuel economy.

## ELECTRONIC GASOLINE INJECTION SYSTEM

In trying to keep up with emissions and fuel efficiency laws, the fuel system used in modern cars has changed a lot over the years. Fuel injection is not a new development. Fuel injection has been around since the 1950s, and electronic fuel injection was used widely on European cars starting around 1980. Now, almost all the cars sold in all over world have fuel injection systems.

## PROBLEMS OF CARBURETOR

There are practical difficulties faced in the carburetors for switching over to electronic fuel injection system. A carburetor is a mechanical device. It is neither totally accurate nor particularly fast at responding to changing engine needs. The intake manifold is also a mechanical device and usually the following problems occur.

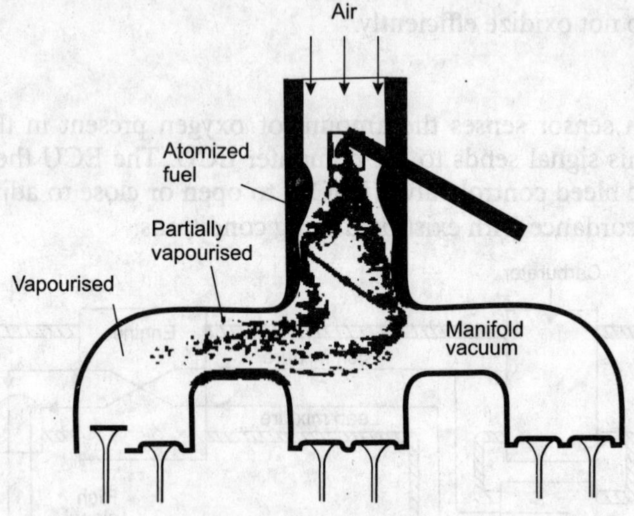


**Fig.11.37:** *Maldistribution*

- Intake manifold distribution problems. Intake manifold supplies different mixture strength to the engine cylinders for the same operating condition called 'maldistribution'.
- Intake manifold is to be designed for the smooth flow of both fuel and air, which is very difficult.

The result, when these two devices are used together, is an engine that does not run efficiently at low and high speeds.

## ADVANTAGES OF EFI

If the air-fuel mixture can be precisely controlled and the distribution to the cylinders can be made more efficient, then today's strict emissions standards can be met, while at the same time the engine will run better and consume less fuel. Fuel injection is one of the best ways to get this precise control.

Fuel injection systems offer several major advantages over carburetors.

- Equal air-fuel ratio to all the cylinders
- Accurate air-fuel ratio can be obtained throughout all engine rpm changes

- Better cold starting
- Reduces fuel consumption with no loss of engine performance
- 10% advance of fuel economy over carbureted systems
- Quick response
- Reduce pollution levels
- Precise control of engine operation
- Reduce design consideration in the manufacture of inlet manifold
- Earlier default detection

## TYPES OF ELECTRONIC FUEL INJECTION SYSTEMS

(According to the method used in sensing the volume of intake air)

EFI systems can be classified into the following types, according to the method used in sensing the volume of intake air.

## D-JETRONIC (MANIFOLD PRESSURE CONTROL TYPE)

This type measures the vacuum in the intake manifold and senses the volume of air by its density.

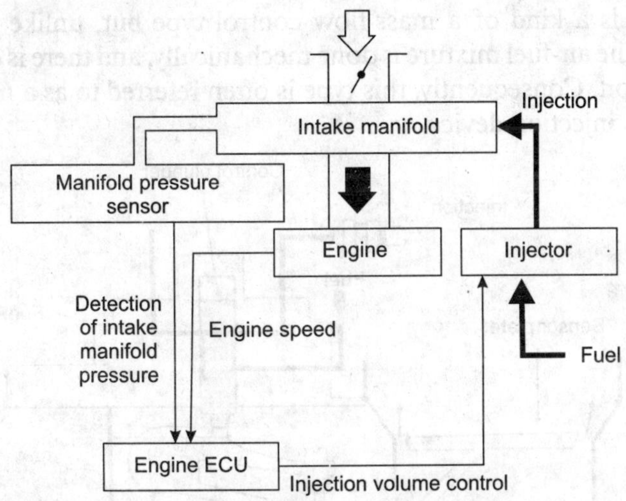


**Fig.11.38:** *D-Jetronic*

D-Jetronic is also referred to as D-EFI. D-Jetronic is a word created from the German word 'Druck' meaning 'pressure' and 'Jetronic', a word coined by Bosch meaning 'injection'.

## L-JETRONIC (AIR FLOW CONTROL TYPE)

This type directly senses the amount of air flowing into the intake manifold by means of an air flow meter.

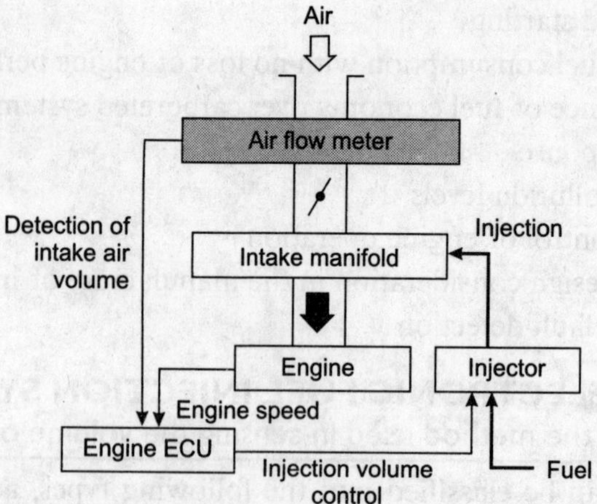


**Fig.11.39:** *L-Jetronic*

L-Jetronic is also referred to as L-Jetronic. The 'L' comes from the German 'Luft' meaning 'air'.

## K-JETRONIC

K-Jetronic is a kind of a mass-flow control type but, unlike L-Jetronic, control of the air-fuel mixture is done mechanically, and there is continuous fuel injection. Consequently, this type is often referred to as a mechanical, continuous injection device.

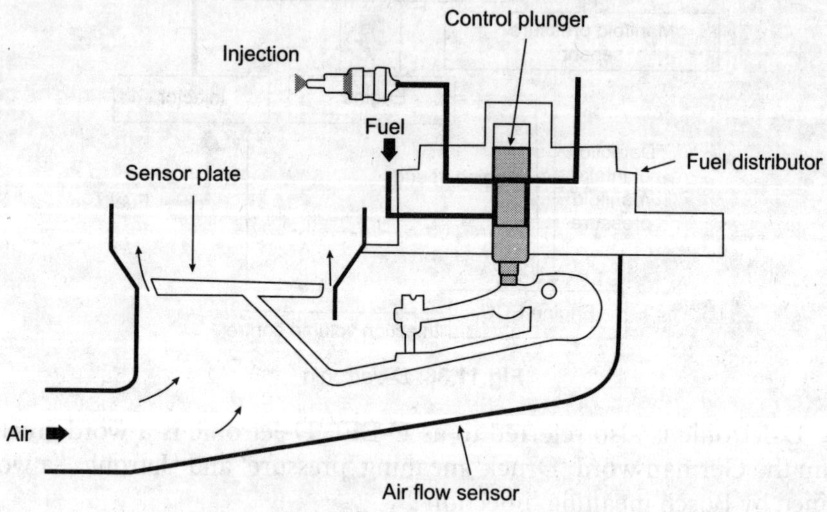


**(At idle & low speeds)**

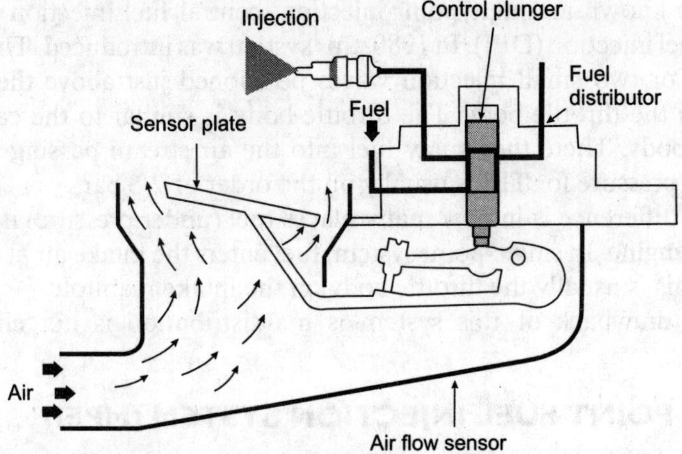


(Under heavy loads & at high speeds)

**Fig.11.40:** *K-Jetronic*

## TYPES OF ELECTRONIC FUEL INJECTION SYSTEMS
(According to the injector location)

There are three basic types of petrol fuel-injection systems, throttle body injection (TBI), multi-point injection (MPFI) and gasoline direct injection (GDI).

## THROTTLE BODY INJECTION SYSTEM (TBI)

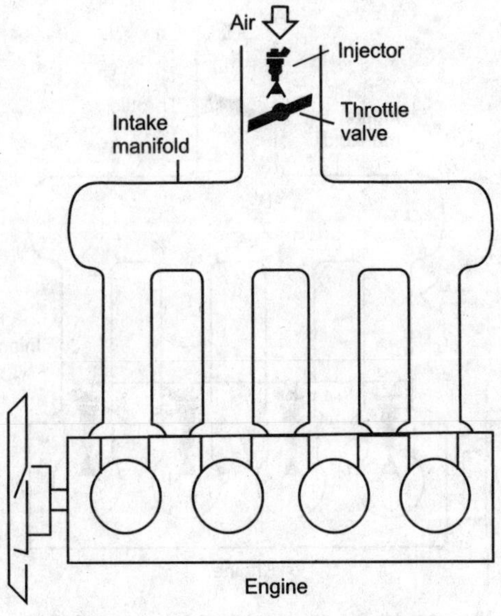


**Fig.11.41:** *Single-point injection*

It is also known as single point injection, central fuel injection (CFI) or digital fuel injection (DFI). In 1980, this system was introduced. The system has one or two small injection valves positioned just above the throttle valves in the throttle body. The throttle body is similar to the carburetor throttle body. There they spray fuel into the air stream passing through. The fuel pressure for TBI is usually on the order of 2-3 bar.

The difference is in how many places fuel (under pressure) is injected into the engine. In single-point system, fuel enters the intake air at only one place. This is usually the throttle body on the intake manifold.

The drawback of this system is maldistribution is not eliminated totally.

## MULTI-POINT FUEL INJECTION SYSTEM (MPFI)

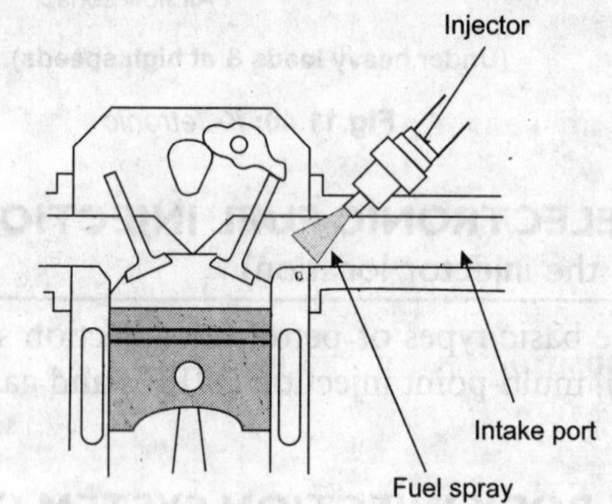


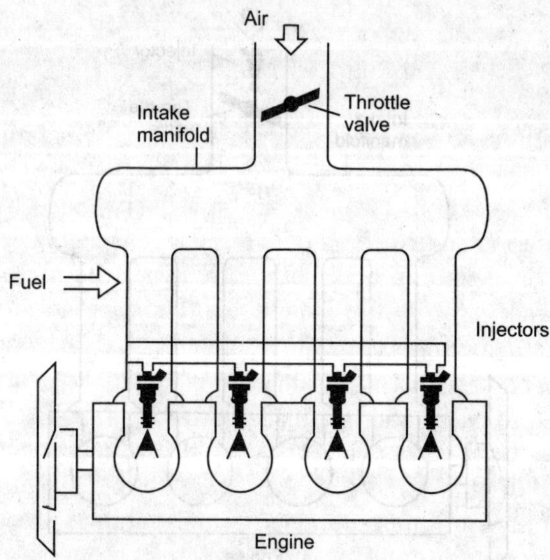


**Fig.11.42:** *Multi-point fuel injection (MPFI)*

In the multi-point system, fuel is injected into the intake air for each cylinder. It is also known as port injection. Multi-point injection systems have a fuel injector for each cylinder. Each injector is placed so that it points at the intake valve for the cylinder. The injectors are connected to fuel rails that are connected through a fuel filter to an electric fuel pump. Only air enters the intake manifold. Then as the air approaches the intake valves, injection valves open to spray fuel into the air stream. The maldistribution problem is completely eliminated with MPFI systems. These systems provide more accurate fuel metering and quicker response. The typical fuel pressure runs between 3-7 bar.

## ENGINE SENSORS IN MPFI SYSTEM

In order to provide the correct amount of fuel for every operating condition, the engine control unit (ECU) has to receive signals from a number of input sensors.

- **Mass airflow sensor**, signals the ECU the mass of air entering the engine.
- **Oxygen sensor**, monitors the amount of oxygen in the exhaust so the ECU can determine how rich or lean the fuel mixture is and make adjustments accordingly.
- **Throttle position sensor**, monitors the throttle valve position (which determines how much air goes into the engine) so the ECU can respond quickly to changes, increasing or decreasing the fuel rate as necessary.
- **Coolant temperature sensor**, allows the ECU to determine when the engine has reached its proper operating temperature.
- **Voltage sensor**, monitors the system voltage in the vehicle so the ECU can raise the idle speed if voltage is dropping (which would indicate a high electrical load).
- **Manifold absolute pressure sensor**, monitors the pressure of the air in the intake manifold. The amount of air being drawn into the engine is a good indication of how much power it is producing. The more air that goes into the engine, the lower the manifold pressure, greater power is the result. Hence this reading is used to gauge how much power is being produced.
- **Engine speed sensor**, monitors engine speed, which is one of the factors used to calculate the pulse width.

### Working principle of MPFI

In MPFI system, a high-pressure electric fuel pump is located in or near the fuel tank. The pump sends fuel from the tank through the fuel-pressure regulator to the injectors at constant high pressure all the time the engine runs. Injector is operated by a small electric solenoid. When the solenoid is connected to the battery, it pulls back on the nozzle needle, opening the valve. Now fuel can spray out of the valve. Various sensors located on the engine detect the operating mode of the engine and signal this condition to the electronic control unit (ECU). It processes the information. Then the ECU computes when and for how long the injectors should be opened. The ECU sends a voltage pulse to the injectors. The voltage pulse causes the injectors to remain open for the required time. To increase engine speed, the pulse width increases so that injectors stay open longer. If less fuel is needed, the pulse width decreases. Therefore the injectors close earlier and engine speed decreases.

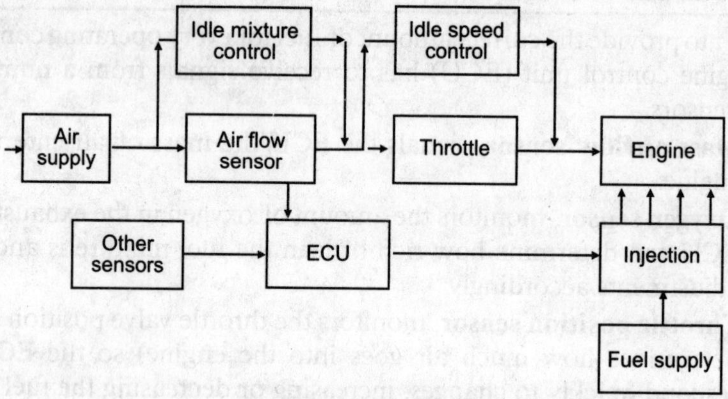


**Fig.11.43:** *MPFI system*

## TYPES OF MPFI

There are three types of MPFI systems: Sequential, Batched and Simultaneous.

### Sequential injection

In sequential injection, injection is timed to coincide with each cylinder's intake stroke. Injectors are driven independently and sequentially by separate driver circuits. Injection is timed to deliver the entire fuel charge just prior to each intake valve opening. This timing method provides optimum engine performance, emissions and fuel economy.

Sequential fuel injection is also called sequential port fuel injection (SPFI) or timed injection. Sequential fuel injection triggers each injector nozzle independently. Timed like spark plugs, they spray the fuel immediately before or as their intake valve opens.

Many modern EFI systems utilize sequential MPFI; however, in newer gasoline engines, direct injection systems are beginning to replace sequential ones.

### Simultaneous injection

Fuel is injected at the same time to all the cylinders in simultaneous injection. All injectors are pulsed simultaneously by a common driver circuit. Injection occurs once per crankshaft revolution just prior to the crankshaft reaching TDC cylinder 1. This means that twice per engine cycle one half of the calculated fuel is delivered by the injectors. This is the simplest and most common injection timing method in use.

### Grouped injection

In group injection, fuel is injected to the cylinders in groups, without precise synchronization to any particular cylinder's intake stroke. Injectors are grouped into pairs. The pairs consist of two consecutive cylinders in the firing order, each pair is driven by a separate driver circuit. Four cylinder engines use two groups, six cylinder engines three groups and the eight cylinder engine uses four groups of injectors.

Injection is timed to deliver fuel immediately preceding the intake stroke for the leading cylinder in the pair. The entire group is pulsed once per engine cycle, delivering the entire calculated charge of fuel. This timing method ensures that fuel does not reluctant to leave behind the intake valve, thereby, reducing emissions, improving fuel economy and throttle response.

## GASOLINE DIRECT INJECTION (GDI)

Gasoline direct-injection engines generate the air/fuel mixture in the combustion chamber. During the suction stroke, only the air flows through the intake valve. The fuel is injected directly into the combustion chamber at high pressure by special fuel injectors. As in a diesel engine, air-fuel mixture formation takes place inside the combustion chamber.

The electric fuel pump delivers fuel to the high-pressure pump at a pre-supply pressure of 3-5 bar. The highly pressurized fuel flows into and is stored in the fuel rail. The high-pressure fuel injectors are mounted on the fuel rail, also known as the "common rail". These injectors are actuated by the engine ECU and spray the fuel into the cylinder combustion chambers. For a GDI engine, fuel pressure is usually on the order of 100-500 bar.

In developing the gasoline direct injection engine, it is aimed to cool the interior of the cylinder as much as possible by promoting fuel vaporization and uniform mixing of atomized fuel and air. This produces a high charging efficiency of the air-fuel mixture and a high compression ratio, which results in significant improvements in both torque and fuel efficiency.

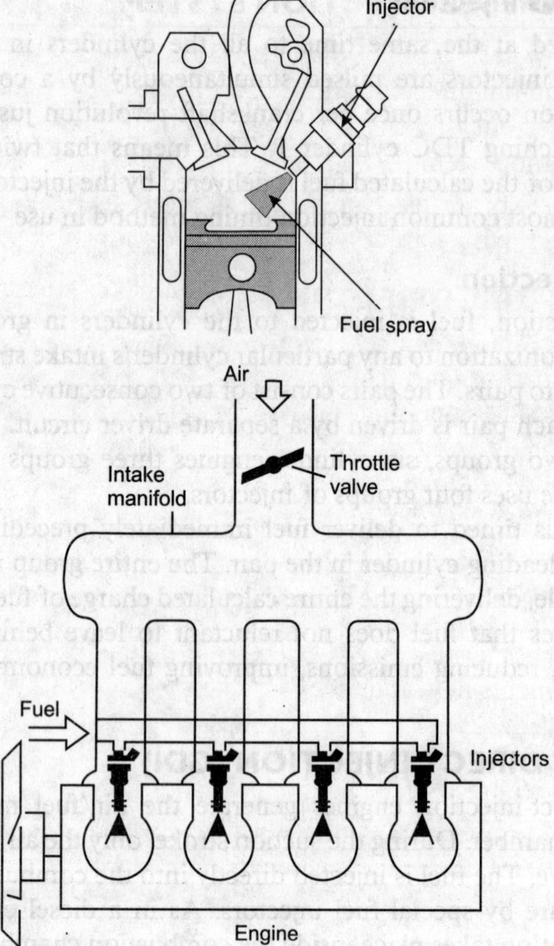


**Fig.11.44:** *Gasoline direct injection*

## Advantages of GDI

1. Fuel is injected from a tiny nozzle into a relatively large cylinder, so it has a high latent heat of vaporization, which efficiently cools the air within (in-cylinder cooling effect).
2. The air temperature in the cylinder decreases, which means:
   - more air may be charged into the combustion chamber, which produces increased torque.
   - the engine is less prone to knocking. This contributes to increased torque, and enables a higher compression ratio.
   - good fuel efficiency.

## MODERN DIESEL INJECTION SYSTEMS

There are three basic types of diesel injection systems:

1. Common rail direct injection (CRDi)
2. Unit injector system
3. Unit pump system

The first two are used on the large diesel engines of sports utility vehicles, trucks, etc.

## COMMON-RAIL DIRECT INJECTION

Conventional diesel engines (non-CRDI engines) are sluggish, noisy and poor in performance compared to a CRDI engine. This technology has turned diesel engines into fuel efficient, clean, high torque and quieter engines. CRDI is an intelligent way of controlling a diesel engine with use of modern computer systems. So it helps to improve the power, performance and reduce harmful emissions from a diesel engine.

### Brand names

CRDI system is also sometimes referred to by many similar or different names. All these systems work on same principles with slight variations and enhancements here and there.

| | | |
|---|---|---|
| CRDe | Common Rail Diesel Engine | Mahindra |
| DICOR | Direct Injection Common Rail Engine | Tata |
| Quadrajet | Multijet Diesel Engine | Tata |
| DDIS | Diesel Direct Injection System | Fiat |
| TDI | Turbocharged Direct Injection | Volkswagen |

### Construction

A common rail diesel fuel injection system consists of a high pressure fuel pump driven by the engine of a vehicle, a thick-walled steel tube known as an accumulator or rail that runs alongside the cylinder head, and electrically operated injectors. The rail creates a reservoir of fuel "common" to all the fuel injectors, hence the name. Common rail diesel injectors are controlled by an Electronic Diesel Control (EDC) unit, which regulates fuel pressure and the timing and duration of fuel injection.

A high-pressure pump increases the fuel pressure in the accumulator up to 1,800 bar. The pressure is set by the engine control unit and is independent of the engine speed and quantity of fuel being injected into any of the cylinders. The fuel is then transferred through rigid pipes to the fuel injectors, which inject the correct amount of fuel into the combustion chambers.

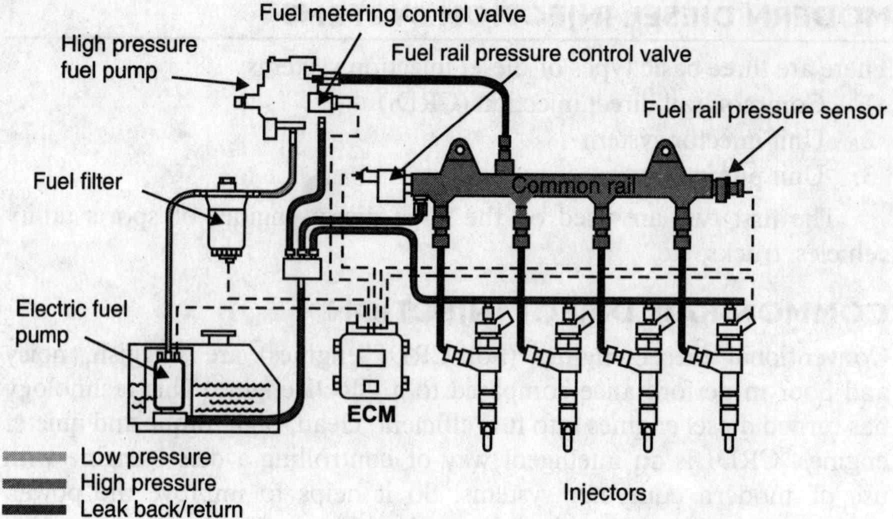


**Fig.11.45:** *Common rail direct injection*

## Operation

To run a CRDI system, the microprocessor works with input from multiple sensors which include throttle position sensor, crank position sensor, pressure sensor, lambda sensor, etc. Based on the input from these sensors, the microprocessor can calculate the precise amount of diesel and the timing when the diesel should be injected inside the cylinder. Using these calculations, the CRDI control system delivers the right amount of diesel at the right time to allow best possible output with least emissions and least possible wastage of fuel.

## Advantages

The electronic diesel control unit precisely meters the amount of fuel injected, and improves atomization of the fuel by controlling the injector pulsations. This results in quieter, more fuel efficient engines, reduced exhaust emissions (upto 30%) and more power output.

CRDI engine fitted cars offer 25% more power than the normal direct injection engine with a superior pickup and torque offering sometimes up to 70% more power than the conventional diesel engines.

## Disadvantages

CRDI engines are at least 25% more costly than the conventional engines. They also require a higher degree of maintenance and spares are not cheap also.

## UNIT INJECTORS

The lines connecting the injection pump to the fuel injector produced problems for early diesel engineers. So in 1905, Carl Weidman got rid of them by combing the injection pump and the injector.

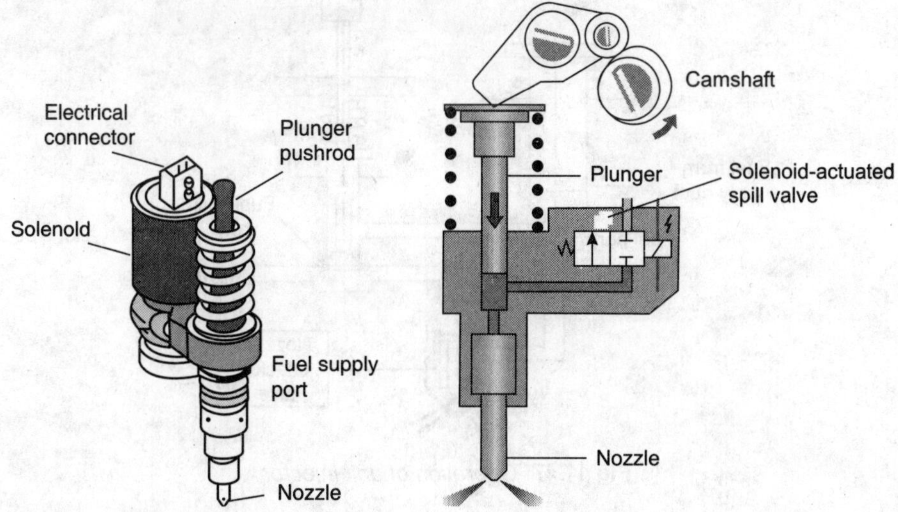


**Fig.11.46:** *Electronic type unit injector*

### Construction

Unit Injector System (UIS) is an integrated direct fuel injection system for diesel engines, combining the injector nozzle and the injection pump in a single component. The pump is usually driven by a shared camshaft. The unit injector is fitted into the engine cylinder head, where the fuel is supplied via integral ducts machined directly into the cylinder head. Each injector has its own pumping element, and in the case of electronic control, a fuel solenoid valve as well.

### Operation

The basic operation can be described as a sequence of four separate phases: the filling phase, the spill phase, the injection phase, and the pressure reduction phase.

The fuel system is divided into the low pressure fuel supply system and the high pressure injection system. A low pressure fuel delivery pump supplies filtered diesel fuel into the cylinder head fuel ducts, and into each injector fuel port of constant stroke pump plunger injector, which is overhead camshaft operated.

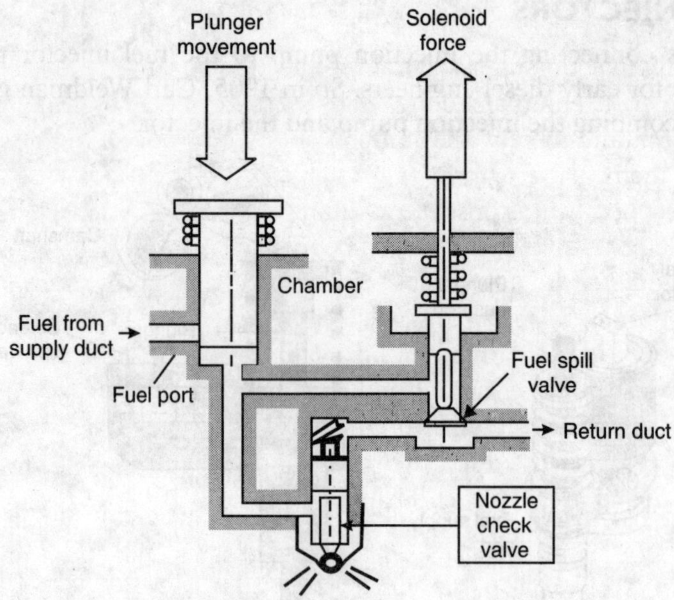


**Fig.11.47:** *Operation of unit injector*

### (i) Fill phase

The constant stroke pump element on the way up draws fuel from the supply duct into the chamber, and as long as electric solenoid valve remains de-energized fuel line is open.

### (ii) Spill phase

The pump element is on the way down, and as long as solenoid valve remains de-energized the fuel line is open and fuel flows in through into the return duct.

### (iii) Injection phase

The pump element is still on the way down, the solenoid is now energized and fuel line is now closed. The fuel cannot pass back into return duct, and is now compressed by the plunger until pressure exceeds specific opening pressure, and the injector nozzle needle lifts, allowing fuel to be injected into the combustion chamber.

### (iv) Pressure reduction phase

The plunger is still on its way down, the engine ECU de-energises the solenoid when required quantity of fuel is delivered, the fuel valve opens, fuel can flow back into return duct, causing pressure drop, which in turn causes the injector nozzle needle to shut, hence no more fuel is injected.

## Advantages

1. As unit injection uses a combined injection pump and nozzle for each cylinder, it eliminates long, high pressure pipelines from pump to injector, and thereby the compressibility and wave effects associated with them.
2. Consequently, it is possible to attain very high injection pressure (up to 2000 bar) so that smoke can be reduced.
3. By reducing the degree of air swirl needed, high pressure improves volumetric efficiency and reduces heat loss to coolant. Therefore, unit injection also offers a reduction in fuel consumption.
4. Allows accurate injection timing, and amount control
5. Servicing is also simplified.
6. The high pressure ducts used in this system are very short, of equal length, and are drilled in the body of the injector. As a result, the rate of delivery of fuel is consistent from cylinder-to-cylinder, chances of leakage is practically eliminated, any wave effects if occur are at very high frequencies and low amplitudes, and injection timing can be more accurate and consistent.

## Application

The most common application for unit injectors are Volkswagens and large diesel engines.

## UNIT PUMP SYSTEM (UPS)

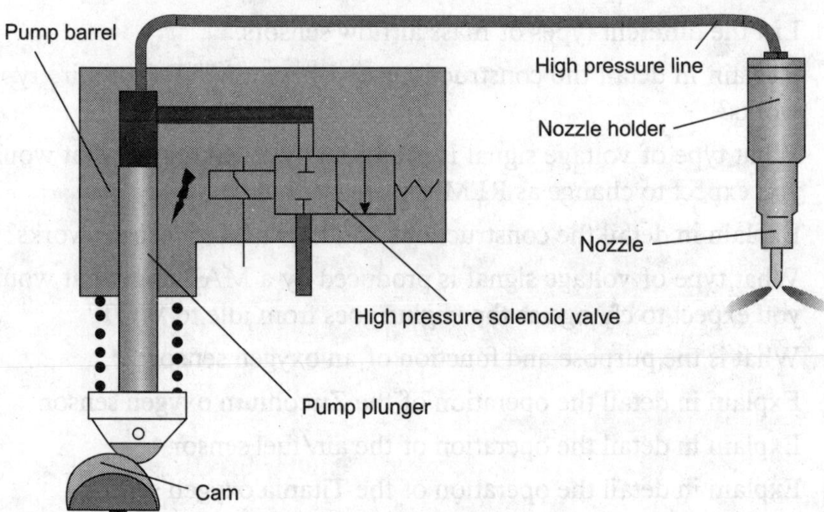


**Fig.11.48:** *Unit pump system*

The unit pump system (UPS) use an individual injection pump mounted on the engine block for each cylinder so it is primarily designed for OHV or cam in the block engines. The fuel pumps are driven by an extra camshaft lobe where each pump unit is connected to the injector via a short precise length high-pressure fuel line.

This system also feature electronically controlled fuel solenoids for precise timing and the injection of fuel quantity is variably adjusted for each cylinder.

## Benefits

- High pressure capability
- Robust design
- Proven reliability
- Low emissions
- Low fuel consumption and less CO2 output
- High torque even at low engine speeds

## Applications

The Delphi diesel electronic unit pump is suitable for heavy duty diesel engines for on- and off-highway applications and premium medium duty diesel engine programs.

## REVIEW QUESTIONS

1. Explain the purpose of a mass airflow sensor?
2. List the different types of mass airflow sensors.
3. Explain in detail the constructions and how a MAF (hot wire type) works?
4. What type of voltage signal is produced by a MAF and what would you expect to change as RPM is increased?
5. Explain in detail the constructions and how a MAP sensor works?
6. What type of voltage signal is produced by a MAP and what would you expect to change as the engine goes from idle to WOT?
7. What is the purpose and function of an oxygen sensor?
8. Explain in detail the operation of the Zirconium oxygen sensor.
9. Explain in detail the operation of the air/fuel sensor.
10. Explain in detail the operation of the Titania oxygen sensor.
11. What is the purpose of a knock sensor?
12. Explain how the ECU uses the knock sensor input signal?

13. Where are knock sensors usually located?
14. Explain the construction of a knock sensor.
15. What are the limitations of carburetors?
16. How does an ECU control the quantity of fuel injected.
17. At what pressure diesel is injected by the CRDi?
18. List out the advantages of EFI system over carburetor?

UNIT

12

# Safety and Warning Systems

## AUTOMOBILE SAFETY

An important design feature of any modern vehicle is its ability to protect the occupants from serious physical harm resulting from a collision. In recent years the regulations governing vehicle construction have been updated, making safety a primary consideration. All vehicles are now required to pass certain technical safety standards.

The terms "active" and "passive" are used in the context of automobile safety. At the most basic level, the terms apply to the vehicle occupant's involvement in the function of the safety device or system. Active safety devices and systems are those which the vehicle occupant must act to make functional, as for example by fastening a seat belt. Passive safety devices and systems are those, such as air bags, which operate without any input or action from the vehicle occupant.

The following important active and passive safety systems are employed in the modern vehicles.

## CRASH PROTECTION DEVICES

Crash protection features provide greater levels of injury protection to drivers and passengers in car crashes, they include:

1. Crumple zones
2. Seat belts
3. Airbags
4. Head rests

## CRUMPLE ZONES

Modern cars protect drivers and passengers in frontal, rear and offset crashes by using crumple zones to absorb crash energy. This means that the car absorbs the impact of the crash, not the driver or passengers.

## SEAT BELT

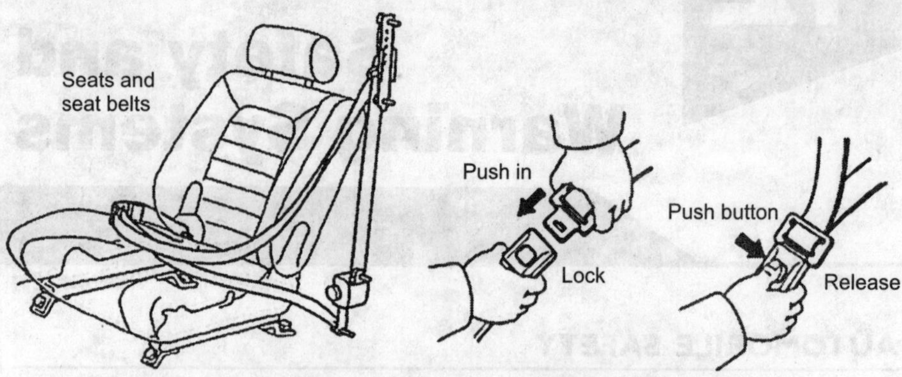


**Fig.12.1:** *Seatbelt*

### Purpose

The purpose of a seat belt is to reduce the risk of injury in an accident. This prevents the wearer from being thrown forwards and possibly being injured during a frontal impact. It is essential to have such an automatic device to restrain the body, because in a collision the time to act is too short-one tenth of a second-and the forces are too great to resist. In a really severe crash a human baby effectively weighs the same as a baby elephant.

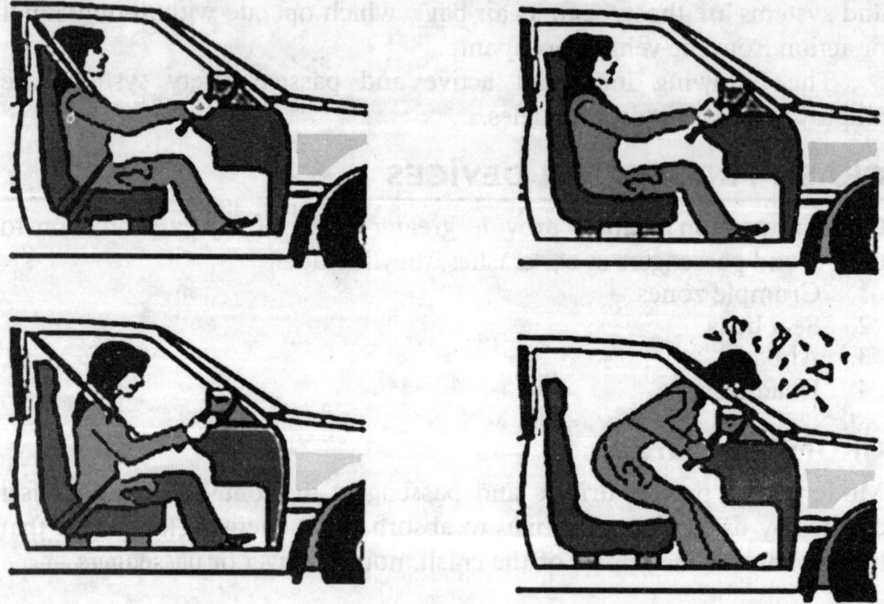


**Fig.12.2:** *Need of a seat belt*

### Operation

A seat-belt is made of a woven fabric usually mounted on a spring-loaded inertia mechanism. Under normal driving conditions the seat-belt can be easily pulled off the spool against spring tension. When the vehicle decelerates rapidly as a result of hard braking or impact, a sensor ball is thrown forwards and by means of the actuator plate prevents the spool from unwinding. The seat-belt is now locked in position. When the vehicle comes to rest the ball falls back and the seat-belt is released.

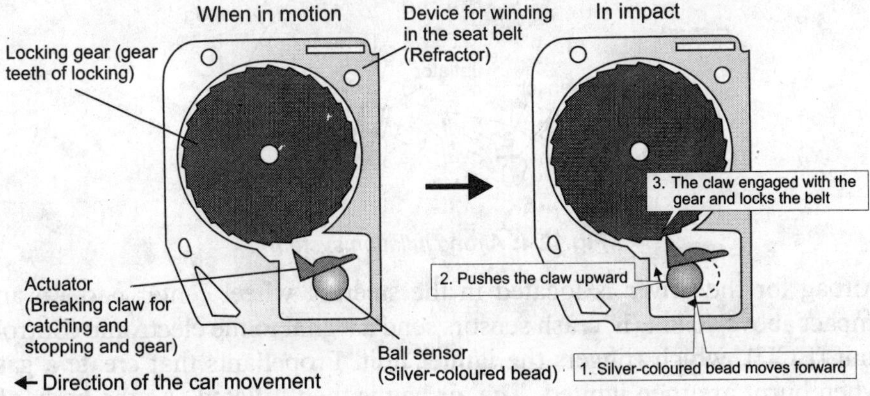


**Fig.12.3:** *Spring loaded inertia mechanism*

## SUPPLEMENTARY RESTRAINT SYSTEM

At speeds in excess of 40 kmph the seat belt alone is no longer adequate in the event of a serious accident. The Supplementary Restraint System (SRS) is basically an air-bag system designed to supplement the seatbelt system to reduce the risk of head and chest injuries in severe collisions.

## COMPONENTS OF SRS

An air bag system consists of the following components.

1. Air bag module
2. Crash sensor
3. Warning lamp

**NOTE:** The Supplemental Restraint System can function only when the ignition switch is in the II position.

## AIRBAG MODULE

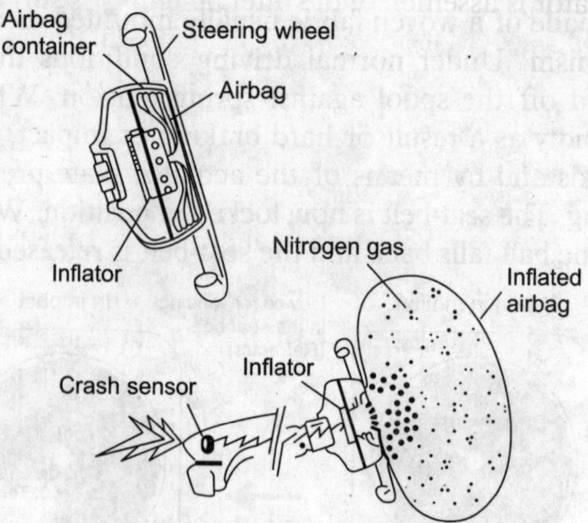


**Fig.12.4:** *Airbag inflation system*

Airbag for the driver is located in the steering wheel center pad. In an impact above 40 km/h, crash sensors send a signal to the electronic control unit (ECU), which triggers the igniter unit. Propellants that create a gas when burnt are then ignited. The air bag is then inflated. As the body of the occupant sinks into the airbag, the gas is forced out of vent holes.

The SRS is designed to function on a one time only basis. In the event the SRS deploys, the airbag module will not function again and must be replaced.

An air bag module has three main parts: the air bag, the inflator, and the propellant.

### Airbag

The airbag is made from a woven nylon fabric and comes in different shapes and sizes depending on specific vehicle requirements. The driver's-side airbag material is manufactured with a heat shield coating to protect the fabric from scorching, especially near the inflator assembly, during deployment. Talcum powder or corn starch is also used to coat the air bag. These substances prevent the fabric from sticking together and make it easier to assemble. Newer silicone and urethane coated airbag materials require little or no heat shield coating, although talcum powder or corn starch will probably still be used as a processing aid.

### Inflator

The inflator canister or body is made from either stamped stainless steel or cast aluminum. Inside the inflator canister is a filter assembly consisting of a stainless steel wire mesh with ceramic material sandwiched in between. When the inflator is assembled, the filter assembly is surrounded by metal foil to maintain a seal that prevents propellant contamination.

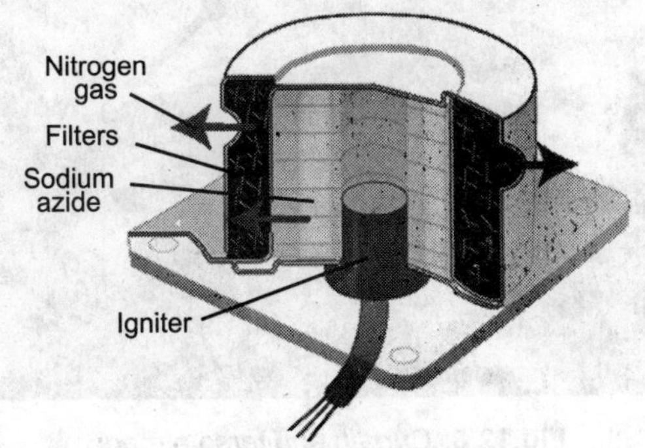


**Fig.12.5:** *Inflator*

### Propellant

The propellant, in the form of black pellets, is primarily sodium azide combined with an oxidizer and is typically located inside the inflator canister between the filter assembly and the igniter. Sodium azide burns so fast like an explosion. It's main product is nitrogen, so that's mostly what fills the bag. Nitrogen is a lot more than an inert gas and is quite harmless.

## OTHER AIRBAGS

Other available airbags in addition to the ones for the front are side airbags, knee airbags, curtain airbags and a torso airbag system.

### Side airbags

Side airbags are an important safety innovation and help protect occupants in side-impact crashes. These often occur at intersections or when a vehicle veers off the road and crashes side-on into a pole, tree or other object. Side airbags also offer protection in roll-over crashes or when two vehicles have an adjacent angle crash.

### Knee airbags

These are extra airbags that are designed to cushion the knees of the driver and front passenger in the event of a crash.

### Curtain airbags

Curtain airbags, also referred to as head airbags, are designed to protect the driver's and passenger's heads in a side crash. When deployed, they form a cushion between the occupant's head and the window. They may also offer protection when vehicles roll over.

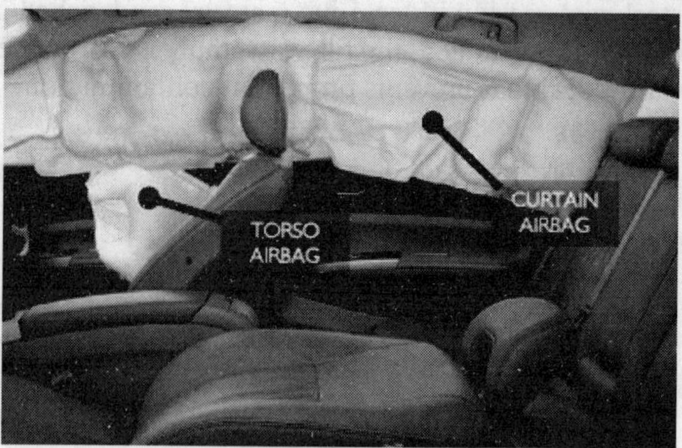


**Fig.12.6:** *Curtain and torso airbags*

## Torso airbags

These protect the body from the shoulders to the hip. They are often fitted for front-seat occupants, but depending on the manufacturer, may also be included for rear seat occupants. They are deployed from the seat or the door to form a cushion between the occupant's torso and the door panels and vehicle body.

## Combination head/torso airbags

This system protects the head and the body from the shoulders down to the hip and is typically provided for front seat occupants only. They are located in a similar position to the torso airbags, but are larger as they protect the head and torso.

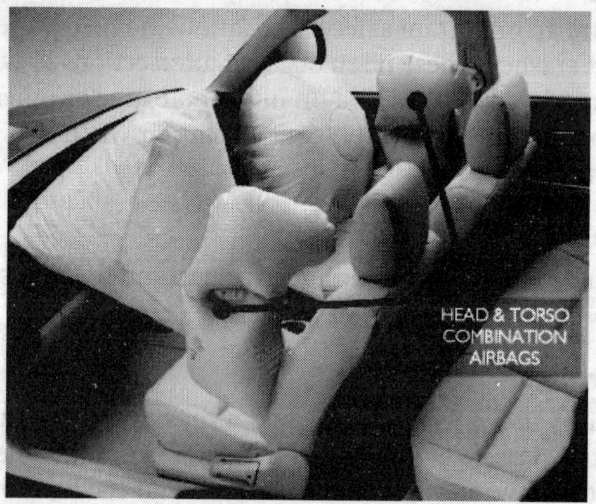


**Fig.12.7:** *Combination head/torso airbags*

## LOCATION OF AIRBAGS

Airbags are an important safety system but, they are usually not visible. The driver side airbag is located in the center of the steering wheel and the passenger side airbag is located inside the dashboard. Side airbags that inflate during side collisions are usually located within the seat. Curtain airbags are installed above the side windows, in the headliner, on both sides. Knee airbags are located in front of the legs of the driver and front seat passenger. They are concealed within the steering column or the lower part of the dashboard and deploy during frontal impacts.

There are also airbags for motorcycles. These are installed in front of the gasoline tank, and, in head-on collisions, this restraint system may help keep the driver on the motorcycle.

## CRASH SENSOR

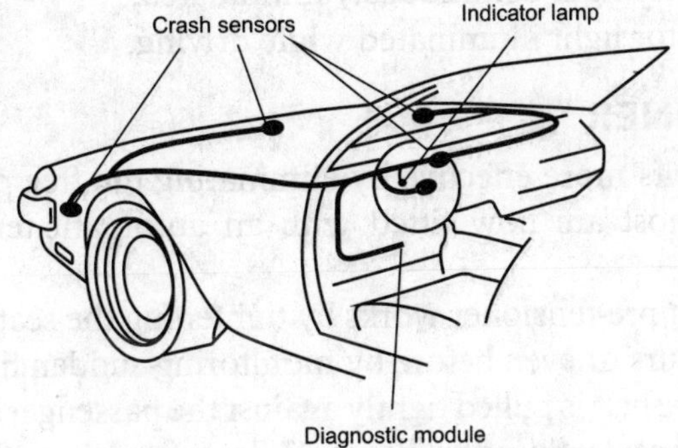


**Fig.12.8:** *Crash sensor*

The crash sensors must be able to detect a collision and convert it into corresponding signals within milliseconds. The airbag control unit receives this signal, deploys the airbags and belt tensioners. Additional, remote crash sensors are also located on the vehicle to assist with crash detection.

The measurement principle employed by the crash sensors is usually based on the effect of inertia. When a vehicle is stopped abruptly by an impact, all bodies or objects which are not firmly fixed to the vehicle will continue to move at the impact speed. The sensors' design enables them to measure this acceleration and relay it to the control unit as usable data.

## WARNING LIGHT

When the SRS warning or indicator light is illuminated on the dash, this means that the airbag control unit sees an issue in the system and has disabled the system. This means that the airbags will not deploy in the event of an accident.

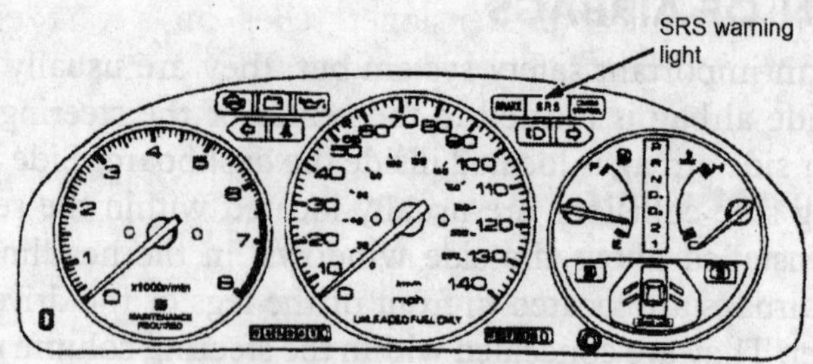


**Fig.12.9:** *SRS warning light*

A malfunction is shown by:

1. The indicator light flashing or flickering.
2. The indicator light failing to illuminate when the ignition switch is first turned to the II position.
3. The indicator light continuously illuminated.
4. The indicator light illuminated while driving.

## PRE-TENSIONER

To make seat-belts more effective by eliminating the free play and stretch of the fabric, most are now fitted with an automatic-tensioner or pre-tensioner.

The seat belt pre-tensioner works by tightening the seat belt as soon as the collision occurs or even before by monitoring sudden hard application of the brakes. The belt is pulled tightly against the passenger and holds them firmly in the soft seats. Some milliseconds later it releases the passenger in a timely manner to hit the airbag at the correct time.

This mechanism, usually housed in the seat-belt reel housing, is commonly operated by a small explosive charge.

## TYPES OF PRE-TENSIONERS

There are three types of seatbelt pretensioners, mechanical, pyrotechnic, and electrical.

Pyrotechnic pre-tensioners are intended to be used once. Once a pre-tensioner has been deployed the entire mechanism including the seatbelts should be replaced.

### Mechanical type

This pre-tensioner is operated mechanically. It is usually connected to the seat belt buckle assembly with a long powerful spring that is compressed in a tube and is kept compressed by a latching mechanism. If an accident occurs the impact causes the latch to release the spring which pulls on the buckle assembly and tightens the seat belt.

The other mechanical pre-tensioner relies on a cam engagement resulting from a sudden acceleration change. The inertia of the mass of the engagement wheel then develops the locking and tensioning forces.

## Electronic pyrotechnic actuated

This pre-tensioner operates with the same ECU as the airbag and a gas generator system in a similar way to the airbag.

A pyrotechnic charge (explosive charge) is used and causes the gas generator to produce a volume of gas (1). This volume of gas produces a pressure which acts on a piston and rack (2) to rotate the pinion (3), which retracts the seat belt. This pre-tensioner is usually located in the B-pillar.

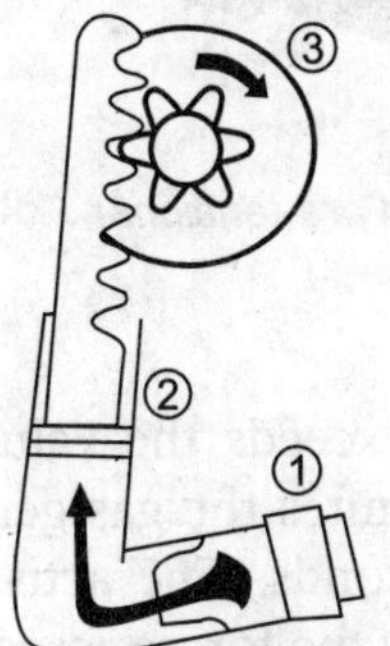


1. The pyrotechnic charge is activated, rapidly filling the chamber with gas.
2. The gas pressure builds up quickly, forcing the piston upwards.
3. The piston rotates the pinion, which retracts the seatbelt.

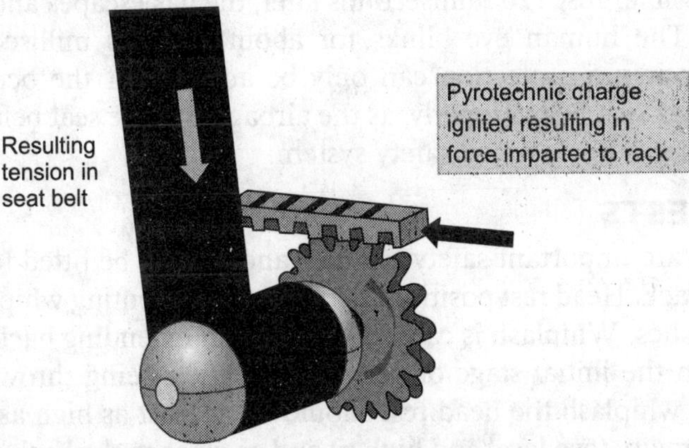


**Fig.12.10:** *Electronic pyrotechnic Pre-tensioner*

## Electric motor actuated

The electric motor actuated system is controlled in the same manner as the pyrotechnic. However, in most common designs a gear connected to the electric motor engages the belt spool. The motor is then controlled to first remove the slack in the belt and provide the torque to the gear equating to the prescribed tension.

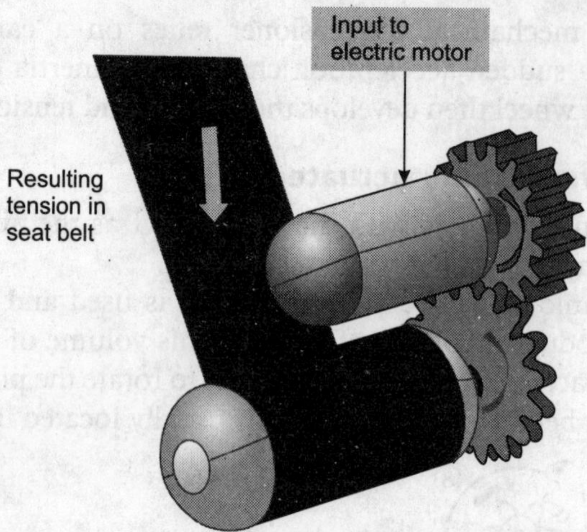


**Fig.12.11:** *Electric motor actuated pre-tensioner*

## WORKING OF SRS

If the crash sensors sense an impact that exceeds the value needed to trigger the airbags, the airbag control unit ignites the gas generator. This inflates the airbags within 35 to 45 milliseconds. The activated airbags cushion the head and upper body, distributing the forces across as large an area as possible. Just 120 milliseconds later, the gas escapes and the airbag collapses. The human eye blinks for about 100-300 milliseconds (0.1-0.3sec). Optimum protection can only be achieved if the occupants are wearing their seat belts properly, as the airbags and the seat belt tensioners together form a finely tuned safety system.

## HEAD RESTS

Head rests are important safety features and should be fitted to all seats - front and back. Head rest position is critical for preventing whiplash in rear impact crashes. Whiplash is caused by the head extending backward from the torso in the initial stage of rear impact, then being thrown forward. To prevent whiplash the head rest should be at least as high as the head's centre of gravity (eye level and higher) and as close to the back of the head as possible.

## CRASH AVOIDANCE FEATURES

Crash avoidance features are safety features that help you avoid being in a crash. They include:

- Electronic Stability Control (ESC)
- Traction control
- ABS

- Auto Emergency Braking (AEB)
- Intelligent Speed Assist (ISA)
- Active Braking Systems
- Handling and stability
- Visibility (seeing and being seen)

## ELECTRONIC STABILITY CONTROL

Electronic Stability Control (ESC) helps drivers to avoid crashes by reducing the danger of skidding, or losing control as a result of sudden directional change, understeer or oversteer during turns on slippery roads, or hydroplaning. ESC becomes active when a driver loses control of their car. It incorporates the functions of anti-lock braking systems and traction control.

Different manufacturers use different names for their electronic stability control systems. Here are a few:

| Other names of ESC | User |
|---|---|
| Electronic Stability Program (ESP) | Hyundai |
| Vehicle Stability Assist (VSA) | Honda |
| Vehicle Stability Control (VSC) | Toyota |
| Vehicle Dynamic Control (VDC) | Nissan |
| Dynamic Stability Control (DSC) | BMW |
| AdvanceTrac with Roll Stability Control | Ford |

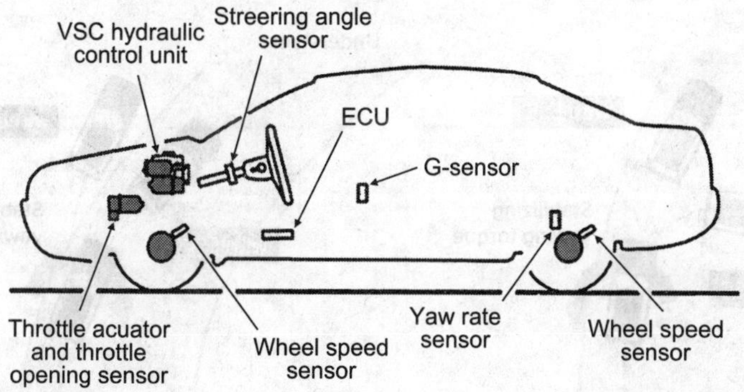


**Fig.12.12:** *Components of ESC*

## SENSORS IN ESC

ESC incorporates yaw rate control into the anti-lock braking system (ABS). Yaw is rotation around the vertical axis i.e. spinning left or right. Anti-lock brakes enable ESC to brake individual wheels. Many ESC systems also incorporate a traction control system, which senses drive-wheel slip under acceleration and individually brakes the slipping wheel or wheels and/ or reduces excess engine power until control is regained. However, ESC achieves a different purpose than ABS or traction control.

The ESC system uses several sensors to determine the driver intention. Other sensors indicate the actual state of the vehicle (response). The most important sensors are:

### Steering wheel angle sensor

Steering wheel angle sensor determines the driver's intended rotation; i.e. where the driver wants to steer.

### Yaw rate sensor

This sensor measures the rotation rate of the car i.e. how much the car is actually turning. The data from the yaw sensor is compared with the data from the steering wheel angle sensor to determine regulating action.

### Lateral acceleration sensor

Lateral acceleration sensor often based on the Hall effect. Measures the lateral acceleration of the vehicle.

### Wheel speed sensor

It measures the wheel speed.

## WORKING OF ESC

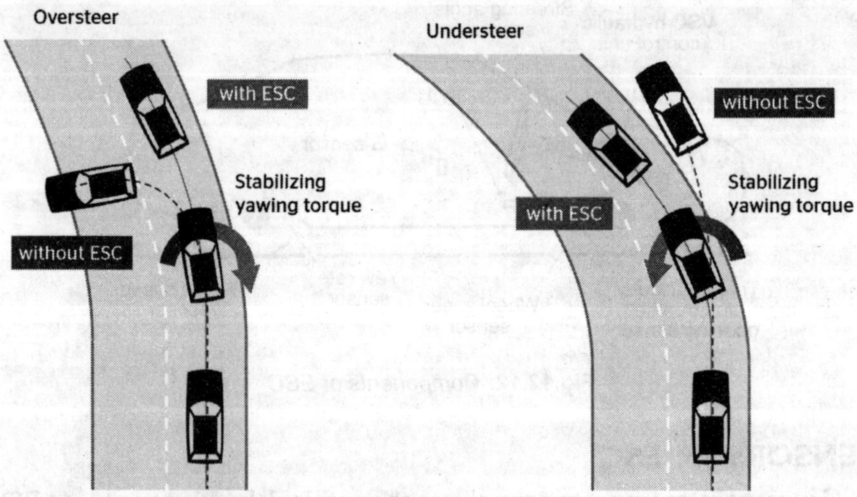


**Fig.12.13:** *Understeer and Oversteer without ESC*

During normal driving, ESC continuously monitoring steering and vehicle direction using sensors to determine which direction the driver wants the car to go, and compares that to which way the car is actually going. If the system senses that a skid is imminent or has already started or in other words, that the car is not going in the direction the driver is telling it to go and it can apply the brakes on individual wheels to bring the car back under control. Because the system can brake individual wheels, whereas the driver can only brake all four wheels at once, ESC can recover from skids that a human driver cannot.

Additionally, the system may reduce engine power or operate the transmission to slow the vehicle down.

## ANTI-LOCK BRAKE SYSTEM (ABS)

ABS (Antilock Braking System) reduces the risk of tyres skidding under heavy braking and allows the driver to maintain steering control of the car and decrease the stopping distance in slippery conditions. ABS only operates under heavy braking or on slippery surfaces.

### Need for ABS

Friction between the tyre and the road is essential for all aspects of vehicle control. The brakes may lock up and control of the vehicle may be lost under emergency braking or when the road surface is slippery.

Ordinary brakes work to slow or stop a vehicle by utilizing two types of resistance. These are the resistance between the brake pads and discs (or brake shoes and drums) and the resistance between tyres and the road surface. Braking can be controlled in a stable manner if the following relationship between the resistance between the road and tyres exists:

Brake system resistance < Resistance between tyres and road

However, if this relationship is reversed, the tyres will lock up and the vehicle will begin skidding:

Brake system resistance > Resistance between tyres and road

As a result, if the front wheels lock up, it will become impossible to steer the car. If the rear wheels lock up, the differences in the coefficients of friction between the left and right sides on the road surface will cause a tail spin to occur.

The ABS controls the hydraulic pressure acting on the wheel cylinders to prevent the wheels from locking up when the brakes are applied on a slippery surface or during panic braking. It also maintains directional stability during braking so that the vehicle can be steered.

## COMPONENTS AND OPERATION

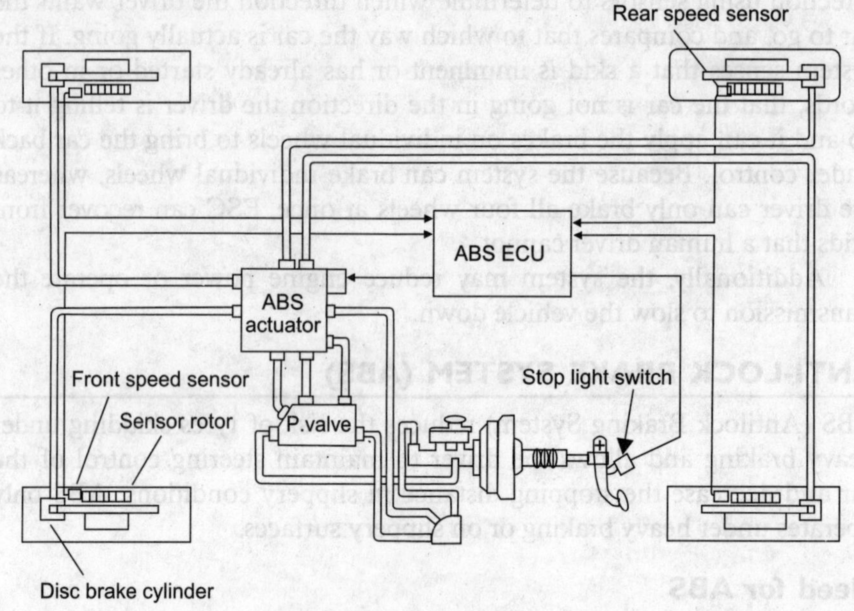


**Fig.12.14:** *ABS braking system*

The heart of the ABS system is an electronic control unit (ECU) which is provided with inputs from sensors and gives outputs to the controls. The sensors are located at each wheel and send pulse generated signals to the ECU. They indicate whether each wheel is revolving and how fast it is turning. The ECU then controls the output of a hydraulic pump assembly. Based on the input signals from the wheel sensors, the ABS system's ECU knows the speed of each wheel. When the brakes are applied, the wheels and the vehicle start to slow down. The ECU monitors the reducing pulses from each of the wheel sensors. If a wheel seems about to lock up, the ECU reduces the pressure at that wheel until the tyre grips again.

## ELECTRONIC BRAKE FORCE DISTRIBUTION

Electronic brake force distribution (EBD) is actually a part of the anti-lock brakes and helps in regulating and applying the braking force needed to each wheel according to the road condition, vehicle's load and its speed. The EBD ensures how much braking force is needed by each wheel and then applies the same in order to achieve the most suitable braking impact. Usually, most of the cars have the front wheels carrying maximum load, therefore, this braking system offers minimal pressure to rear wheels in case of deceleration so that the car does not skid at all.

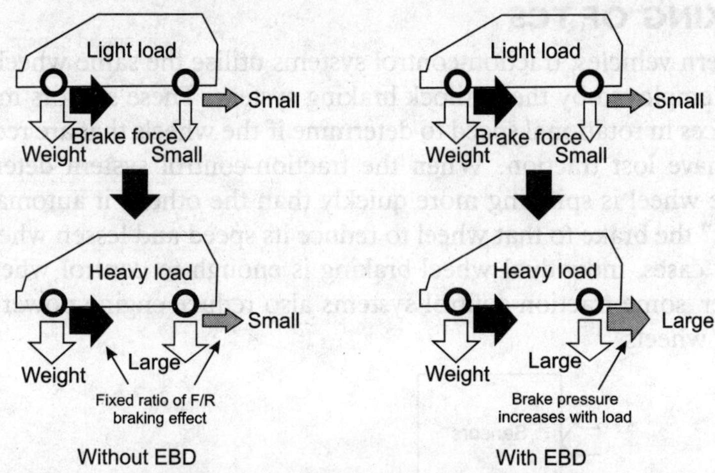


**Fig.12.15:** *Electronic brake force distribution*

A number of vehicles, such as Honda, Volkswagen and Nissan cars have both these features as standard.

### Benefits

EBD helps ensure a stable vehicle posture and braking performance when there is a change in the number of occupants or load condition, and even when the vehicle is cornering and the vehicle center of gravity is likely to move.

## TRACTION CONTROL SYSTEM

Traction control system (TCS) optimises grip and stability of the car on the road during acceleration by measuring wheel rotation. It stops wheel spin by reducing engine power or temporarily applying the brakes to that wheel, allowing the car to accelerate smoothly, even on slippery surfaces.

### Need for TCS

On roads with a low coefficient of friction, such as snowy, icy or wet roads, the drive wheels spin if the vehicle is started or accelerated quickly, resulting in loss of torque and possibly causing vehicle skidding.

Maintaining the torque at a level appropriate to the coefficient of friction at such times is sometimes difficult for the driver. The traction control system lowers the engine's torque, regardless of the driver's intentions, when the wheels start to spin, while at the same time controlling the brake system, thereby reducing the torque transmitted to the road surface to appropriate levels. In this way, stable and rapid starting and acceleration can be achieved.

## WORKING OF TCS

In modern vehicles, traction-control systems utilise the same wheel-speed sensors employed by the antilock braking system. These sensors measure differences in rotational speed to determine if the wheels that are receiving power have lost traction. When the traction-control system determines that one wheel is spinning more quickly than the others, it automatically "pumps" the brake to that wheel to reduce its speed and lessen wheel slip. In most cases, individual wheel braking is enough to control wheel slip. However, some traction-control systems also reduce engine power to the slipping wheels.

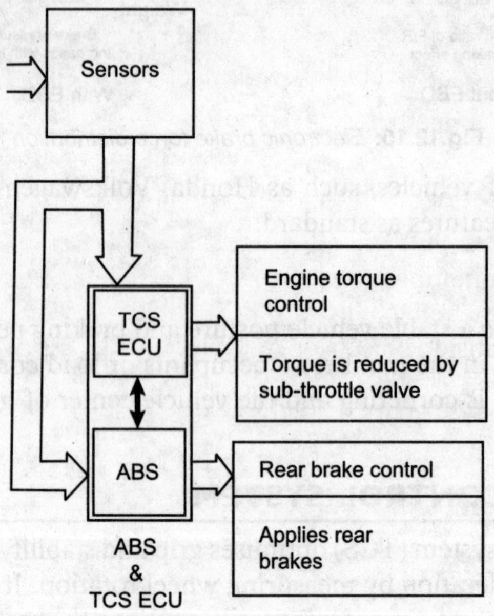


**Fig.12.16:** *Traction control system*

Traction control is basically an anti-lock braking system in reverse. Anti-lock brakes prevent the wheels from locking up when the car is decelerating. Traction control keeps the wheels from locking up or skidding when the car is accelerating.

## AUTO EMERGENCY BRAKING

Auto Emergency Braking (AEB) can:

- Alert the driver to an imminent crash and help them use the maximum braking capacity of the car and
- Apply the brakes independently of the driver if the situation becomes critical

It is important to note that AEB systems are designed to support the driver only in emergency situations and that the driver remains responsible for the vehicle at all times.

AEB systems use sensors, radar, laser or cameras to monitor for risk and detect potential collisions with other vehicles, pedestrians or hazards.

Although AEB systems vary in their functionality, most systems will provide a warning (audible and/or visual) to the driver. AEB systems will also intervene and brake the vehicle automatically if the driver does not respond. Some systems will also charge the brakes in order to provide the most efficient braking and prepare the vehicle for collision by pre-tensioning the seatbelts.

Some AEB systems deactivate if they detect avoidance action being taken by the driver.

## TYPES OF AEB

AEB can be classified into three main categories:

### Low speed system

This system targets city driving where crashes often occur at low speeds but can cause debilitating injury such as whiplash injuries. Typically, these systems look for the reflectivity of other vehicles and are not as sensitive to pedestrians or roadside objects.

### Higher speed system

This system typically utilises long range radar to scan further ahead of the vehicle (up to 200 metres) at higher speeds.

### Pedestrians

These systems use a camera combined with radar to detect vulnerable road users through their shape and characteristics. The way in which pedestrians move relative to the path of the vehicle is calculated to determine whether they are in danger of being struck.

## VOICE WARNING SYSTEM

Voice warning systems use computer synthesized speech to alert the driver to certain conditions in the vehicle. Human speech can be broken down into simple, basic sound called phonemes. There are about 50 for the English language. Synthesized speech is speech produced by combining phonemes together. An electronic circuit called a phoneme generator can produce sounds which duplicates these basic sounds. The phoneme generator is connected to a small loudspeaker. Human speech can be simulated by causing the generator to produce certain combinations of phonemes. The generator can do nothing more than produce sounds. A computer must be programmed to control the generator to produce phonemes in the right combinations to form words and sentences.

## Principle

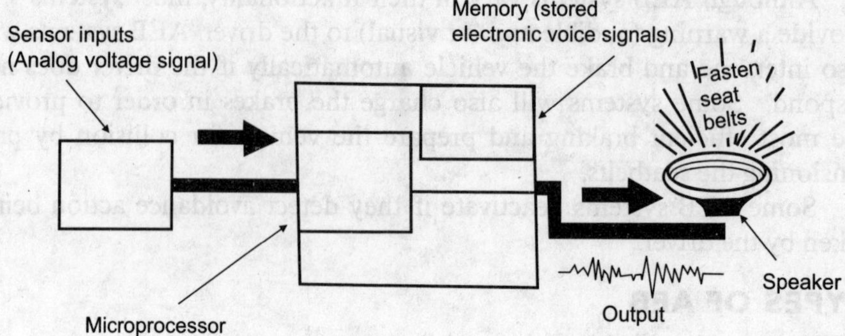


**Fig.12.17:** *Principle of voice warning system*

The electronic voice alert system uses a voice synthesizer type microprocessor which stores the electronic voice signals in its memory. Many different functions are monitored by the microprocessor. When one of these functions is inoperative, the microprocessor senses the condition from the inputs and then selects the appropriate audible voice message from its memory. This message is heard on the radio speaker nearest to the driver.

## 6-warning voice-alert system

A voice warning system used by Nissan is shown in Fig.12.13. If the ignition switch is ON and the right door is open, the voice warning will be, 'Right door is open'. This system has six warnings. Table lists the warnings and describes the conditions which trigger them. Some voice warning systems say 'thank you' when the computer receives the signal that the driver has remedied the fault.

| Item | Condition | | | Voice warning |
|---|---|---|---|---|
| Left door | Ignition switch ON | Left door switch is ON (left door is open) | Speed switch is ON Car speed is more than 10 km per hour. | Left door is open |
| Right door | | Right door switch is ON (right door is open) | | Right door is open |
| | | Parking brake switch is ON | | Parking brake is ON |
| Parking brake | | | | |
| Fuel level | | Fuel level is less than 10 litres | — | Fuel level is low |
| Ignition key | Ignition switch OFF | Left door switch is ON (left door is open) | Steering lock switch is ON | Key is in the ignition |
| Light | | | Lighting switch is ON | Lights are ON |

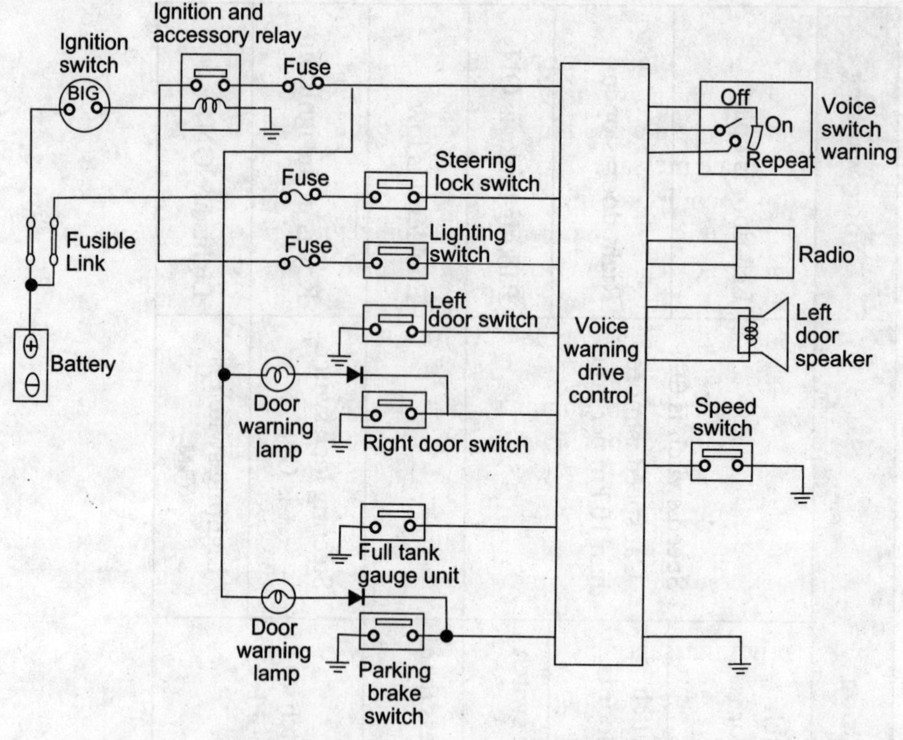


**Fig.12.18:** *6-voice warning system*

## 24-warning voice-alert system

The Fig.12.14 shows a voice warning system used by Chrysler that has 24 warning functions. This voice warning is part of an electronic monitor system which consists of a microcomputer, a vacuum fluorescent display, and a voice-alert module.

## Sensors

1. Four different types of sensors are used with this system to:
2. Sense headlight, taillight, and stoplight failure.
3. Sense a low level of engine oil.
4. Measure charging system voltage.

Sense door ajar, fluid level, oil pressure, and brake pad wear.

In addition, the microcomputer receives information on coolant temperature from a temperature sensor, vehicle speed from a speed sensor, and engine speed from the ignition system.

## Working

If a sensor detects a failure or an unsafe condition, a warning is activated. When this occurs, a warning message is indicated on the display and the voice-alert module produces a short tone to alert the driver and then delivers the voice message.

The microcomputer may require more than one input signal to activate a warning. For example, the engine temperature warning is a two-part message. The first voice warning is , 'your engine temperature is above normal'. To activate this message, engine coolant temperature must be too high, and engine speed must remain above 300 rpm for longer than 30 seconds. Therefore, the microcomputer must receive temperature information, engine speed information, and time information (from its internal electronic clock). On the basis of these three inputs, the microcomputer activates the warning message if the failure conditions are met. If the engine temperature is too high for 30 seconds and if the engine speed is greater than 300 rpm for 60 seconds, then the microcomputer activates a second message 'your engine is overheating', prompt service is required'.

The other messages provided by this system are:

- Your keys are in the ignition switch
- Your headlights are on
- Please fasten your seatbelts
- Your parking brake is on
- Please close your left rear door (driver's door) (right rear door) (trunk lid) (rear hatch)
- Your washer fluid is low
- Please check your fuel level (engine coolant level) (brake fluid level) (gearbox oil level) (engine oil level)
- Please check your disc brake pads
- Please check your headlights (taillights) (brake lights)
- Your engine oil pressure is critical. Engine damage may occur.
- Your charging system is malfunctioning. Prompt service is required.

## COLLISION AVOIDANCE RADAR WARNING SYSTEM

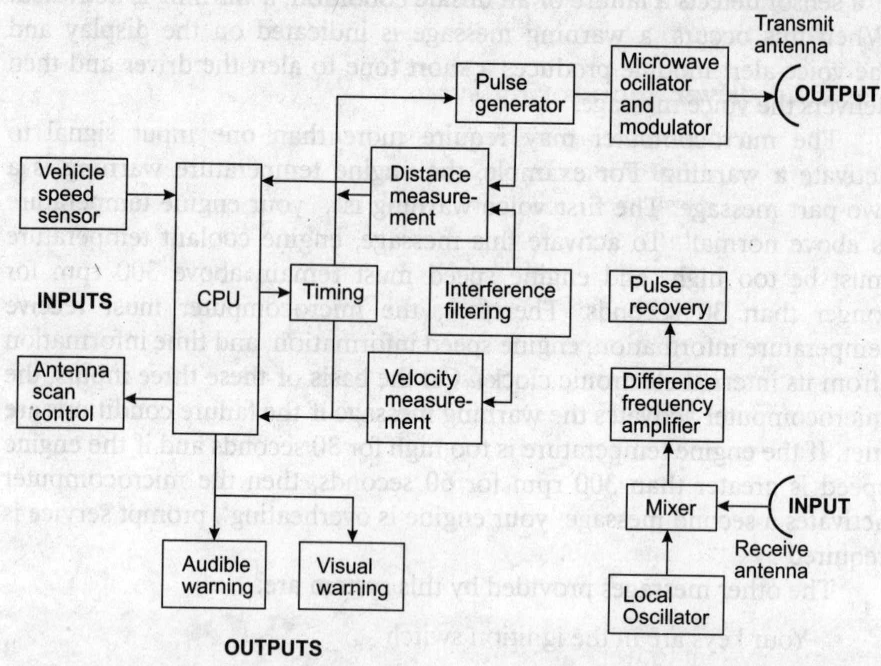


**Fig.12.19:** *Collision avoidance radar warning system*

A collision warning system, also known as pre-collision system, collision avoidance system, forward collision warning system or collision mitigating system, is an automobile safety system designed to reduce the severity of an accident. It uses radar (all-weather) and sometimes laser and camera to detect an imminent crash.

Once the detection is done, these systems either provide a warning to the driver when there is an imminent collision or take action autonomously without any driver input (by braking or steering or both).

Cars with collision warning system are also equipped with adaptive cruise control.

### Pre-collision system

Pre-collision system is the radar-based system that uses a forward-facing millimeter-wave radar. When the system determines that a frontal collision is unavoidable, it pre-emptively tightens the seat belts, removing any slack, and pre-charges the brakes using brake assist to give the driver maximum stopping power instantly when the driver depresses the brake pedal.

### CCD camera

A charge coupled device (CCD) camera on the steering column monitors the driver's face to determine where the driver is looking. If the driver's head turns away from road and a frontal obstacle is detected, the system will alert the driver using a buzzer, and if necessary, pre-charge the brakes and tighten the safety belts.

### Updated driver monitoring system

Updated driver monitoring system detects whether the driver's eyes are properly open. It monitors the driver's eyes to detect the driver's level of wakefulness. This system is designed to work even if the driver is wearing sunglasses, and at night.

### Driver monitoring system

Driver monitoring system includes a rearward-facing millimeter-wave radar mounted in the rear bumper. This system adjusts the active head restraints by moving them upward and forward to reduce the risk of whiplash injuries if an imminent rear collision is detected.

### Advanced pre-collision system

Advanced pre-collision system has a twin-lens stereo camera located on the windshield and a more sensitive radar to detect "soft" objects such as animals and pedestrians. A near-infrared projector located in the headlights allows the system to work at night.

### Lane departure warning system

The lane departure warning system will make automatic steering adjustments to help ensure that the vehicle maintains its lane in case the driver fails to react.

### Collision warning system with GPS-navigation

Collision warning system with GPS-navigation is designed to determine if the driver is late in decelerating at an approaching stop sign, will then sound an alert and can also pre-charge the brakes to provide optimum braking force if necessary.

The further advanced pre-collision systems have additional a front-side millimeter-wave radar to detect potential side collisions primarily at intersections or when another vehicle crosses the center line. The latest version tilts the rear seat upward, placing the passenger in a more ideal crash position if it detects a front or rear impact.

### Higher speed collision warning system

Higher speed collision warning systems enable deceleration of up to 60 kmh (37 mph) as compared to the current system of 40 kmh (25 mph).

This higher speed system uses the same technologies with the increased braking force up to twice that applied by average drivers.

### Pre-collision system with pedestrian-avoidance steer assist and steering bypass assist

Pre-collision system with pedestrian-avoidance steer assist and steering bypass assist can help prevent collisions in cases where automatic braking alone is not sufficient, such as when the vehicle is travelling too fast or a pedestrian suddenly steps into the vehicle's path. An on-board sensor detects pedestrians and issues a visual alert on the dashboard immediately in front of the driver if the system determines that there is a risk of collision. If the likelihood of a collision increases, the system issues an audio and visual alarm to encourage the driver to take evasive action, and the increased pre-collision braking force and automatic braking functions are activated. If the system determines that a collision cannot be avoided by braking alone and there is sufficient room for avoidance, steer assist is activated to steer the vehicle away from the pedestrian.

## DAYTIME RUNNING LIGHTS

Daytime Running Lights (DRLs) are headlights that are illuminated during the day in order to make vehicles more visible and thus reduce their involvement in crashes. This system is implemented in India recently. It is possible to fit vehicles with a device that will automatically activate DRLS when the ignition is switched on but is overridden by full strength headlights.

DRLS have been found to increase driver's peripheral perception of vehicles. It is also easier for drivers to estimate the distance to vehicles with DRLs.

## VEHICLE CRUISE CONTROL

## CRUISE CONTROL

Cruise control is employed to retain a steady speed under varying load conditions and relieves the driver from constant foot throttle manipulation. The system is a servomechanism that takes over the throttle of the car to maintain a steady speed as set by the driver.

The greatest feature of this system is to improve the fuel efficiency by controlling throttle positions gradually. The key input signals for this system are vehicle speed signal, throttle position signal, and engine speed signal.

## ADAPTIVE CRUISE CONTROL

Adaptive cruise control (ACC) is a cruise control system in some modern vehicles which maintains a safe distance from the vehicle in front in the same lane.

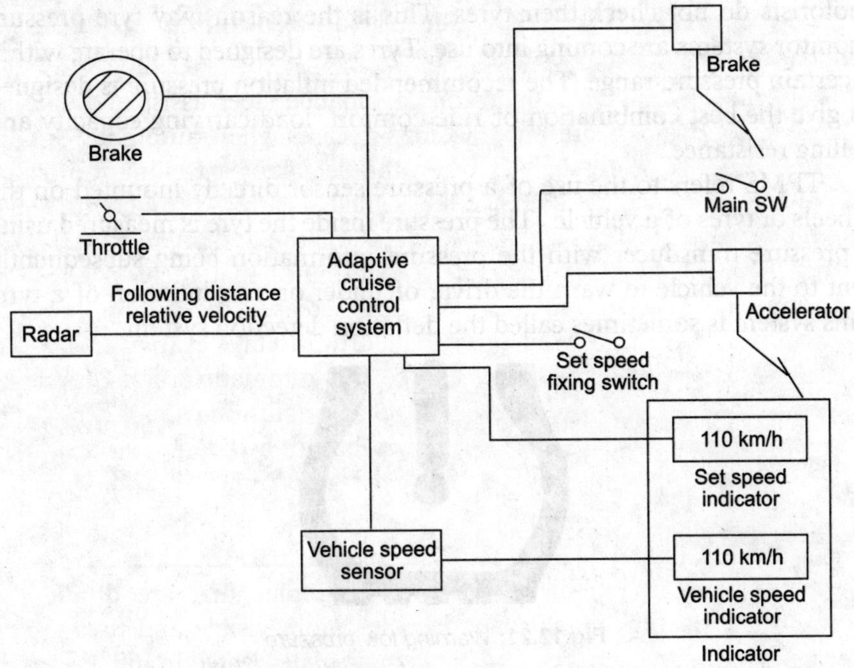


**Fig.12.20:** *Adaptive cruise control system*

The system also called active cruise control (ACC) or intelligent cruise control (ICC). This system uses either a radar or laser setup to monitor the distance between the vehicle and the vehicles ahead, and regulates speed. The ACC system decelerates vehicle when necessary to maintain a safe distance between vehicles. When the way is clear, the ACC system resumes preset cruising speed.

Although ACC systems automatically take action to control the vehicle's speed, they may not sufficiently decelerate the vehicle in all circumstances to avoid a crash. Depending on the vehicle load, road grade, and vehicle performance parameters, the level of deceleration control typically used by the system is limited to a range between 0.1 and 0.2 g, where g is a unit of force equal to the force exerted by gravity. As a result, drivers remain responsible for the safe operation of their vehicles, and they should use steering and apply the service brakes to avoid a crash.

## TYRE PRESSURE MONITORING SYSTEM (TPMS)

Tyre pressure monitors detect when a tyre drops below the designated pressure and alerts the driver.

## NEED FOR TPMS

The inflation pressure of the tyres should be checked regularly, but many motorists do not check their tyres. This is the reason why tyre pressure monitor systems are coming into use. Tyres are designed to operate within a certain pressure range. The recommended inflation pressure is designed to give the best combination of ride comfort, load carrying capacity and rolling resistance.

TPMS refers to the use of a pressure sensor directly mounted on the wheels or tyres of a vehicle. The pressure inside the tyre is measured using a pressure transducer with the pressure information being subsequently sent to the vehicle to warn the driver of under or over inflation of a tyre. This system is sometimes called the deflation detection system.

**Fig.12.21:** *Warning low pressure*

**Note**: The TPMS system must warn the driver if the pressure inside any tyre on the vehicle is 25 percent or more under the recommended inflation pressure.

## TYPES OF TPMS

There are essentially two basic ways to monitor tyre pressure electronically: direct method and indirect method.

### Direct method

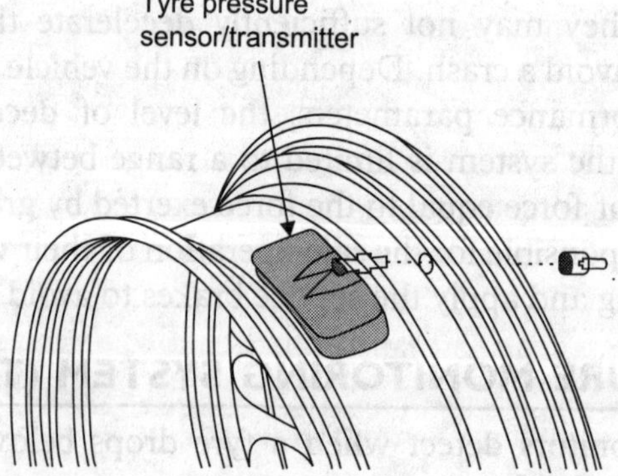


**Fig.12.22:** *Pressure sensor inside a wheel*

In direct method, a small pressure sensor is located inside each wheel. The sensor has a built-in transponder that broadcasts a radio signal to an external module. The module identifies the signal from each wheel and keeps an eye on pressure. If pressure drops below a predetermined threshold, the module turns on a light or displays a message to warn the driver.

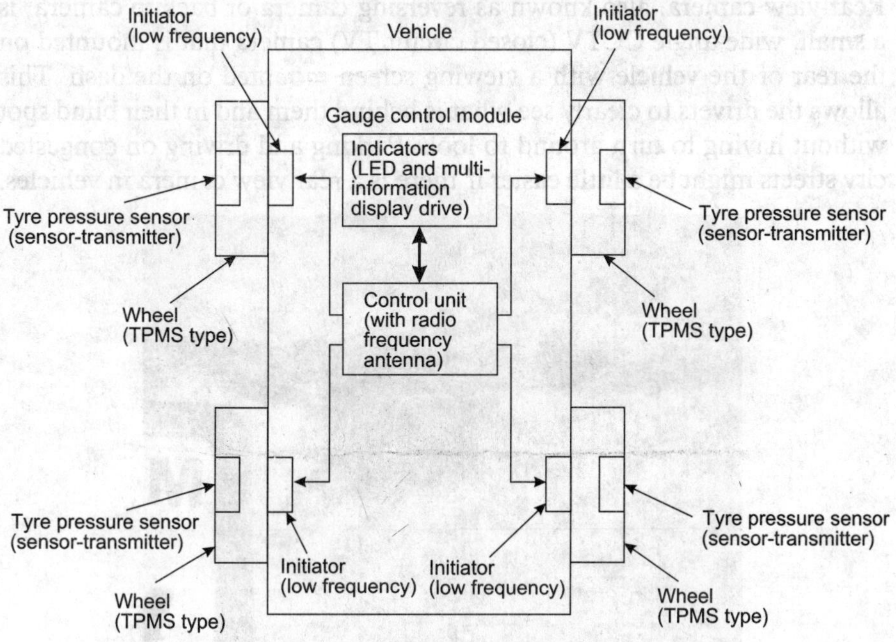


**Fig.12.23:** *Direct type TPMS*

The pressure sensors are very accurate but they have their own limitations. One is battery life. The pressure sensor inside each wheel contains a small battery to power the sensor and transponder. Most are long-life lithium batteries that may last from 5 to 10 years.

## Indirect method

The other method of monitoring tyre pressure does not require any additional hardware inside or outside the wheel. This approach makes use of the existing wheel speed sensors in the antilock brake (ABS) system. By modifying the ABS operating software, the ABS system learns the average speed at which each wheel rotates when traveling straight at a constant velocity, and detects low tyres by comparing relative wheel speeds. It is a simple, and cost-effective way to monitor tyre pressure.

In the indirect ABS-based TPMS, if a tyre loses air, the diameter of the tyre shrinks slightly causing it to turn at a slightly faster rate than the others. Most ABS systems cannot detect changes smaller than 1 mm. For this reason, indirect ABS-based tyre pressure monitoring systems are not as sensitive as direct TPMS systems that have a pressure sensor inside each wheel.

## REARVIEW CAMERA

Rear-view camera, also known as reversing camera or backup camera, is a small, wide-angle CCTV (closed circuit TV) camera that is mounted on the rear of the vehicle with a viewing screen mounted on the dash. This allows the drivers to clearly see what is behind them and in their blind spot without having to turn around to look. Parking and driving on congested city streets might be a little easier if there is a rear view camera in vehicles.

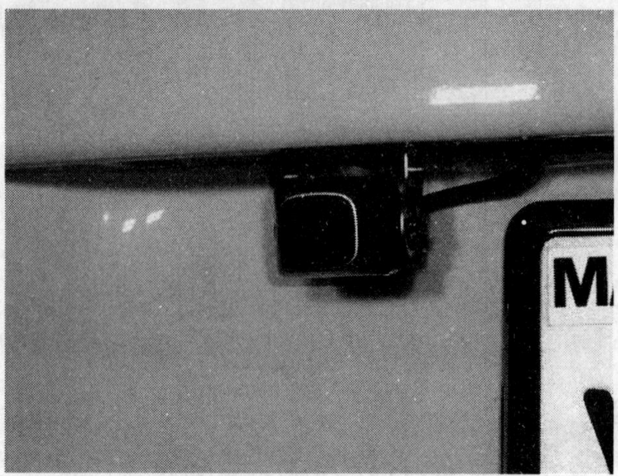

**Fig.12.24:** *Rear view camera*

### Working of a rear-view camera

The structure of a rear-view camera is different from other cameras. The reflection of a rear-view camera is horizontally flipped so that it produces a mirrored image. This is essential because the driver and the rear-view camera are in opposite positions. The mirrored image is also necessary because it causes the direction of the display to be constant with the mirrors mounted on the motor vehicle. The rear-view camera display is usually wired to automatically sense when the gearbox is set in reverse, and it shows the rear view while the car is in reverse. The other items in the rear of the vehicle are also shown at all times. The rear-view camera usually consists of a wide-angle or fish-eye lens. Although the lens doesn't allow the camera to view distant objects, it does permit the camera to view a continuous horizontal view from one corner, behind the car, to another.

Rear-view cameras are also generally mounted at a downward angle. This allows the camera to show potential barriers on the ground in addition to the location and position of walls and docks that may be looming around the corner.

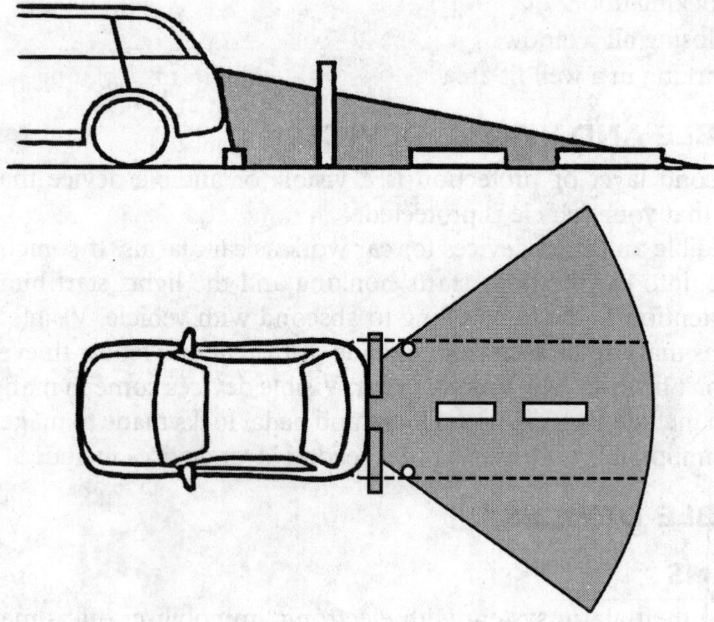

**Fig.12.25:** *Range of view*

## Benefits

The benefits of rear-view cameras in vehicles are:

1. Eliminate blind zone
2. Keep kids and pets safe
3. Parking
4. Traffic

## ANTI-THEFT SYSTEM

Motorists with high-risk vehicles should also consider using several "layers" of protection. The more layers of protection on vehicle, the more difficult it is to steal.

There are three basic types of anti-theft devices for cars: audible and visible devices (deterrents), immobilizing devices and vehicle tracking or recovery systems. Deterrents warn or scare off potential thieves, immobilizers make it difficult or impossible to drive a stolen vehicle away, and tracking systems facilitate the process of finding vehicles after they have been stolen.

The common sense approach is the first layer of protection in the anti-theft process. The following are the simplest and most cost-effective first layer protection.

1. Removing keys from the ignition
2. Locking doors
3. Closing all windows
4. Parking in a well-lit area

## AUDIBLE AND VISIBLE DEVICES

The second layer of protection is a visible or audible device that alerts thieves that your vehicle is protected.

Audible anti-theft devices for car work as car alarms. If someone tries to break into car the horn starts honking and the lights start blinking to draw attention to the thief trying to abscond with vehicle. Visible devices are items that can be seen from outside of the car and let car thieves know they won't have an easy time with car. Visible devices come in many forms and can include steering wheel locks and pedal locks made to make the car almost impossible to drive. Popular second layer devices include:

## AUDIBLE DEVICES

## ALARMS

The anti-theft alarm system with electronic immobiliser offers maximum protection against vehicle theft. Audible and visual warning signals attract the attention of passers-by and deter potential thieves. An acoustic alarm is triggered when unauthorised entry is detected by contacts on the doors, engine compartment and luggage compartment lid or by ultrasonic motion sensors. The ultrasonic system also detects anyone trying to get into the car through a broken window.

### Operation

Car alarms consist of at least three components, which include:

- Sensors
- Siren
- Control unit

The sensors are connected to different parts of the vehicle, and they are designed to trip whenever a thief attempts to gain access. When one of the sensors is tripped, it sends a signal to the control unit. The control unit then activates the siren, which will call attention to the vehicle and may scare off the thief.

### Types of sensors

### Door

The most common car alarm sensors are simple switches that are installed in the doors. When a door is opened, the switch sends a signal to the control unit.

### Movement

These sensors send a signal to the control unit if the vehicle is jostled in any way.

### Microphones and pressure sensors

Microphones detect the ambient sound level, which allows the control unit to monitor for sounds like breaking glass that indicate a theft in process. Pressure sensors work on the same basic principle that microphones do, but they are set off when the pressure in the vehicle changes. Since breaking a window or opening a door will trigger a pressure change, this type of sensor can be relatively effective.

## VISIBLE DEVICES

### Steering wheel lock

Prevents rotation of the steering wheel.

### Brake pedal lock

Prevents depression of the brake pedal.

### Wheel locks

Similar to the circular steel "boots" used by many large city police departments, tyre locks prevent the vehicle from being driven.

### Tyre locks/Tyre deflators

Attaches to the tyre valve stem and causes the tyre to go flat if the tyre rotates before they are removed.

### Theft deterrent decals

Theft deterrent decals visually warn thieves the vehicle is protected by an alarm or other antitheft devices.

### Identification markers in or on vehicle

Security labels marking various vehicle parts enabling the part to be identified if removed from the vehicle.

### Window etching

Etching the vehicle identification number or other traceable number onto the vehicle's windows makes it difficult for thieves to resell the vehicle or its parts.

### Laminated glass

Impact resistant protective glazing applied to vehicle windows designed to fortify the windows against penetration or splintering.

## IMMOBILIZING DEVICE

The third layer of protection is a device that prevents thieves from bypassing ignition and hot-wiring the vehicle. Some electronic devices have computer chips in ignition keys. Other devices inhibit the flow of electricity or fuel to the engine until a hidden switch or button is activated. Popular third layer devices include:

### Smart keys

Smart keys contain specially coded computer chips or radio frequencies. Without the exact key, the vehicle's engine cannot be started.

### Fuse cut-offs

Short circuits the electrical system, preventing the vehicle from starting.

### Kill switches

Kill switches inhibit the flow of electricity or fuel to the engine until a hidden switch is activated.

### Starter, ignition and fuel disablers

Short circuits the electrical or fuel system, preventing the vehicle from starting, or causes the vehicle to stop running within a short time.

## VEHICLE TRACKING SYSTEM

The final stage of the car security is tracking. Professional thieves can steal any car. After a vehicle has actually been stolen, it can be very difficult to successfully track it down and recover it. If it has some type of tracking system installed, the process is streamlined, and the recovery rate increases exponentially.

Most vehicle tracking systems use GPS technology, and some also make use of cellular or radio transmitters hidden in the vehicle which emit signals to the police or monitoring stations when the vehicle is reported stolen.

### Types of vehicle tracking

There are two types of vehicle tracking, each of which is useful in specific situations.

#### Passive

The simplest vehicle tracking systems are passive in nature. These trackers typically use a GPS device to record the position of a vehicle over time. When the tracker is removed, the data can be transferred to a computer and analyzed. These tracking systems are useful for fleet management, but they also have other applications.

#### Active

More complex tracking systems also include the ability to transmit the location of a vehicle in real time. For fleet management and dispatch purposes, this data is typically monitored from a central location. This type of system can also be used for stolen vehicle recovery.

## KEYLESS ENTRY SYSTEM

Most cars today come with the following keyless entry systems that can be used to turn on cars remotely.

| Type of keyless entry | Features |
|---|---|
| Programmable keypad | Access car without by using a passcode |
| Key fob | Unlock and lock doors; open trunk with touch of button; alarm activation; unique to each car |
| Smart key | Unlock and lock doors; open trunk, and start car without a traditional key; unique to each car; each driver can have unique key with setting programmed |

## BENEFITS

The Benefits of key fobs and remote entry systems in vehicles are:

1. Convenience
2. Safety
3. Customizable Options

## PROGRAMMABLE KEYPAD

Programmable keypad permits the driver to unlock the car door and/or release the trunk lid (luggage compartment) without a key. It consists of five touch buttons located on the driver's door belt molding.

Each button has two numbers, 1-2, 3-4, 5-6, 7-8, 9-10 and all are connected to a microcomputer for 'code' and 'sequence' identification.

**Fig.12.26:** *Keyless entry system keypad*

Two five-digit codes are available, one code is programmed at the factory and cannot be changed, the other code is for owner programming and can be changed at any time. Entry into the car is gained by pressing the proper buttons in the proper sequence. For night time use, pushing any of the five buttons will illuminate all of the buttons, the interior car lighting system and the key lock light. The driver can then unlock the door by the use of a key, in the conventional manner, or by using either of the two codes-factory or owner programmed.

## KEY FOB

**Fig.12.27:** *Key fob*

A key fob is a palm-sized portable device capable of controlling various parts of a car. Their primary function is to manage the remote-entry system installed in a car.

Aside from locking and unlocking car doors, some key fobs can:

- Pop the trunk latch
- Remotely start the vehicle ignition
- Arm/disarm a security system
- Activate the panic alarm
- Control automatic windows

## SMART KEY

SmartCARD keyless entry with Radio Frequency Identification (RFID) technology and encryption presents the safer car alarm system. One can access a car without even a need to press the button on the remote control or take out any key from bag or pocket.

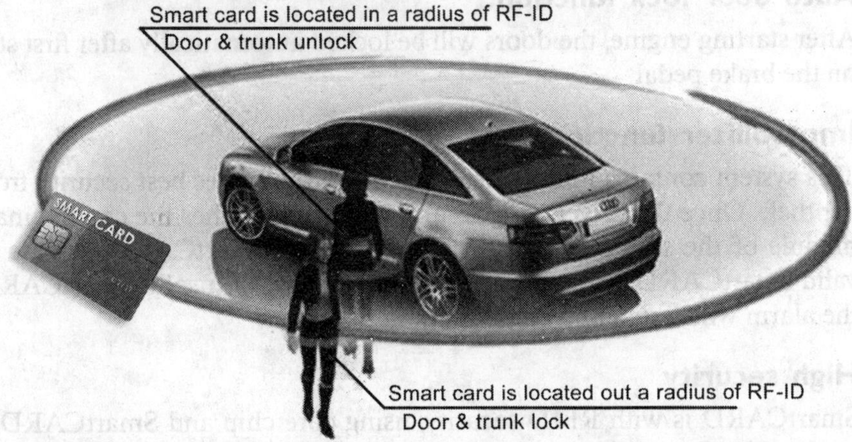


**Fig.12.28:** *Smart key*

## MAIN FEATURES

The important features of this system are:

### AUTO Door Close

If driver with SmartCARD is apart 1.5~2.0m from the car, the door will be closed automatically and it will enter into security mode. At this time, LED indicator will flicker in blue and red light in order.

### AUTO Door Open

If driver with SmartCARD approaches to the car within 1.5~2.0m, the system recognises the validity of the card and door will be opened if its code matches. At this time, LED indicator will flicker in BLUE light.

### Door open alarm

If the SmartCARD is apart and door is opened, the system will trigger the alarm sound and hazard light will flicker to notify that door is opened.

### Slim smart key

SmartCARD is very thin and size is about two third of general credit card size. It is convenient to carry and keep. Therefore it is safe and convenient to no risk to lose.

### Battery replacement time notification

If the battery of the SmartCARD is low, the system will notify the driver by alarm sound and flickering. At this time, battery should be replaced. (It can last for up to two more weeks after notification).

### Auto door lock function

After starting engine, the doors will be locked automatically after first step on the brake pedal.

### Immobilizer function

This system contains immobilizer function. It provides best security from car theft. Once the driver with SmartCARD approaches the car, the main module of the system will check the ID of the SmartCARD. In case of valid SmartCARD, driver can start the car. In case of invalid SmartCARD, the alarm will sound and engine cannot be started.

### High security

SmartCARD is with RFID security, using core chip and SmartCARD is made of hoping code technology making it very difficult to copy.

## DOOR LOCKS

Different types of door locks used in cars are:

1. Electric door locks
   (a) Solenoid operated door locks
   (b) Motor-operated door locks
2. Automatic door locks

## ELECTRIC DOOR LOCKS

Electric door locks often use electric solenoids to move the linkage to the door locks. When the ignition key is turned or a door lock button is pressed, current flows to the solenoids in the doors. The solenoids convert the electrical energy into motion. The solenoid motion moves the lock linkages and latches. this locks or unlocks the doors.

## AUTOMATIC DOOR LOCKS

Automatic door lock system facilitates automatic locking of all doors providing,

- all doors are closed with courtesy lamps OFF
- the ignition switch is ON
- the driver is seated
- the gear selector is in drive

When the gear selector lever is moved to Park, Neutral or Reverse, all doors will automatically unlock if an inside door handle is pulled. Also if the car is stopped and the gear selector is in Drive, an individual door can be unlocked by pulling its inside door handle, or all doors can be unlocked from the main control switch. The door, or doors, will automatically relock when the door is closed.

The automatic door lock system consists of an electronic logic module, a permanent-magnet motor lock actuator relay, a back-up lamp switch, right and left front door lock remote control handle switches and a driver's seat sensor switch. This system is interconnected with the regular electric door lock harness and the permanent-magnet motor lock actuators and switches.

## REVIEW QUESTIONS

1. What are the features incorporated for vehicle safety?
2. What do you understand by active safety and passive safety?
3. What is meant by cruise control?
4. Discuss briefly the construction of seat belts.
5. With relevant sketches explain the construction and operation of air bags.
6. Give a note on rear view camera.
7. Indicate the design feature of electronic stability control.
8. Give the importance of adaptive cruise control.
9. What is the need of emergency brake assist system?
10. Discuss briefly the tyre pressure monitoring system.

UNIT

# 13 Microcomputer Controlled Devices

Computers are devices that will store datas, manipulate them through arithmatic and logic operations and gives an output accordingly, as per the instructions already given to it. In automobiles also computers do the same functions, but the datas are, real time datas collected through various sensors in the automobile systems and the output is directly employed to control the functions of the systems. A brief review is presented in this chapter about the application of computers in automobiles.

## ELECTRONIC DASH BOARD INSTRUMENTS

All the information displayed as pointer movements on graduated scale (analog instruments) are being replaced by digital instruments. Apart from typical instruments like fuel gauge, speedometer, odometer, coolant temperature gauge, ammeter, clock, etc., the following warning systems may also be included in thc dash board.

- Low fuel
- Malfunctioning of changing system
- Low pressure of the braking system
- Burning out of any of the lamps (head lamp, tail lamp, fog lamp, etc)
- Non closing/locking of any of the doors
- Low lubricating oil pressure
- High engine temperature

Additional information that may be obtained from the dash board are:

- Distance travelled
- Time elapsed after the start
- Average speed
- Distance covered/distance to destination
- Estimated time of arrival at destination
- Distance that may be covered with the fuel remaining in the tank
- Most economical route selection

## TRAVEL INFORMATION SYSTEM

Travel information systems mainly used for:

1. Highway networks, including freeways and toll way
2. Major city streets

Travel information systems provide the driver with information such as the time of day, fuel economy, and estimated time of arrival.

In this system various sensors provide information on vehicle operation to a microcomputer. The microcomputer provides the following travel information:

- Month, date, day, and time of day
- Distance remaining before fuel tank is empty
- Estimated time of arrival at a given destination
- Average vehicle speed
- Average fuel economy
- Instantaneous fuel economy
- Distance remaining to a given destination
- Elapsed time since the system was last reset
- Elapsed distance since the system was last reset

Travel information systems also update drivers on current roadway conditions including delays, incidents, weather-related messages, travel times, emergency alerts, and alternate routes. Providing this information to drivers before and during trips allows them to make more effective travel decisions about changing routes, modes, departure times, or even destinations. More informed drivers result in more efficiently utilised roadway capacity. This means less traffic congestion and better traffic flow.

Travel information is generated by sensors reporting to a traffic management center or through private entities using data from in-vehicle location devices, or from smart phones communicating location and speed. This information is then disseminated by means of traditional broadcast media, internet, mobile devices, or roadside messaging. Personalised travel messages and alerts enable individuals to get trip-specific information on demand, or have it pushed to them via email or text message subscription services. Once familiar with these services, nearly 80% of drivers use traveler information to make daily decisions about route or departure time.

### Benefits

1. Maximise efficiency and capacity by providing current transportation system information to drivers.

2. Reduce the impacts of congestion.
3. Increase safety by alerting drivers of upcoming hazards.

## ELECTRONIC POWER STEERING

The electronically controlled power steering improves steering ability and performance. To reduce the steering effort when driving at low speeds and to provide a corrective action when driving at high speeds.

It consists of wheel speed sensor, a steering sensor (torque, angular velocity), an electronic control unit, a drive unit and a motor. Signal output from each sensor are input to ECU and necessary steering assistance is calculated and applied by the drive unit to control the operation of the motor.

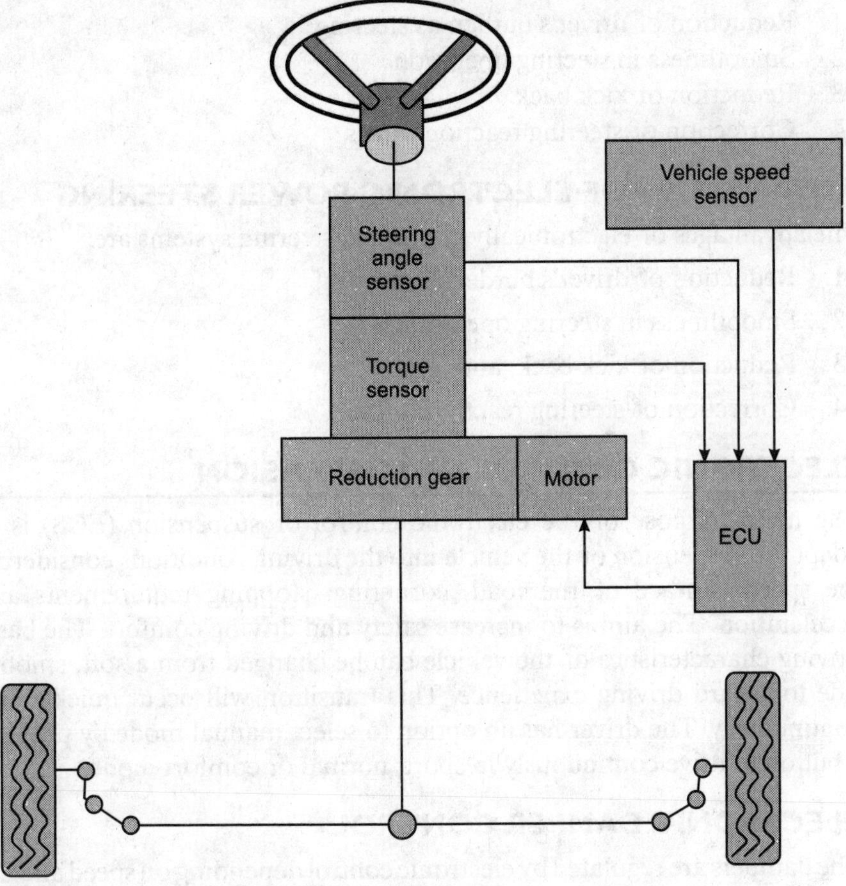


**Fig.13.1:** *Electronically controlled power steering*

### (a) Power steering ECU

The power steering ECU calculates assisting power based on steering torque signals from the torque sensor and vehicle speed signals from the skid control ECU. For vehicles without ABS, the power steering ECU receives vehicle speed signals from the speedometer.

### (b) Torque sensor

The torque sensor detects the steering effort generated when the steering wheel is turned and converts it to an electrical signal.

### (c) EPS motor

The EPS motor is activated by the current from the power steering ECU and generates torque to assist the steering effort.

### Advantages of electronic power steering

1. Reduction of driver's burden as steering
2. Smoothness in steering operation
3. Reduction of kick back
4. Correction of steering reaction forces

## ADVANTAGES OF ELECTRONIC POWER STEERING

The advantages of electronically controlled steering systems are:

1. Reduction of driver's burden as steering
2. Smoothness in steering operation
3. Reduction of kick back, and
4. Correction of steering reaction forces.

## ELECTRONIC CONTROL OF SUSPENSION

The main purpose of the electronic control of suspension (ECS) is to adapt the suspension of the vehicle into the driving conditions considering the speed, surface of the road, cornering, stopping requirements and acceleration. The aim is to increase safety and driving comfort. The basic driving characteristics of the vehicle can be changed from a soft, smooth ride to a hard driving experience. This transition will occur quickly and continuously. The driver has an option to select manual mode by pushing a button to drive continuously in sport, normal or comfort mode.

## ELECTRONIC DAMPER CONTROL

The dampers are regulated by electronic control depending on speed and the condition of the road. The control unit uses information from acceleration sensors on the wheels and body to calculate the most appropriate damper settings. Electronic Damper Control (EDC) automatically or manually adjusts each damper to suit the driving conditions.

Sensors constantly monitor all factors influencing the vehicle's behaviour and occupants' comfort, in order to precisely adjust the damper control. In a fraction of a second, the signals are analysed by the EDC microprocessor and signals are sent to the actuators on the shock absorbers, which, with the help of magnetic valves, are variably adjusted to provide optimal suspension. With EDC, the tendency for the nose to dip when braking is practically eliminated. The influence of potholes and unevenness on the road surface is reduced to minimum.

In addition to increased driving comfort and improved roll characteristics on the tyres, EDC also contributes to vehicle stability and safety. By reducing the nose's tendency to dip when braking and improving the tyre traction, EDC shortens the braking distance when braking heavily. The damper adjustment means that even when braking with ABS the vehicle chassis remains upright on the road and all four wheels have the largest possible contact to the surface.

## ADVANTAGES

Adding electronic damper control to the air suspension system provides the following advantages when driving over loose surfaces or on off-road terrain.

1. Extra ride comfort
2. Driving pleasure
3. Safety
4. Improved handling capabilities
5. Compensate for any body movements
6. Dampen the pitching vibrations produced when towing a trailer very effectively.

## ON-BOARD DIAGNOSTICS

On-Board Diagnostics (OBD) is a computer-based system originally designed to reduce emissions by monitoring the performance of major engine components.

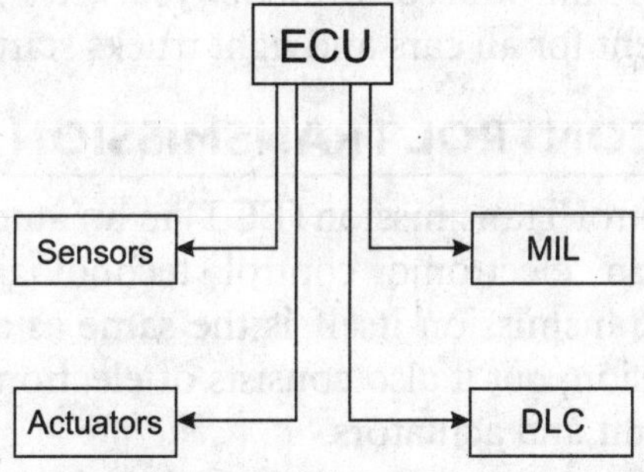


**Fig.13.2:** *On-board diagnostics*

A basic OBD system consists of an ECU, which uses input from various sensors to control the actuators (fuel injectors) to get the desired performance. The Check Engine light, also known as the Malfunction Indicator Light (MIL), provides an early warning of malfunctions to the vehicle owner. A modern vehicle can support hundreds of parameters, which can be accessed via the Diagnostic Link Connector (DLC) using a device called a scan tool.

There are two kinds of on-board diagnostic systems: OBD-I and OBD-II.

## OBD-I

OBD-I refers to the first generation OBD systems which were developed throughout the 1980s. These early systems use proprietary connectors, hardware interfaces, and protocols. A mechanic who wanted to access diagnostic information typically had to buy a tool for every different vehicle make. OBD-I scan tools that support multiple protocols are supplied with an array of different adapter cables.

## OBD-II

In the early 1990s, Society of Automotive Engineers (SAE) and International Standardization Organization (ISO) issued a set of standards which described the interchange of digital information between ECUs and a diagnostic scan tool. All OBD-II compliant vehicles were required to use a standard diagnostic connector, and communicate via one of the standard OBD-II communication protocols.

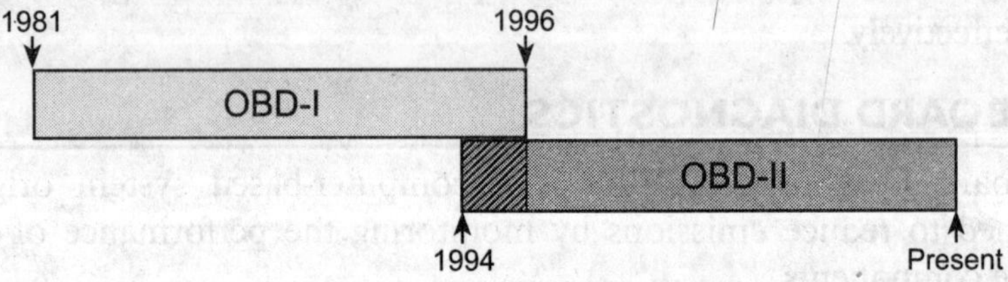


OBD-II was first introduced in model year (MY) 1994 vehicles, and became a requirement for all cars and light trucks starting with MY1996.

## ELECTRONIC CONTROL TRANSMISSION

The Electronic Control Transmission (ECT) in an automatic transmission which uses modern electronic control technologies to control the transmission. The transmission itself is the same as a fully hydraulically controlled transmission, but it also consists of electronic parts, sensors, an electronic control unit and actuators.

The electronic sensors monitor the speed of the vehicle, gear position selection and throttle opening, sending this information to the ECU. The

ECU then controls the operation of the clutches and brakes based on this data and controls the timing of shift points and torque converter lock-up.

## SENSORS IN ECT

### Throttle position sensor

This sensor is mounted on the throttle body and electronically senses how far the throttle is open and then sends this data to the ECU. The throttle position sensor takes the place of throttle pressure in a fully hydraulic control transmission. By measuring the throttle position, it gives the ECU an indication of engine load to control the shifting and lock-up timing of the transmission.

### Water temperature sensor

The water temperature sensor monitors engine coolant temperature and is typically located near the cylinder head water outlet.

When the engine coolant is below a predetermined temperature, the engine performance and the vehicle's drivability would suffer if the transmission is shifted into overdrive or the converter clutch is locked-up. The engine ECU monitors coolant temperature and sends a signal to the ECT ECU. The ECU prevents the transmission from upshifting into overdrive and lock-up until the coolant has reached a predetermined operating temperature.

### Speed sensor

The speed sensor is used in place of governor pressure in the conventional hydraulically controlled transmission. To ensure that the ECT ECU is kept informed of the correct vehicle speed at all times, vehicle speed signals are input into it by two speed sensors. For further accuracy, the ECT ECU constantly compares these two signals to monitor whether they are the same.

## SOLENOID VALVES IN ECT

Solenoid valves are electro-mechanical devices which control hydraulic circuits by opening a drain for pressurised hydraulic fluid. Of the solenoid valves No.1 and No.2 control gear shifting while No.3 controls torque converter lock-up.

### No.1 and No.2 Solenoid valves

These solenoid valves are mounted on the valve body and are turned on and off by electrical signals from the ECU, causing various hydraulic circuits to be switched as necessary. By controlling the two solenoids on and off sequences, four forward gears are provided as well as upshifts into third or fourth gear is prevented. The No.1 and No.2 solenoids are normally closed. The plunger is spring loaded to the closed position, and when energised, the plunger is pulled, allowing line pressure fluid to drain.

### No.3 Solenoid valve

This solenoid valve is mounted on the transmission exterior or valve body. It controls line pressure which affects the operation of the torque converter lock-up system. This solenoid is either a normally open or normally closed solenoid.

### No.4 Solenoid valve

This solenoid is a normally closed solenoid which controls the shift to low 4-wheel drive. It is controlled by the ECT ECU when low 4-wheel drive has been selected at vehicle speeds below 18 mph with light throttle opening.

## OPERATION OF ECT ECU

The ECU controls No.1 and No.2 solenoid valves based on vehicle speed, throttle opening angle and mode select switch position.

The ECT ECU has lock-up clutch operation pattern for each driving mode (normal and power) programmed in its memory. The ECU turns the No.3 solenoid valve on or off according to vehicle speed and throttle opening signals. The lock-up control valve changes the fluid passages for the converter pressure acting on the torque converter piston to engage or disengage the lock-up clutch.

## OPERATION OF ECT SHIFT VALVE

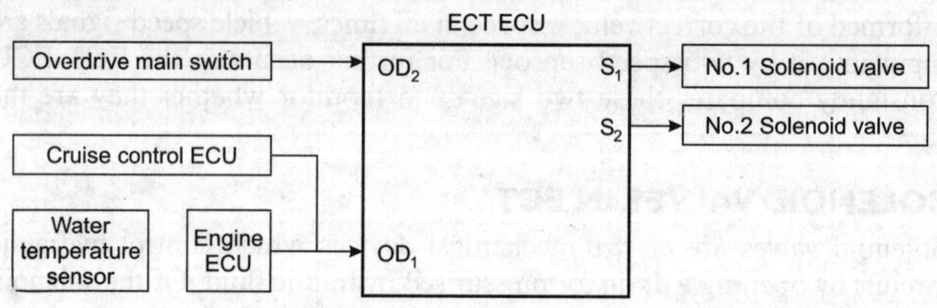


**Fig.13.3:** *ECT ECU*

Two electrically operated solenoids control the shifting of all forward gears in the electronic control four speed automatic transmission. These solenoids are controlled by an ECU which uses throttle position and speed sensor input to determine when the solenoids are turned on. The solenoids normal position is closed, but when it is turned on, it opens to drain fluid from the hydraulic circuit. Solenoid No.1 controls the 2-3 shift valve. It is located between the manual valve and the top of the 2-3 shift valve. Solenoid No.2 controls the 1-2 shift valve and the 3-4 shift valve.

## First gear

During first gear operation, solenoid No.1 is on and solenoid No.2 is off. With line pressure drained from the top of the 2-3 shift valve by solenoid No.1, spring tension at the base of the valve pushes it upward. With the shift valve up, line pressure flows from the manual valve through the 2-shift valve and on to the base of the 3-4 shift valve.

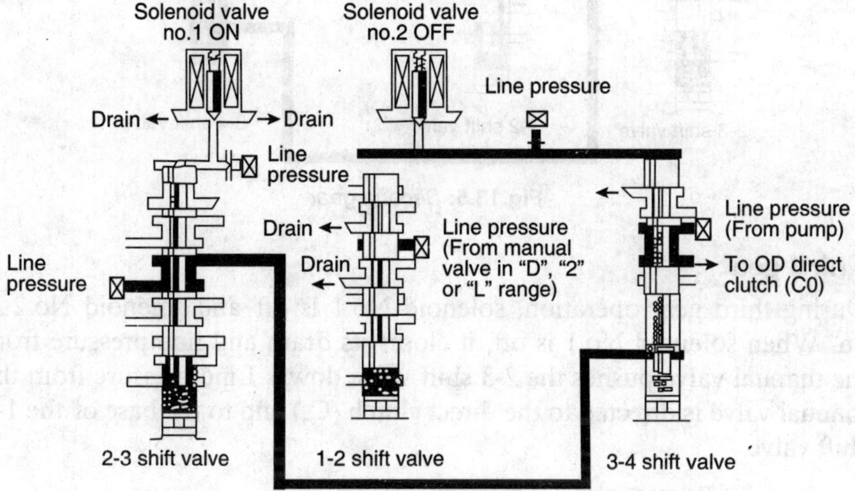


**Fig.13.4:** *ECT shift valve operation*

With solenoid No.2 off, line pressure pushes the 1-2 shift valve down. In this position, the 1-2 shift valve blocks line pressure from the manual valve. Line pressure and spring tension at the base of the 3-4 shift valve push it upward.

## Second gear

During second gear operation, solenoid No.1 and No.2 are on. Solenoid No.1 has the same effect that it had in first gear with the 2- shift valve being held up by the spring at its base. Pressure from the manual valve flows through the 2-3 shift valve and holds the 3-4 shift valve up.

With solenoid No.2 on, line pressure from the top of the 1-2 shift valve bleeds through the solenoid. Spring tension at the base of the 1-2 shift valve pushes it upward. Line pressure which was blocked, now is directed to the second brake ($B_2$), causing second gear. The 3-4 shift valve maintains its position with line pressure from the 2-3 shift valve holding it up.

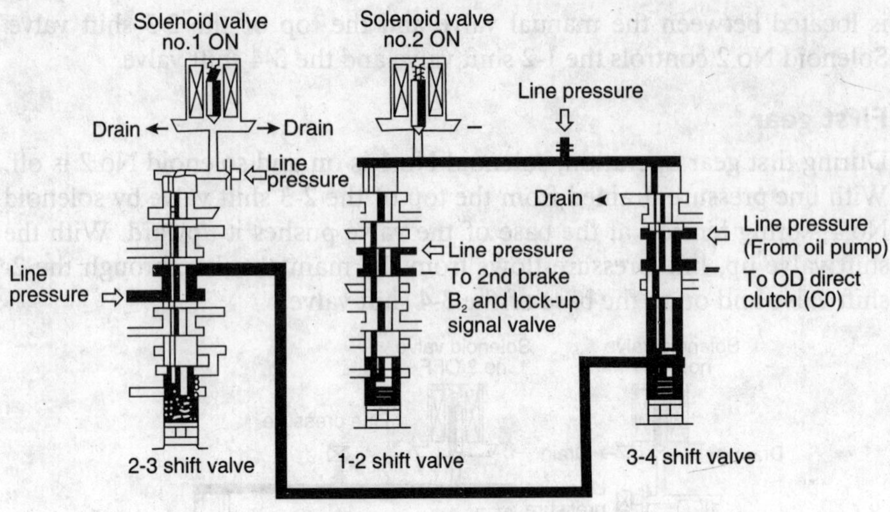


**Fig.13.5:** *Second gear*

## Third gear

During third gear operation, solenoid No.1 is off and solenoid No.2 is on. When solenoid No.1 is off, it closes its drain and line pressure from the manual valve pushes the 2-3 shift valve down. Line pressure from the manual valve is directed to the direct clutch ($C_2$) and to the base of the 1-2 shift valve.

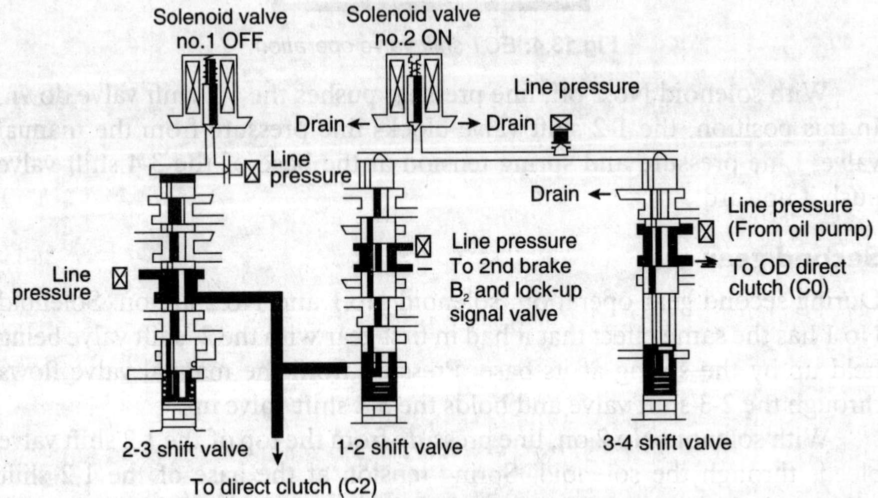


**Fig.13.6:** *Third gear*

With solenoid No.2 on, it has the same effect that is had in second gear. Pressure is bled at the top of the 1-2 shift valve and spring tension pushes it up. Line pressure is directed to the second brake ($B_2$). However in third gear, the second brake ($B_2$) has no effect since it holds the one-way clutch No.1 (F1) and free wheels in the clockwise direction. The second coast brake is ready in the event of a downshift when the OD direct clutch ($C_2$) is released.

### Fourth gear

During fourth gear operation, both solenoids are off. A third solenoid controls lock-up operation.

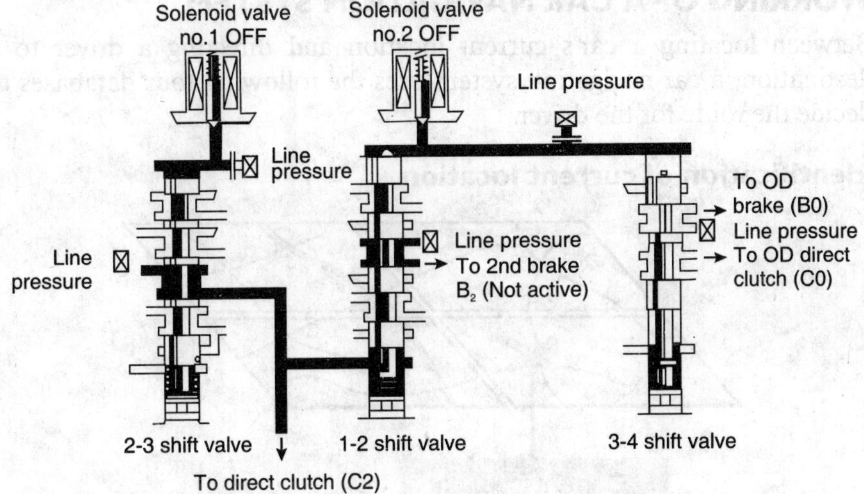


**Fig.13.7:** *Fourth gear*

## ROAD NAVIGATION SYSTEM

Car navigation systems receive signals from satellites and identify the vehicle's position and direction by combining that data with information obtained from various onboard sensors.

### Radio positioning

This is the process of special signals from one or more radio transmitters at known locations to determine the position of the receiving equipment.

### Distance sensor

This is an electronic odometer which provides discrete signals from a rotating shaft or wheel.

### Distance road map using vector encoding

It provides mathematic modelling concepts to represent geometrical features such as road way and boundaries in abstract form with a minimum of data. By considering each road or street as a series of straight lines and intersection as a node, a map may be viewed as a of interrelated nodes, lines and enclosed areas.

### Map matching

It will recognises a vehicle's location by matching the pattern of its apparent path using radio positioning with the road patterns of digital maps and stored in computer memory.

## WORKING OF A CAR NAVIGATION SYSTEM

Between locating a car's current location and directing a driver to a destination, a car navigation system uses the following four databases to decide the route for the driver.

### Identification of current location

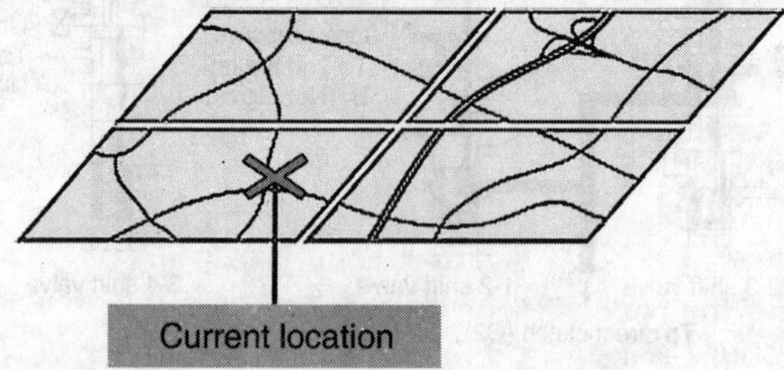


**Fig.13.8:** *Identification of current location*

Initially, the current location is identified and indicated on the map display. At that time, the configurations and the locations of nearby roads, connecting routes between roads, as well as attributes such as one-way streets and intersections are ascertained from the system's database.

### Finding and setting the destination

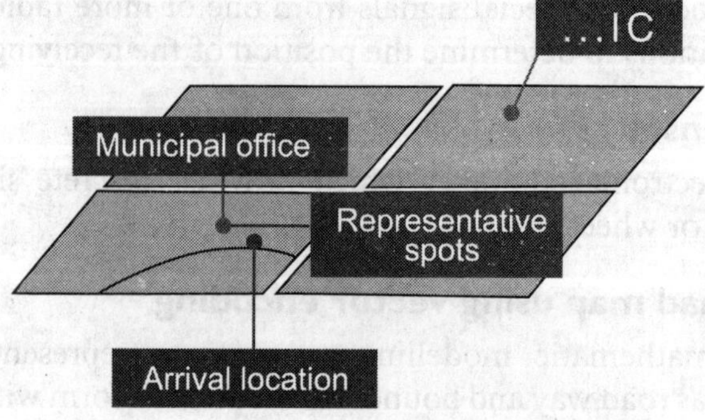


**Fig.13.9:** *Finding and setting the destination*

Next, the destination is located. Information on destination address, properties (hotel, park, etc.) and other representative features is obtained.

## Route search

After confirming the current location and the vicinity of the destination, the system confirms the route that connects those two locations. Road connection status, intersections, left/right turn conditions, etc. are determined.

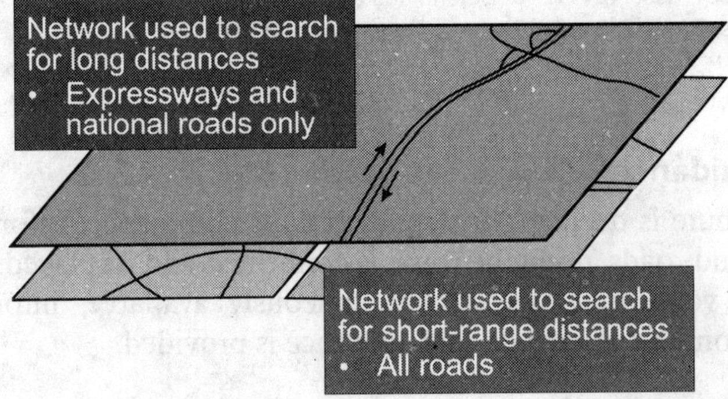


**Fig.13.10:** *Route search*

## Searching for recommended route

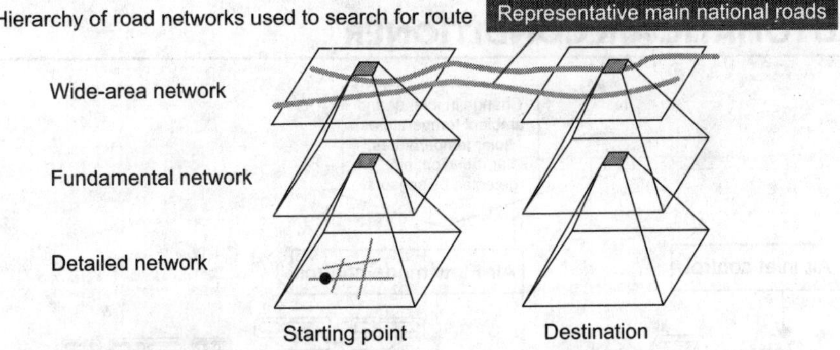


**Fig.13.11:** *Searching for recommended route*

The system determines various possible routes from the current location to the destination, then verifies the optimum route.

In order to identify the recommended route, the following decisions are made. Factors such as one-way streets, narrow roads, etc. are recognized as costs in terms of route selection, and the route with the least cost is recommended.

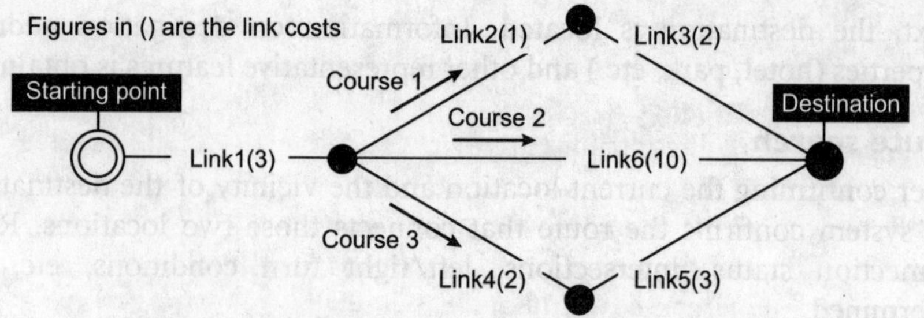


**Fig.13.12:** *Selection of route*

## Route Guidance

After the route is decided, driving directions commence. Information on buildings and roads along the route is confirmed and displayed.

When voice guidance is simultaneously available, information is obtained from the voice data and guidance is provided.

## Map display

While route directions are being provided, information on the locale through which the car is driving is displayed. In this way, roads, buildings, topographical features, etc., along the way are identified.

# AUTOMATIC AIR CONDITIONER

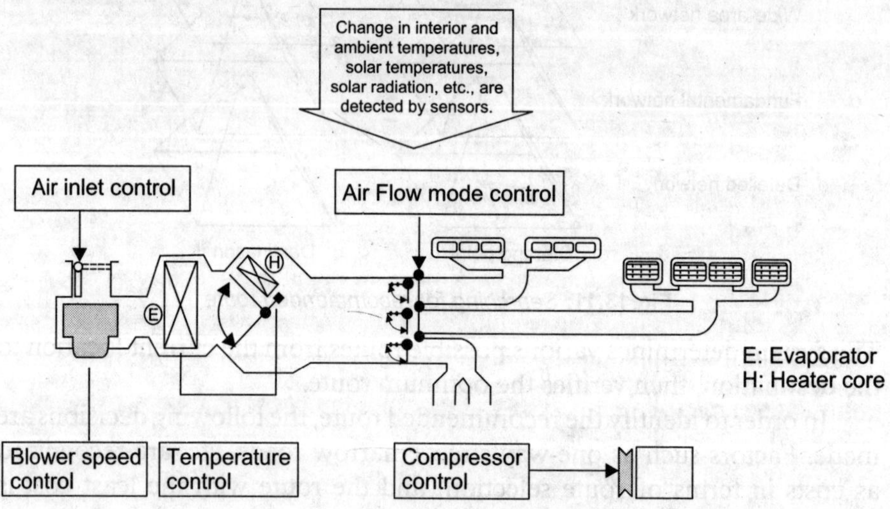


**Fig.13.13:** *Automatic air-conditioner*

The manually controlled air-conditioner always operates at the blower air temperature and blower speed set by the driver. However, such factors as solar radiation, engine heat, exhaust pipe heat, heat generated by the passengers, etc., will contribute to the interior temperature over time.

It is therefore necessary with an air conditioner of this type to readjust the blower air temperature, blower speed, or both, as the need arises. The automatic air conditioner has been developed to eliminate the necessity for such annoying manual adjustments.

The automatic air conditioner detects the interior and ambient temperatures, solar radiation, etc., and adjusts the blower air temperature and blower speed automatically according to the temperature set by the driver, thus maintaining the interior temperature at the set temperature.

These days, some models also provide such controls as air inlet control, air flow mode control (air outlet control) and/or compressor control in addition to temperature control and blower speed control.

## FUNCTIONAL COMPONENTS

An automatic air conditioner is a manual air conditioner equipped with (i) sensors which detect changes in temperature and solar radiation, (ii) controllers which determine the operating conditions according to signals arriving from the sensors, and (iii) actuator which are driven by controllers, and physically move dampers and other components.

## SENSORS

The typical sensors employed in an automatic air conditioner use the following semiconductors:

- Sensors using thermistors and inputting changes in temperature as changes in electrical resistance to the controllers:
  1. Interior air temperature sensor
  2. Ambient air temperature sensor
  3. Evaporator sensor
  4. Water temperature sensor
- A solar sensor using a photodiode and inputting changes in solar radiation as changes in electric current to the controllers.

## Interior air temperature sensor

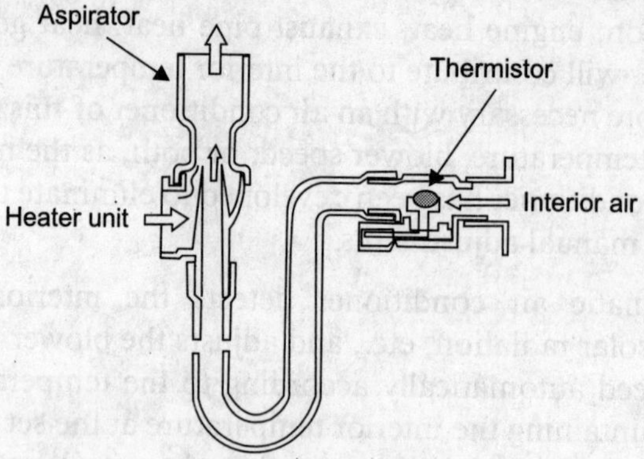


**Fig.13.14:** *Aspirator*

An interior air temperature sensor draws in interior air to determine the mean temperature of the air in the passenger compartment. Formerly, a motor type interior air temperature sensor, which used a motor to draw in air, was used, but these days an aspirator type which uses the air flow through the heater unit is commonly used.

## Ambient air temperature sensor

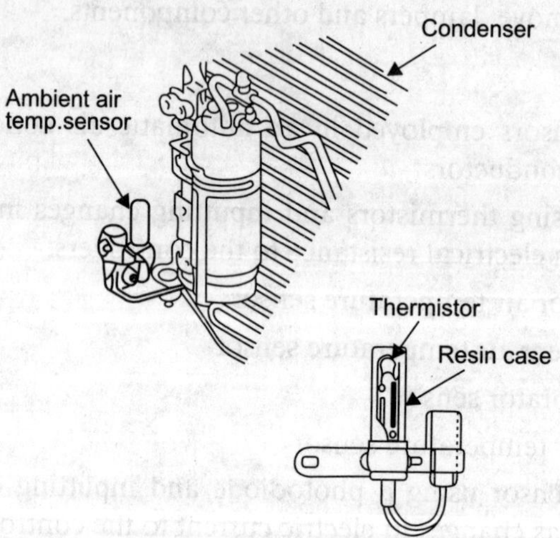


**Fig.13.15:** *Ambient air temperature sensor*

The ambient air temperature sensor is enveloped in a moulded resin casing designed not to react to abrupt changes in temperature. This allows it to accurately detect the mean ambient temperature.

## Evaporator sensor

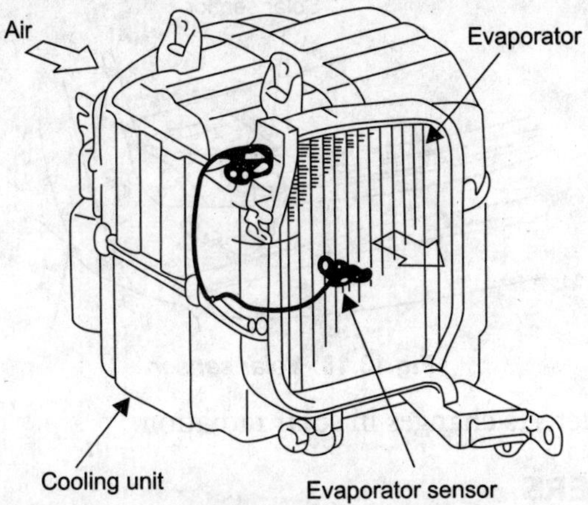


**Fig.13.16:** *Evaporator sensor*

The evaporator sensor detects the temperature of the air passing through the evaporator. This is only for the microcomputer-controlled automatic air conditioners.

## Water temperature sensor

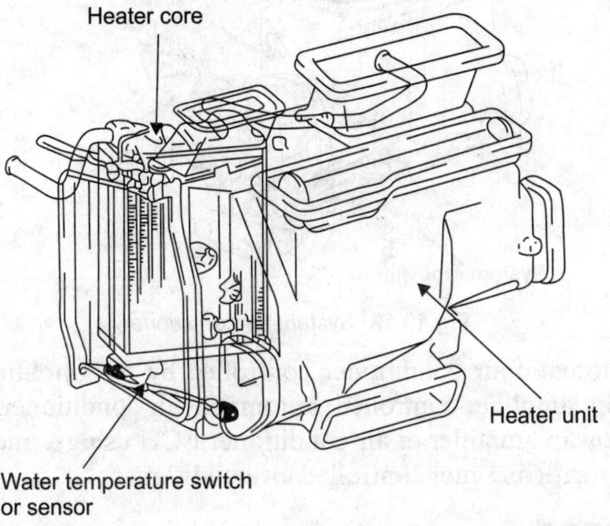


**Fig.13.17:** *Water temperature sensor*

The water temperature switch or sensor detects the temperature of the engine coolant.

## Solar sensor

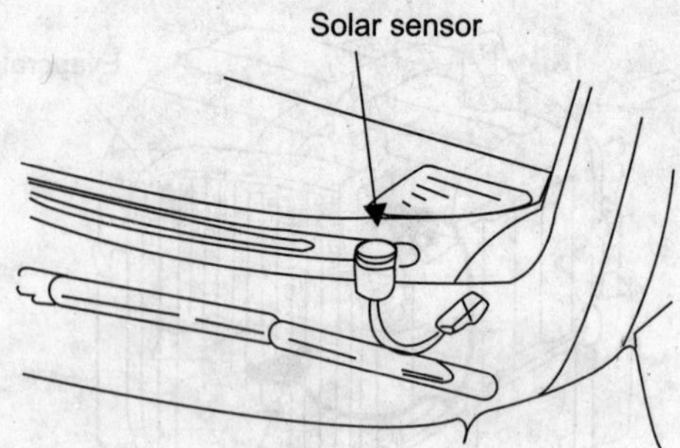


**Fig.13.18:** *Solar sensor*

A solar sensor detects changes in solar radiation.

## CONTROLLERS

Controllers can be divided into two types: those using ICs and those using a microcomputer. They are called system amplifiers, automatic air-conditioner amplifiers or air conditioner ECUs.

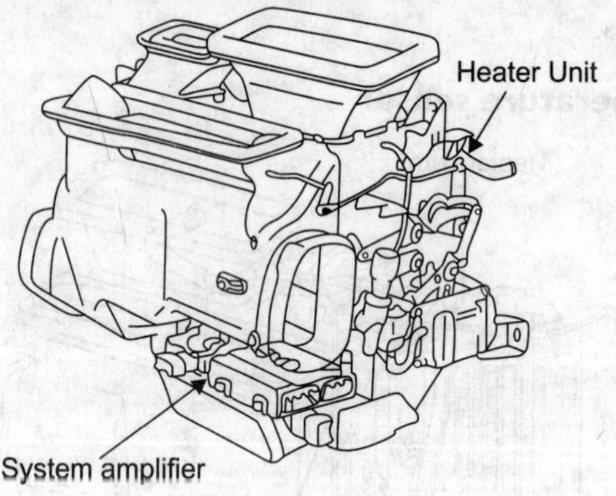


**Fig.13.19:** *System air conditioner*

An automatic air conditioner controlled by an amplifier using ICs is called an amplifier-controlled automatic air-conditioner, while one controlled by an amplifier or air-conditioner ECU using a microcomputer is called a microcomputer controlled air-conditioner.

## ACTUATORS

Actuators include blower motor, compressor, and servomotors.

# INDEX

# AUTOMOTIVE ELECTRICAL AND ELECTRONICS

## A K BABU

## FEEDBACK FORM

1. Which topics from your examination point of view are inadequately or not covered in this book?

2. Is there any misprints/ mistakes/ factual inaccuracy in this book? If so, please specify.

3. What is your assessment of this book as regards the presentation of the subject-matter, expression, precision and price in relation to the other recommended books available on this subject?

4. Which competing books you regard as better than this? Please mention their author and publishers.

5. Any further suggestions/comments you would like to make for the improvement of the book?

***PS:** If needed, please attach a separate sheet.*

**Name** : ______________________

**College** : ______________________

**Address** : ______________________

**Email** : ______________________

**Mobile** : ______________________

*Please mail the feedback form at:*

**KHANNA BOOK PUBLISHING CO. (P) LTD.**

4C/4344, Ansari Road, Daryaganj, New Delhi-110002

You can also email this feedback form at:

contact@khannabooks.com or Whatsapp at +91 – 99109 09320

**BEST FEEDBACK WILL BE REWARDED SUITABLY**